Ageing, Health and Pensions in Europe

Also by Lans Bovenberg

PUBLIC FINANCE AND THE ENVIRONMENT IN AN IMPERFECT WORLD (co-authored)

ECONOMIC SCIENCE AND PRACTICE: The Roles of Academic Economists and Policymakers (co-authored)

ECONOMIC SCIENCE: Art or Asset? The Case of the Netherlands (co-authored)

DEALING WITH NEW GIANTS: Rethinking the Role of Pension Funds (co-authored)

Also by Asghar Zaidi

MAINSTREAMING AGEING: Indicators to Monitor Sustainable Policies (co-authored)

WELL-BEING OF OLDER PEOPLE IN AGEING SOCIETIES

NEW FRONTIERS IN MICROSIMULATION MODELLING (co-authored)

Ageing, Health and Pensions in Europe

An Economic and Social Policy Perspective

Edited By

Lans Bovenberg
University of Tilburg, The Netherlands

Arthur Van Soest
University of Tilburg, The Netherlands

and

Asghar Zaidi
European Centre for Social Welfare Policy and Research, Vienna

First published 2010 by
PALGRAVE MACMILLAN

Palgrave Macmillan in the UK is an imprint of Macmillan Publishers
Limited, registered in England, company number 785998, of Houndmills,
Basingstoke, Hampshire RG21 6XS.

Palgrave Macmillan in the US is a division of St Martin's Press LLC,
175 Fifth Avenue, New York, NY 10010.

Palgrave Macmillan is the global academic imprint of the above
companies and has companies and representatives throughout
the world.

Palgrave® and Macmillan® are registered trademarks in the United States,
the United Kingdom, Europe and other countries

ISBN 978–0–230–28290–2   hardback

This book is printed on paper suitable for recycling and made from fully
managed and sustained forest sources. Logging, pulping and manufacturing
processes are expected to conform to the environmental regulations of the
country of origin.

A catalogue record for this book is available from the British Library.

A catalog record for this book is available from the Library of Congress.

10 9 8 7 6 5 4 3 2 1
19 18 17 16 15 14 13 12 11 10

Printed and bound in Great Britain by
CPI Antony Rowe, Chippenham and Eastbourne

Ageing, Health and Pensions in Europe

An Economic Perspective

Contents

Part II: Well-being of the Elderly 105

Chapter 5
Socioeconomic and Psychosocial Determinants
Johannes Siegrist / Morten Wahrendorf

Chapter 6
Martin Kohli / Harald Künemund

Chapter 7
Subjective Well-being in Older Adults:
Dieter Ferring / Thomas Boll

Chapter 8
Alberto Holly

Chapter 11
Arthur van Soest

Chapter 12
Policy and Research Challenges for Ageing,
Arthur van Soest / Lans Bovenberg / Asghar Zaidi

List of Tables, Figures and Box

Tables

Figures

Box

1

Ageing, Health and Pensions in Europe: An Economic Perspective

Arthur van Soest / Lans Bovenberg / Asghar Zaidi

The ageing of populations is one of the main economic and social developments shaping the 21st century. Population ageing affects welfare states, as it creates the need for reforms of pension-, healthcare- and social care systems, which will have reverberations for European welfare states and their labour markets, and the health and well-being of European citizens. Yet, ageing also yields new opportunities and benefits, raising important issues such as how societies can benefit longer from the talents of populations enjoying rising longevity, and how human resources can be maintained better during the entire life course. Since population ageing is proceeding more rapidly in Europe than in other continents, Europe can be expected to take the lead in social innovations aimed at adopting suitable policies for successful ageing.

This book discusses population ageing and the main associated research challenges in Europe for economists and social scientists concerning welfare regimes, pension provision, public health, employment, income security and well-being. It develops a research agenda that exploits the diversity of European pension-, welfare- and health systems to study causal links between institutional arrangements, individual decision-making, labour-force transitions, financial security, health, and general well-being of older age groups.

Most of the topics relevant to the economics of ageing, health and pensions fit into one of three broad policy and research themes: income security, the well-being of older age groups, and labour-market issues. These three broad themes are the basis for the three main parts of the book.

1

The first theme, income security, is covered in chapters 2, 3 and 4. Broadly speaking, it involves many features of state- and occupational pension systems, risk sharing between generations, individual decisions on voluntary retirement savings (such as pension portfolio choice, housing assets), consumption patterns before- and after retirement, financial knowledge and planning during the life course. The most important ways to provide old-age financial security in Europe are government systems of old-age social security provision (first-pillar pensions), occupational pensions provided mainly through private pension funds (second-pillar pensions), and voluntary household savings for retirement – for example, through life insurance (third-pillar pensions). Also other forms of household saving may be used. A specific example is owner-occupied housing (which offers a vehicle for savings for retirement and may reduce housing costs in retirement, while also a reverse mortgage or the proceeds of selling the house can be used to finance [part of] the cost of living in old age). In some countries, financial support by family members also plays a role. The main question is how public policy can help to guarantee an adequate combination of these various ways of obtaining financial security in old age. On the one hand, people's preferences and opportunities vary, which is an argument for giving individuals their own responsibility and the freedom to make the best choices in their own situation. On the other hand, recent research has shown convincingly that people have great difficulty understanding financial products and making long-term plans, and seemingly do not make the decisions that are in their own best interests.

Chapter 2 discusses how European pension reforms are having an impact on the adequacy of retirement incomes. The authors also address the question of how household financial preparation for retirement can be improved by suitable public policies and by development of a suitable financial market. They hold the view that reforms are reducing future replacement rates, and that there is thus a need for additional discretionary savings. Reforms are also reducing annuitization, which leaves individuals more exposed to longevity risks. On the positive side, in general, the reforms have reduced incentives towards early retirement, introduced greater freedom of choice, and moved towards a better-diversified pension design. There is as yet not enough evidence to support the conclusion that workers are working longer and saving more in response to a reduction in mandatory savings. It is also not clear whether the policy reforms will have (negative) unintended consequences: in the form of rising numbers of poor or – contrary to the intentions of pension reforms – in the form of larger future fiscal costs of supporting pensioners by income-related benefits.

The authors point to the scarcity of data for European countries in measuring the extent and distribution of a saving gap for an adequate retirement income. Their policy recommendations include not only adoption of financial education programmes and carefully designed default options for complex choice situations, but also a strengthening of safety nets for the less fortunate. Policy measures should vary according to age, given the different constraints faced by the young and the old. The young should be provided with the instruments to make the best possible – and informed – saving decisions. For the elderly, there should be instruments to make housing equity more liquid – and this is a matter for policy-making as well as for markets. The main role of policy would therefore be to help people, through education and incentives, to accumulate and use up their pension capital and not adopt more paternalistic programmes of mandatory savings of one sort or another.

Chapter 3 discusses the optimal management of financial portfolios during the life cycle in individual and collective pension plans. It explores whether an individual should reduce risk exposure when growing older, and whether one should adjust this exposure if market circumstances change. The optimal portfolio appears to depend, among other things, on longevity insurance in pay-as-you-go pension plans and the flexibility of the labour market as well as the housing market. The role of collective pension plans in sharing risks across generations is also investigated, including the problem of how to commit future generations to these plans. The chapter identifies various gaps in the literature on optimal portfolio management – for example, the impact of model uncertainty, small-probability catastrophic events and household psychology.

Chapter 4 studies the optimal design of pension systems. It presents a benchmark model for thinking about optimal life-cycle financial planning, and analyses how researchers have recently extended this model and explored how people actually behave in investing, saving and insuring longevity risk during the life cycle. The chapter investigates also the remaining research challenges, including the documentation of shocks in human capital and health, the modelling of financial-market and inflation risks, and the implications of alternative preference structures (such as habit formation and loss aversion). The chapter discusses the roles of institutions that can help financially unsophisticated consumers to delegate complex financial decisions to professionals they can trust. These institutions include solvency regulations and the choice architecture and governance of pension plans. The chapter concludes by investigating the research infrastructures that would be required to address the research challenges most efficiently.

The second theme, the well-being of older age groups (chapters 5, 6, 7 and 8), involves not only financial status but also other factors affecting wider aspects of quality of life. The theme covers psychological well-being, family contacts and other social networks and inter- and intra-generational transfers, time use and satisfaction with daily activities (including paid work and volunteer work), physical and mental health (and health behaviour and prevention), availability of formal and informal long-term care, and other aspects of welfare regimes and the healthcare system. For this purpose, a distinction can be made between various stages of the life course of older age groups. For the younger ones (until age 70 or perhaps 75; obviously, no strict threshold applies here), formal work and other socially productive activities play a major role. Good working conditions and opportunities to remain socially productive after (early) retirement can contribute to a higher sense of well-being. Intrinsic and extrinsic rewards for these activities are crucial. At older ages, physical and mental health problems often lead to an increasing dependency on formal and informal healthcare and social care. Mobility tends to diminish, which increases the importance of suitable housing and living arrangements. Social networks play an important role at all ages, and these networks tend to become smaller and more family-centred at older ages. The role of family networks changes with age, from giving support to parents and to children and grandchildren, to receiving support from younger generations. The importance of social networks also varies with the national context determining the intergenerational balance of income and wealth, and the nature of the housing market.

Chapter 5 starts from the notion that healthy ageing is determined by the ability to achieve self-defined or other-defined goals by own agency, implying that welfare institutions and (more generally) societal opportunity structures that support the capability of individuals to achieve goals by means of autonomous activity can contribute to healthy ageing at the population level. By comparing different European countries, the chapter considers the potential of societal opportunity structures to provide options of goal-oriented agency to early old-age populations. Examples include specific policy programmes that favour participation in productive activities after labour-market exit, such as insurance protection or tax relief for volunteers.

Moreover, the chapter analyses how different types of goal-oriented agency of older people affect their health status and well-being, reviewing the available evidence and identifying gaps of knowledge. It addresses two policy issues: what needs to be done to retain older people in work and to better protect and improve older workers' health and well-being? And what could be done to stimulate retired people to do socially productive activities such as volunteering, or informal care for a sick or disabled person?

Chapter 6 focuses on the social networks – the social connectedness – of older individuals: on their relationships with others (family members, friends, neighbours, colleagues or service providers), and on the content of their exchanges with these others. Social networks are crucial for the well-being of older individuals, offering a range of benefits for ageing societies. They are a source of support for persons in need, act as a focal point for productive activities of the elderly, and serve as catalysts for social participation in community affairs. They also contribute to a reduction in public expenditure. The chapter discusses how three conceptualisations of social connectedness (focusing on activities, networks, and social capital) are interrelated and can be fruitfully used in combination to keep older people socially connected. It emphasises the importance of policy issues such as how intergenerational transfers in family networks can help to provide financial security in case of unemployment or divorce when the coverage provided by public transfers is decreasing due to current welfare state retrenchments, how grandparents can play a role in reconciling employment and parenthood, particularly in countries with a weak provision of public childcare, or how informal care by family members or friends can replace formal care for the dependent elderly.

Chapter 7 focuses on subjective well-being (SWB) in older adults. SWB essentially refers to an evaluation of an individual's life from his or her own perspective, and not to external evaluations based upon objective criteria such as health, education or income. SWB encompasses life satisfaction as well as the presence of positive- and absence of negative feelings. It can refer to one's life as a whole or to specific life domains (including health, material wealth, social relationships). The chapter discusses measurement issues, theoretical models, and reviews the main factors that drive SWB in older adults (life circumstances including income, social indicators, and functional limitations), and the effectiveness of interventions like psychotherapy and mental health services for older people.

Chapter 8 first addresses two main health-related issues in the economics of ageing: health and wealth, and housing and living arrangements. It then examines issues related to the impact of population ageing on health and long-term healthcare expenditures in European countries. The author also emphasises the relevance of 'epidemiological transition', alongside the commonly discussed demographic transition. The former is characterised by the shifting burden of illness toward non-communicable diseases and injuries. He points to the fact that with changing age structure of the society the incidence of heart problems, hip fracture and dementia will be increasing and there are also greater risks of people experiencing multiple chronic diseases. The author concludes that ageing as such is not a main driving force in health expenditure growth, but the main contributing factor for future health- and care-related expenses is in fact the diffusion of medical technology. Thus, studies of patterns of diffusion of medical technology, and of the incentives at work, form a major research issue with important policy implications. If economic and regulatory incentives matter for technological change, then national health policies may have dynamic, long-term consequences for the costs and quality of care – particularly for the treatment of conditions in the older population.

The final theme, labour-market issues for older workers (chapters 9, 10 and 11), involves a key policy concern of contemporary European policymakers: How to keep older people at work and how to increase the average retirement age? Previous research has demonstrated that retirement decisions depend not only on economic incentives such as generous early-retirement benefits, but also on non-financial factors such as job characteristics and quality of work, job satisfaction, health status, social networks, attitudes of employers and co-workers, and peer-group behaviour. In many countries, generous early retirement benefits have been abolished and workers are seeking to retire later. The crucial issue then becomes whether firms are willing to employ and retain these older workers, and also organise the work in such a way that the job remains doable and financially attractive. This may involve accommodation of the workplace, reorganising tasks, providing opportunities for part-time work and phased retirement, and so forth. Furthermore, investments in human capital and in new skills required by changing technologies through on-the-job training are crucial to keep older workers productive enough in relation to their costs. But also employer attitudes and perceptions of older workers play an important role.

Improving the demand for older workers is key to the objective of increasing the labour-market participation of older workers, and of reducing the potential negative effect on long-term growth of shrinking working-age population. This key area of research and policy action has not been fully acknowledged by policymakers and researchers in Europe. *Chapter 9* provides a review of key theoretical models that explain the factors underlying demand for older workers as well as the empirical evidence on a number of actual factors that condition the demand for older workers. The factors identified are the impact of wages and productivity variations, macroeconomic conditions, discrimination in recruitment processes and labour-market regulations. The authors emphasise that more research is required on developing an understanding of the relationship between age, wage and productivity. Also required is a better understanding of the impact of business cycle fluctuations and technological innovations and labour-market regulations on the demand for older workers, and on the extent and impact of age discrimination in recruitment processes. They point to the benefits of an age-diverse workforce and recommend changing attitudes and reducing age discrimination to achieve this diversity. One key factor to promote employment of older workers is to reconsider the weight of seniority components in wages to bring wages more in line with productivity, to update the skills of workers when they age and provide incentives in collective agreements to recruit older unemployed workers. Based on an assessment of the current state of the research infrastructures and networks on this topic, the authors identify where investment is to be made to improve research output in this area of high policy importance.

Chapter 10 develops a theoretical framework to explore the life-cycle interactions between human-capital investment, retirement decisions and pension saving in order to understand individual behaviour and to make policy recommendations on the tax treatment of retirement saving, retirement systems and labour-market institutions, more generally. The chapter explores the underlying assumptions and the empirical content of this framework, while also considering various competing theories to explain the data. Future research challenges include discerning the impact of imperfect labour markets on human-capital investments and developing econometric models to identify non-observable investment in human capital. Moreover, micro panel data are needed to estimate life-cycle models, while quasi-experiments are required to estimate the impact of institutions on life-cycle interactions between human capital and saving.

Chapter 11 studies retirement decisions from the point of view of workers. It focuses on economic factors such as the generosity of early retirement benefits and pensions, as well as psychological and social factors such as quality of work and work satisfaction, social networks and retirement decisions of family members and peer groups, and health and work disability. Public policy affects the economic environment under which retirement decisions are made, through eligibility rules and levels of state pensions, and through taxation of occupational pensions and other savings. The chapter emphasises the consequences of institutional differences across countries in pension systems, disability benefits, other social security arrangements, taxation, and gradual retirement opportunities for the observed differences in retirement patterns.

Structure

Chapters 2 to 11 all have a similar structure. They first introduce their research topic of interest, which is part of one of the three themes introduced above, and formulate various policy questions related to that topic. Each chapter then presents an overview of the current state of the art concerning the topic of the chapter:

- What are the most important research results obtained thus far, and what are the remaining gaps in scientific knowledge?
- Which research questions can be addressed in the near future, and what is needed to fruitfully address these questions in terms of, for example, developing a theoretical framework, data collection, international cooperation, or institutional knowledge? and
- How will answering the research questions contribute to development of socioeconomic policy?

Each chapter is followed by one or two brief discussions, which emphasise the strengths, weaknesses, opportunities and threats for research on the topics addressed in the chapter.

Together, the chapters show that the socioeconomic, cultural and policy diversity of Europe can be exploited to learn more about causal links between ageing, pensions, welfare regimes, retirement decisions, health outcomes and well-being. Our understanding of the impact of population ageing benefits substantially from a cross-national comparative point of view, since comparing different policy approaches and exploring the consequences of various policy changes in different institutional environments and business cycle stages helps enormously to learn about the impact of policy. We focus on understanding how individuals and firms make their decisions and on gaining insight into the mechanisms that drive individual well-being. Understanding the mechanisms at this micro level is crucial for analysing the functioning of institutions and markets and for obtaining insight in the potential consequences of economic and non-economic policies for a country's economy or for Europe as a whole.

The topics covered are relevant not only for science but also for public policy: the economics of ageing, pensions and health are at the top of the policy agenda in Europe and in almost all individual European countries.

Part I:
Income Security of the Elderly

2

Adequacy of Savings for Old Age in Europe

Elsa Fornero / Annamaria Lusardi / Chiara Monticone

1   Policy questions

Household saving rates and wealth levels are very heterogeneous – both across and within countries, varying with respect not only to age, cohort, and time, but also to education, family size, health and so forth. Given these differences, in what sense is it possible to look into the *adequacy* of retirement savings? And on what grounds is the question *relevant*?

The concept of (retirement) savings *adequacy* combines two dimensions: a well-structured institutional design for an efficient sharing and diversification of risks, and sensible individual behaviour with respect to the time allocation of resources, in a given market and institutional context.

The first aspect is crucial, because even rational individuals will accumulate wealth poorly or inefficiently if they are forced to participate in ill-designed pension schemes or if they lack proper instruments and markets for accumulation. Institutional features are extremely important but difficult to characterise in a single model. In Europe, for example, a wide variety of retirement provisions are in place, with countries varying according to the degree of state intervention, the provision of inter- and intra-generational insurance, the amount of redistribution, and other characteristics. Moreover, reforms are modifying – in some cases rather radically – the playing field.

Given this diversity, individual saving behaviour is expected to vary – not only because of heterogeneous preferences and constraints, but also because of the different levels of mandated saving, its characteristics, and its substitutability with respect to 'discretionary' accumulation. In particular, the pension reform process that started in the 1990s in most, if not all, European countries substantially increased workers' uncertainty with regard to their replacement rates, typically by shifting from defined-benefit (DB) to defined-contribution (DC) formulae. Reforms made future pensions not only less generous and more 'self-made', but also more uncertain and difficult to understand, thus imposing greater costs upon planning ahead.

The issue of retirement savings adequacy is *relevant* for Europe on three grounds: first, because there are differences among countries in the level of mandatory provisions and in the institutional setting for voluntary complementary pensions; second, because reforms are increasing the extent to which individuals are responsible for their own retirement; and third, because 'inadequate' decisions can be improved through suitable policies, possibly *targeted* to specific goals / groups.

In this perspective, a few questions deserve to be addressed:

- *What consequences will reforms have on the adequacy of retirement provisions?* While a cutback on past promises would seem to undermine the adequacy of pension systems, it could indeed reinforce their adequacy, at least for young and future generations, by restoring conditions of financial sustainability and by attaining both a better diversification of risks and a better incentive structure.
- *How will the discretionary savings of households respond to changes in pension provisions?* As reforms reduce public generosity and redistribute risks, will workers be sufficiently encouraged / prepared to fill the gap? The answer depends on how individual decisions are made. Economic theory typically *assumes* that individuals are rational, farsighted and well informed – and able both to plan ahead and to correctly interpret market products as well as the rules and incentives provided by policy.
- *Are conventional models really able to capture individual behaviour?* A vast amount of empirical literature has indeed documented significant departures from such an ideal standard because of myopia, inertia, and / or lack of financial literacy. Accordingly, the degree of savings inadequacy depends on how distant individuals are from the rationality paradigm, and in what respects.
- *What can policy do to improve retirement saving choices?* Although at any particular moment personal characteristics are given, they can be influenced by both the market (promoting the demand for private insurance products) and by public policies (trying to strengthen attitudes conducive to saving behaviour). In particular, the provision of information on

the mechanisms regulating the pension system and the improvement of financial literacy among the population are instruments that can stimulate individuals' *preparedness* for retirement and strengthen the adequacy of their decisions (for example, by acknowledging the *status quo* bias in individual behaviour).

In dealing with these rather broad issues, this chapter has chosen to concentrate on *income risk in retirement* (and therefore on pensions, as the main source of security in retirement), and thus to disregard other risks – in particular those connected to participation in the labour market (such as unemployment, disability, and so on), even though lack of resources in retirement is typically the direct consequence of a poor working career. Further, the chapter does not explicitly consider the consequences on household savings of heterogeneous health- and Long-Term Care (LTC) provisions even if, due to incomplete public coverage and imperfect insurance markets, both health- and LTC risks add to the precautionary motive of savings; as a matter of fact, the health risks of the elderly are, in Europe, generally provided for by national healthcare systems, and out-of-pocket expenditures are less relevant than in more pro-market countries, like the United States. Finally, the chapter pays no attention to the way in which fiscal incentives might affect retirement saving choices. Even though a vast literature suggests that tax-exempted instruments for wealth accumulation mainly attract savings previously held in other forms, rather than spurring the creation of 'new' saving, the issue is far from being settled (Börsch-Supan and Brugiavini, 2001).

The chapter is organised as follows. After an overview of the most relevant policy questions, section 2 reviews the major contributions of theoretical and empirical economic analysis in addressing such questions. Section 3 describes the main remaining gaps in knowledge. Sections 4 and 5 discuss respectively the current conditions of existing infrastructures and the requirements in terms of data and methodological innovations. Section 6 suggests what can and should be delivered on the initial policy questions.

2   Major progress in understanding

2.1   *Adequacy from the perspective of the pension system*

Differences in European retirement provisions are reflected in differences in the age-saving profiles. According to Börsch-Supan and Lusardi (2003), for example, these are pronouncedly hump-shaped in the Netherlands, moderately hump-shaped in Germany, almost flat in Italy, and increasing at all ages in the UK. In none of these cases does there seem to be dissaving in old age, as predicted by simple versions of the life-cycle model (LCM). Without ignoring the possibility that other driving forces may have a role in explaining these differences (such as the stringency of borrowing constraints, as measured, for example, by the average down payment necessary to buy a house[1]), part of the variations are the direct effects of the different pension set-ups: the more generous social security provisions in both Italy and Germany reduce the need to save for retirement during working ages, while the Dutch flat-rate pension benefits – with rather low replacement rates – are at the root of the marked hump-shaped profile in that country. Thus, a proper understanding of the adequacy of retirement savings cannot but start from pension provisions, which are the main vehicle for the accumulation of retirement wealth, substituting for discretionary savings and creating various (dis)incentives.

In any examination of pension systems from an adequacy perspective, more important than benefit levels *per se* is the government's role in promoting/delivering a good *ex ante* allocation/diversification of risks. This entails an institutional framework that, under a financial sustainability constraint, does the following:[2]

a.   provides *efficient* ways to broaden the scope for risk pooling and sharing, not only through public pensions (and other benefits for the elderly, such as survivor benefits), but also through good regulation/supervision of market provisions;

b.   reduces poverty among the elderly;

c.   encourages individuals' *awareness* of retirement needs, and their capacity to make informed and farsighted decisions, by means of financial literacy programmes and appropriate choice designs.

With regard to efficiency (point a), in overlapping-generations models a source of market incompleteness comes from the impossibility of individuals to engage in *intergenerational* risk sharing with yet-unborn generations: in the absence of such markets, governments substitute for them by establishing – as a vehicle

---

1   This appears to drive up savings at young ages in both Germany and Italy, and to increase aggregate savings in general with respect to Anglo-Saxon countries and the Netherlands.

2   Adequacy should always be viewed within a context of financial sustainability, given that it is always possible to increase benefit levels by ignoring – at least for a certain period – the government's intertemporal budget constraint. Financial sustainability, however, 'does not imply fully funded pensions, but only that unfunded obligations are not growing excessively relative to the contribution base' (Barr and Diamond, 2008: 10).

to set up an intergenerational contract – a Pay-As-You-Go (PAYG) method of financing (Shiller, 1998; Ball and Mankiw, 2007). Risk diversification, however, demands more than just PAYG financing; it requires a good combination of public and private choices as well as good regulation/supervision of market provisions, thus offering a rationale for a *mixed system* (Lindbeck and Persson, 2003; Castellino and Fornero, 2007). Moreover, the choice between DB or DC formulae should also be carefully considered in pension system design, as it has important implications in terms of social welfare (Gomes and Michaelides, 2003).

Some degree of state intervention is also justified by *intragenerational* risk sharing, with poverty prevention as another way to look at the provision of 'adequate' pensions (point b). Even though the scope for intragenerational risk pooling might be reduced by issues such as moral hazard and prior income inequality (which cannot be entirely 'cured' by the pension system), there are many practical limitations to the ability of the elderly to diversify their incomes by themselves, which emphasises the government's role in providing this kind of risk sharing, as a way to combat poverty in old age (Shiller, 1998; Barr and Diamond, 2008).

Finally, given that public provisions typically do not (and should not, given the efficiency considerations of point a) fully cover financial needs in retirement, governments should also aim at increasing and improving the ability of individuals to make sensible choices, concerning both the age of retirement and the accumulation/investment of personal savings (point c). This can be done not only by fostering individual preparedness, but also by reducing the distortions embedded in pension formulae, and/or by choosing an enhanced choice structure – for example, through an appropriate design of pension default options (Madrian and Shea, 2001; Holzmann and Pallarès-Miralles, 2005; Lusardi, 2008; OECD, 2008).

Assessing the adequacy of pension systems is very difficult. Suitable indicators, capable of offering a 'benchmark degree' of efficient risk diversification against which to compare actual data are simply absent; moreover, they would be difficult to implement – not only because of the mix of insurance and redistributive features that typically characterises pension systems (which requires that other redistributive programmes are included in the assessment), but also because these features are never in a steady state, and the transition costs imposed by reforms would have to be taken into account. In general, reforms imply high costs and long-term adjustments, since, for most countries, the problem is not to decide whether to create *ex novo* a well-diversified and well-balanced system, but how to favour the birth, or the growth, of funded schemes alongside an already existing, and possibly mature, PAYG one.

All this applies in particular to European countries, whose welfare systems are quite complex and almost invariably undergoing transitions, with uncertain consequences on adequacy (Castellino and Fornero, 2006). First, since most of the recent reforms negatively impact (or will in the future) the replacement rates offered by the public scheme (the first pillar), they have been accompanied by measures aiming at encouraging the development of occupational and personal pension plans (the second- and third pillars). Indeed, most recent pension reforms are designed with the implicit idea that household saving is too scarce, at least for a large segment of the population (Börsch-Supan and Brugiavini, 2001). As the growth of funding is seen as an offsetting measure for the reduction of the PAYG coverage, the transition problem can be very severe: indeed, when the young are told that they will receive lower pensions for the same payroll tax rate, and are encouraged to contribute to a funded pillar to compensate the gap, they are asked to save more for the same replacement ratio. Although it would seem to shrink the adequacy of the pension system, this retrenchment of past promises could serve to reinforce it – by restoring financial sustainability, because all future generations would benefit from a system that does not pile up additional (implicit) public debt.

Second, another important feature of pension reforms is the move from DB to DC type of formulae, which implies both a stronger dependence (at the individual level) of benefits on contributions and a closer proximity (when not a strict correspondence, as in a Notional Defined Contribution, NDC, system) of the internal rate of return to the equilibrium rate represented by the growth of the wage bill (countries such as Italy, Sweden, Poland and Latvia have adopted actuarially fair types of formulae). Pensions based on actuarial principles are in sharp contrast to the history of many PAYG schemes, where workers had been accustomed to higher pay-offs. Moreover, the shift from DB to DC is also occurring in the private sector, induced by the increasing, and in some instances destabilising, cost of DB plans to employers. The expansion of DC formulae within both PAYG- and pension funds clearly implies an increase in the uncertainty surrounding the replacement rate at any given age of retirement and a transfer of risks resting on workers. Again, these greater risks would seem to undermine adequacy – but if the overall design should attain a better diversification of risks, the opposite could be true. Having said this, it should be noted that the incidence of transition and redistributive costs should not be ignored.

Third, reforms are also, in general, aimed at instilling greater flexibility of retirement, instead of reinforcing the traditional 'mandatory' retirement ages, differentiated rather arbitrarily by gender and/or working categories. This introduces an important adjustment margin, as workers are not forced to leave at a certain age, nor are they induced to leave as soon as they reach the minimum requirements by pension formulae, which contain high implicit taxes on the continuation of the activity.

2.2   Adequacy from the individual/household perspective

The *normative benchmark* of the economic analysis of household savings assumes rational individuals who make consistent intertemporal plans over their lifetime (Scholz and Seshadri, 2008: 4). Starting from this premise, on a positive level: *'a household is said to be saving adequately if it is accumulating enough wealth to be able to smooth the marginal utility of consumption over time in accordance with the optimizing model of consumption'* (Engen et al., 1999: 70). The stylised version of the LCM, in which individuals save during their working life to provide for consumption in old age, allows for a neat conceptualisation of retirement savings adequacy (the annuity value that, under the constraint of life-cycle resources, can sustain the preferred consumption path). Starting from the model:

$$\max E_t \sum_{j=0}^{T-t} \beta^{j} U(C_{t+j}, Z_{t+j})$$

$$s.t. A_{t+j+1} = A_{t+j}(1 + R_{t+j}) + Y_{t+j} - C_{t+j}$$

where $\beta = 1/(1+\delta)$ and δ indicates the individual intertemporal discount rate, $U(.)$ is a utility function depending on consumption C, as well as on the household demographic characteristics Z (household composition), and where A represents the stock of assets that grows from one period to the next through savings and the gross interest rate R, the following Euler equation is obtained:

$$U'(C_t, Z_t) = E_t \left[U'(C_{t+1}, Z_{t+1}) \beta (1 + R_{t+1}) \right]$$

where the marginal utility of consumption at time t is equal to the expected (discounted present value of the) marginal utility of consumption in period $t+1$. The inclusion of household composition is relevant both because of the particular risk placed on singles (survivors) in retirement and because changes in family composition help to explain why lower consumption in retirement can be optimal (Scholz and Seshadri, 2007).

Much richer versions of the model have subsequently been exploited by including real-life features such as labour-supply decisions and retirement choices; the timing of income receipts; uncertainty over future earnings, rates of return, length of life, and health conditions, all generating scope – even in old age – for *precautionary savings*. Borrowing constraints – although less important in retirement, as household wealth reaches its peak – may explain the timing of an individual's decision to draw from pension wealth. Other motives for saving, such as bequests, have also been added to the model. Apart from predicting the smoothing of marginal utilities, an important feature of the model is its ability to distinguish between *'inadequate'* and low levels of saving/wealth. For instance, facing an upward-sloping income profile, young people may save little or be borrowing. Similarly, older individuals may have little 'discretionary' saving because the amount of mandatory saving is already providing for their retirement needs.

It is then almost natural to take the LCM model as a benchmark for assessing savings adequacy. However, the empirical evidence, implicitly or explicitly based on the LCM, is mixed (and largely concentrated on data from the United States). Some studies use reduced forms to project households' lifetime asset- and income paths, and derive from them implications for savings adequacy. Results are varied: Kotlikoff et al. (1982), Love et al. (2008) and Hurd and Rohwedder (2008) all find there is no generalized under-saving. On the contrary, according to Haveman et al. (2006), about half of retirees will not have enough resources in retirement to meet their pre-retirement consumption level. Moore and Mitchell (1997) also find that the median household needs to increase its saving rate substantially until the age of retirement in order to obtain an 'adequate' level of wealth for retirement (an additional 16% saving rate to retire at age 62, or 7% to retire at age 65).

Other models are more sophisticated, as optimal household consumption and wealth accumulation profiles are simulated from a structural model and compared to actual data (Bernheim and Scholz, 1992; Engen et al., 1999; Munnell et al., 2006; Scholz et al., 2006; Scholz and Seshadri, 2008). These papers find that saving is adequate (or even *more* than adequate, suggesting some over-saving) for the vast majority of the population, and that *under-saving* is concentrated among households without a college degree (Bernheim and Scholz, 1992) or in the lower part of the wealth/earnings distribution.

More precisely, Scholz et al. (2006) find that only 15.6% of older households in 1992 are below the optimal level of accumulation. This is at odds with the results found by Munnell et al. (2006), who argue that 43% of households are at risk of not being able to maintain their standard of living in retirement. Although much of the difference seems to be attributable to data and methodological differences (Engen et al., 1999; Scholz et al., 2006), discrepancies remain.[3]

A different strand of literature – which also provides some evidence on European countries such as Germany, Italy and the UK – studies the 'consumption drop': in other words, whether or not consumption falls around the time of retirement, and for what reason. This finding would provide *prima facie* evidence against the theoretical predictions of the life-cycle model, and could be interpreted as evidence of inadequate retirement saving.

Optimal saving, however, does not necessarily mean *smooth consumption*; the drop itself could thus be 'optimal' (Banks et al., 1998; Bernheim et al., 2001; Miniaci et al., 2003). Retirement is typically an anticipated (or even chosen) event that does not come as an unforeseen shock, and there are reasons that justify a fall in consumption (a decrease in work-related expenses, for example). The occurrence of unexpected shocks inducing earlier-than-expected retirement, and the possibility of non-separability between consumption and leisure within the per-period utility function, provide other explanations for the drop in consumption that are consistent with the standard life-cycle framework (Haider and Stephens, 2007; Hurd and Rohwedder, 2006; Smith, 2006). Finally, other works claim that the proper entity on which to focus is not 'expenditure' but true 'consumption' (as measured, for instance, by food intake), since retirement provides ample opportunities to economise (Aguiar and Hurst, 2005). The existence of the drop itself is thus uncertain. On the whole, this kind of evidence points again to particular groups at risk, rather than to a common problem of inadequate savings in the general population.

Biswanger and Schunk (2009) take a different approach to assess what is an adequate standard of living; instead of relying on the life-cycle model, they use survey questions to try and elicit individual preferences about adequate old-age provision (in the absence of risk) and about the minimal level of monthly spending that people do not want to fall below. Neglecting any risk associated with retirement spending, they find that *ex-ante* desirable old-age- to

---

3   The work by Scholz et al. (2006) is based on the Health and Retirement Study (HRS), which covers Americans 51 to 61 years of age in 1992, while the study by Munnell et al. (2006) computing the National Retirement Risk Index (NRRI) is based on the 2004 Survey of Consumer Finances (SCF). A comparable sample is constructed by calculating the NRRI on the population age 51–61 in 1992 surveyed in the SCF. In this case, the NRRI takes on the value of 19%, meaning that 19% of households are at risk of not being able to maintain their standard of living in retirement. This is to be compared with the result by Scholz et al. (2006) that 16% of US households had less wealth than their optimal targets.

working-life spending ratios exceed 80% for a majority of respondents in the two countries involved in the survey (the US and the Netherlands).

Other studies concentrated on specific topics, again relevant to the issue of retirement savings adequacy. *Family formation* and, more specifically, *household composition,* have always been recognised as important ingredients of savings adequacy – not only because of the time-varying number of household members and of economies of scale in consumption, but also because differently structured households have a different risk tolerance, and thus a different propensity to save (Skinner, 2007; Hurd and Rohwedder, 2008; Scholz and Seshadri, 2007).

Also *wealth (il)liquidity* is an important element of savings adequacy, as, apart from social security wealth, housing wealth is often the major component of wealth in old age and is rather illiquid. Even though financial markets have developed instruments for transforming housing wealth into more liquid forms, these new instruments are rarely used, as households maintain a preference for living in the home (at least until a health shock forces them to move), and only infrequently plan to use their housing wealth to finance consumption in retirement (Lusardi and Mitchell, 2007a).

The availability of *annuities* is another essential aspect of retirement savings adequacy. Even though economic theory suggests that individuals should annuitize all of their retirement wealth to remove both the risk of outliving their resources and the risk of leaving unintended bequests (Yaary, 1965), full annuitization is not the best strategy when markets are incomplete (Davidoff et al., 2003). Annuity markets are thin, as many problems limit the propensity of individuals to annuitize (including the potential need to pay for uninsured medical expenses or for a nursing home), thereby providing a rationale for preferring lump sums (Turra and Mitchell, 2004; Sinclair and Smetters, 2004; Kifmann, 2008), while risk pooling within a couple / family decreases the value of annuitization for married couples (Brown and Poterba, 2000; Dushi and Webb, 2004). Although the complexity and the risk inherent in the product may act as a further disincentive, it seems hard to explain the lack of demand by remaining within the boundaries of rational models. Psychological factors (a preference, perhaps, for lump sums, *as such*) may very well be at work (Brown, 2009). This opens the way for a different analysis of savings decision-making – one not necessarily consistent with optimising behaviour and rational choices.

2.3 Retirement planning, information about pensions, and financial literacy

A growing literature has documented significant departures from the model and pointed to behavioural and psychological factors that limit the planning ability of individuals. One simple and direct way to examine whether individuals (consistent with the predictions of theoretical models of saving) look ahead and make plans for the future is to study the extent of retirement planning. Lusardi (1999) looked for such evidence using data from the 1992 Health and Retirement Study (HRS). She found that merely one-third of respondents had thought about retirement. While part of this behaviour may be perfectly rational, it is nevertheless surprising that the majority of older respondents had not given any thought to retirement – even when they were only five- to ten years away from it. Lack of planning is concentrated among specific groups of the population, such as those with low educational attainment, African-Americans, Hispanics and women. These are potentially vulnerable groups, who are less likely to save for retirement. These findings, which are not specific to a particular time period or a specific survey, were reported in various studies for the US (Lusardi and Mitchell, 2007a; Yakoboski and Dickemper, 1997) and the UK (Clery et al., 2007; Hedges, 1998).

Another way to examine whether and to what extent individuals prepare for retirement and plan for the future is to look at how much they know about crucial components of a saving plan, such as pension and Social Security wealth. In the United States, and in some European countries such as the Netherlands (Gustman and Steinmeier, 1999; Alessie et al., 1995), pension and Social Security savings account for about half of total wealth accumulation.

Earlier studies indicated that workers were woefully uninformed about their pensions and the characteristics of their pension plans (Mitchell, 1988; Gustman and Steinmeier, 1989). Given that most pensions in the past were DB pensions, and that workers had to make few or no decisions at all about their pension contributions, lack of knowledge is perhaps not surprising. However, recent data from the HRS show that workers continue to be uninformed about the rules and the benefits associated with their pensions (despite the shift from DB to DC pensions) and about the rules governing Social Security (Gustman and Steinmeier, 2004; Gustman et al., 2009).

Findings from the UK seem less worrisome, albeit not entirely comforting. Results from the English Longitudinal Study of Ageing (ELSA) show that 40% of individuals aged 50-59 with a DB employer pension do not know the accrual rate of their pension plan, 30% cannot tell how much they expect to receive from this pension, and 30% do not know whether their pension benefit will go up by more or less than prices after their retirement. These individuals, however, do not seem to perceive any major lack of information, as about 70% report having received enough information (Banks and Oldfield, 2006).

One reason individuals do not engage in planning or are not knowledgeable about pensions or the terms of their financial contracts is that they lack financial literacy. Bernheim (1995, 1998) was one of the first to emphasise that most individuals lack basic financial knowledge and numeracy. Several surveys covering the US population or specific subgroups have consistently documented very low levels of economic and financial literacy. Lusardi and Mitchell (2006) devised a special module on financial literacy for the 2004 HRS. Adding these types of questions to a large US survey is important – not only because it allows researchers to evaluate levels of financial knowledge but also and, most importantly, because it makes it possible to link financial literacy to a very rich set of information about household saving behaviour. The module measures basic financial knowledge related to the working of interest rates, the effects of inflation and the concept of risk diversification. Only 50% of respondents in the sample were able to correctly answer two simple questions about interest rates and inflation, and only one-third of respondents were able to answer correctly these two questions and another about risk diversification (Lusardi and Mitchell, 2006). Financial illiteracy is particularly acute among the elderly, African-American and Hispanics, women, and those with low education (Lusardi and Mitchell, 2007b).

Similar modules on financial literacy have been added to some European surveys, such as the Italian Survey of Household Income and Wealth (SHIW), the Dutch DNB Household Survey and the German SAVE. On average, 60% of Italian families correctly answered a basic question about inflation, and only 39% were able to cope with the compound-interest question (Monticone, 2010). Dutch and German households did somewhat better, as about 80% correctly answered basic questions about interest compounding and inflation (van Rooij et al., 2007; Bucher-Koenen and Lusardi, 2009).

Most importantly, lack of financial literacy has important consequences for wealth accumulation. Those who are not literate are less likely to plan and less likely to accumulate wealth (Lusardi and Mitchell, 2006, 2007a). Similarly, Stango and Zinman (2007) show that those who are not able to correctly calculate interest rates out of a stream of payments end up borrowing more and accumulating lower amounts of wealth. Moreover, those who are less literate are more likely to borrow using high-cost instruments and are more likely to have problems with debt (Lusardi and Tufano, 2008). Hilgert et al. (2003) also document a positive link between financial knowledge and financial behaviour. Van Rooij et al. (2007) and Kimball and Shumway (2006) find that financially sophisticated households are more likely to participate in the stock market. Agarwal et al. (2009) show that financial mistakes are most prevalent among the young and the elderly, who are also among those displaying the lowest amount of financial knowledge. Several programmes and tools have been developed to facilitate saving and to overcome the problems of lack of literacy and information and the complexity of saving decisions. Box 2.1 illustrates some of them.

Box 2.1:   Tools to make life easier
Acknowledging that saving for the long term is often problematic has led economists to devise ways to help individuals perform complex calculations or commit to saving plans. Below, we present some examples:
Planners. Devising optimal saving plans requires complex and lengthy computations. Several tools have been developed to make this task less cumbersome. Some software programmes combine advice on life-cycle planning and portfolio choice (Morningstar and Financial Engines). Some are very simple (Ballpark E$timate and Morningstar, which compute the target saving rate using as inputs only age, the amount of retirement savings, and annual income), while others are more detailed (Financial Engines). One notable example is ESPlanner – developed by Laurence Kotlikoff – which accounts for not only Social Security benefits and pension plans, but also savings accounts, housing and other real estate, and taxation.
Planning aids. Lusardi et al. (2009) devised a seven-step planning aid that describes to new hires in a large not-for-profit institution what they have to do to open a supplementary retirement account. In addition to breaking down the enrolment process into simple steps, the aid provides information about the pension scheme, such as the minimum and maximum amount that employees can contribute, the three pension providers employees have to choose from, and the rules of the on-line enrolment process. Consistent with the fact that many employees lack even basic information about pensions and often claim they do not know where to start when considering retirement saving decisions, this programme resulted in a sharp increase in supplementary retirement accounts.

Automatic enrolment and 'locking in'. One way to stimulate participation and contribution to pensions is to automatically enrol workers in employer-provided pension plans. Thus, rather than let workers choose whether or not to opt in, employers enrol workers and let them choose whether or not to opt out of a pension plan. This simple but ingenious method proved to be very effective in increasing pension participation. For example, according to Madrian and Shea (2001), after a company implemented a change in its 401(k) plan and automatically enrolled its new hires in the plan, pension participation went from 37% to 86%. Not only has the increase been very large, but participation rates have remained high for several years (Choi et al., 2004, 2006). Even legislators took notice of this remarkable success, and the 2006 Pension Protection Act made it much easier for firms to automatically enrol their workers in pension plans. Automatic enrolment with an opt-out option was adopted in the UK by the Pensions Act 2008 and in New Zealand (KiwiSaver schemes).

Save More Tomorrow. Similar to the automatic enrolment programme described earlier, in this scheme workers commit themselves to automatic increases of their pension fund contributions every time they obtain a pay rise (Thaler and Benartzi, 2004). As in automatic enrolment, the idea behind this mechanism is to overcome self-control problems and inertia faced by many workers. The increase in contribution is usually set to be slightly smaller than the increase in earnings, so that workers do not suffer from a reduction (in absolute terms) in their paychecks.

3   Remaining gaps in knowledge: main challenges

3.1   *Understanding saving behaviour*

Inadequate outcomes for whom? A few remarks can be made based on the survey of the literature carried out in section 2 on (in)adequate outcomes. The *first* has to do with the empirical evidence. According to most (US) studies, the issue of inadequate retirement saving does not appear to be as serious as one might expect, since the majority of households behave more or less according to theoretical prediction. Those who do not are not only *under-savers*, but also *over-savers*. Moreover, the under-savers – which Scholz et al. (2006) estimate (as of 1992) at about one-sixth of older Americans – are quite easily identified, being largely concentrated among the less educated and those at the bottom of the wealth or earnings distribution. This would suggest that *targeted actions* could be sufficient to tackle the problem.

The *second* remark has to do with methodology. Some optimisation models tend to underrate the great heterogeneity in saving behaviour, which could explain why they tend to yield fairly similar results. In this respect, the most ambitious attempt comes from the far-reaching methodology experimented with in Scholz et al. (2006), which combines the rigour of an optimisation framework with the full distribution of household saving behaviour. However, its implementation is fairly complex (and it requires strong assumptions about the functional forms and parameter values of preferences), and similar results have been obtained with much simpler techniques, such as the one used in Hurd and Rohwedder (2008).

A *third* observation has to do with the results, which seem to be characterised by a stark dichotomy: whereas models point to groups at risk of inadequate resources in retirement, the most recent literature on financial literacy suggests that ignorance is widespread, and that even college graduates often fail to answer basic questions correctly. Lack of knowledge could be overcome by resorting to financial advice external to the family, but the evidence on planning, again, shows that merely 30% of the US close-to-retirement population have thought about retirement, and that only 18% were able to develop a saving plan and stick to it (Lusardi and Mitchell, 2007a). Not surprisingly, lack of planning is present in the same vulnerable groups that display poor ability to save.

Adopting a more pragmatic research strategy. Rather puzzlingly, then, many households manage to plan adequately, even knowing very little about their own finances. Further evidence is therefore needed to reconcile these findings and to obtain a unified and consistent message from different strands of research. The basic life-cycle model has been successfully extended to account for various real-life features, and has contributed to a greater understanding of saving behaviour; progress has been made in both the modelling of intertemporal choices and the methodological strategy (thus, looking at the entire life cycle rather than just at the few years around retirement; using an optimisation criterion to establish adequacy targets; and simulating life-cycle patterns for each household, rather than looking at mean/median households). The model, however, still suffers from severe limitations (Stiglitz, 2008).

Perhaps the best strategy for understanding why (a fraction of) households appear to save 'inadequately' is to reckon that conventional models are only part of the story, and to continue the investigation of documented behavioural and psychological factors: in other words, household saving behaviour can be better understood if we acknowledge the diversity of decision-making mechanisms and commit to more deeply integrating into the current theoretical framework the suggestions coming from other disciplines (particularly, psychology). For instance, while conventional analysis interprets 'inadequacy' as the gap between actual accumulation and that predicted by theory, behavioural economists can help in understanding why this gap occurs. *'Economists must account for the unaccountable irrationality of real people'* (Pratt, 1995): overconfidence, lack of self-control, inertia, mental accounting, dynamically inconsistent time preferences, reliance on 'rules of thumb' and other so-called 'anomalies' can all rather arbitrarily enter an otherwise rational plan (Thaler, 1994; Laibson, 1998) and explain specific aspects of saving behaviour that are still poorly understood.

Moreover, as it has become clear that individuals have a hard time making long-term financial decisions, it is worth exploring the role of intelligent engineering of choice options. Psychological research could play a major role in this.

3.2   A 'smorgasbord' of open issues

i.   *Why are annuities so uncommon?* Annuitization should be a dominant choice, but is rarely *preferred*: why do the elderly tend to deplete their wealth at a slower pace than the theory predicts would be optimal, and why is the voluntary take-up of annuities so low, despite their hedging characteristics against longevity risk? One possible explanation is that annuities are too complex, entailing numerous margins of choice (when to start drawing down, how to draw down, what type of annuity) that individuals, particularly the elderly, do not seem to appreciate.

ii.   *How are health- and long-term-care (LTC) risks valued by the elderly?* Health- and LTC risks are paramount and burdened with uninsurable components. Important differences across countries in the public coverage of health- and LTC needs in old age determine the differences observed in the amount of precautionary savings accumulated to face unforeseen medical expenses, as well as in the importance of intra-family informal insurance arrangements (with consequences for participation of women in the labour market). The uncertainty related to the arrival of health shocks might be one of the reasons (in addition to altruism and a bequest motive) why individuals tend to prefer lump sums rather than annuities, and to run down their resources slowly.

iii.   *Is housing equity an obstacle to consumption smoothing?* According to the standard LCM, homeowners (particularly older ones) enjoy a positive wealth effect after an increase in the price of houses, and should thus increase their consumption of both housing services and other commodities by some fraction of the increase. This does not seem to happen in Europe (Calcagno et al., 2009), with the possible exception of the UK. Why?

iv. *What is the role of household composition in regard to saving?* As was anticipated in section 2.1, the analysis of consumption patterns often takes into
account family composition and its changes. More attention, however,
should be given to the relationship between demographics and the path
of wealth accumulation, not only by including an altruistic or strategic
bequest motive in the theoretical framework, but by truly integrating
saving- and fertility decisions.

v. *What is the link between micro- and macro dimensions?* As shown in section 2, savings adequacy has two dimensions: the ability of public social
protection systems to deliver adequate benefits (and services) in old age,
and the ability of an individual to save for his or her own future. These
two aspects should be further integrated, especially by looking at the link
between efficient risk diversification and individual optimisation.

4 Required research infrastructures, methodological innovations and consequences for research policy

In the last couple of decades, while remaining firmly in the wake of the life-cycle
theory, research on household savings has made substantial progress. Given
that, in the life-cycle framework, retirement represents the main rationale for
saving, one would expect that any important questions on the adequacy of
retirement savings have already been squarely posed and consensus answers
given. On the contrary, the issue has received scant attention in Europe, and
most of the empirical research on the topic refers to the US. Yet, Europe has a
number of universities with research expertise in the study of ageing, household saving and consumption choices, and pension economics. Moreover, in
addition to the already existing research centres (including CASE, CeRP, IFS,
MEA, NIESR and OEE), important European networks have recently been
established to study pension-related issues (Netspar) and to collect data on
European countries (SHARE; EU-SILC). Although this has induced more comparative studies across countries, more needs to be done in terms of research
infrastructure, data collection and methodological innovation, as these will be
crucial in addressing knowledge gaps.

These gaps relate specifically to the extent of possible saving shortfalls and
their distribution among the population – particularly as a consequence of the
phasing in of recent pension reforms. While lack of suitable data is likely to be
the single most important factor in the gap, this could also reflect, particularly
in Continental Europe, a lack of political/institutional concern, given the strong
paternalistic approach of traditional welfare states.

As for data, Campbell (2006) designs the ideal dataset for household financial analysis. Adapting Campbell's list to the study of household saving and adding some elements, the main characteristics of an ideal dataset for evidence-based policy research on the issue of retirement savings adequacy could be listed as follows:

- *Coverage*: the dataset should relate to a representative sample of the entire population, disaggregated by age and wealth.
- *Variables*: for each household, the dataset should measure consumption, income, wealth, transfers (from family and friends as well as from the government) and bequests. Consumption, income, and wealth measurement should be sufficiently disaggregated to distinguish among major categories. It should also include information regarding expectations, preferences (particularly concerning risk aversion) and the extent of financial literacy.
- *Quality*: the data should be reported with a high level of accuracy. This raises two issues: first, the achievement of a high level of accuracy concerning wealth data might entail some degree of over-sampling of the wealthier group – both because they hold most of the population wealth and in order to acquire sufficient information about the tails of the distribution; second, the lower the quality of the data collected, the heavier the imputation effort needed. Therefore, eliciting complete and reliable answers – especially as far as monetary values are concerned – is key. Enormous effort to improve data quality has been put into recent surveys (e.g., HRS and SHARE), for instance, by means of unfolding brackets (using wealth ranges when the exact wealth level is not provided). These methods have proved successful and could be adopted in all future data collection efforts.
- *Temporal dimension*: the dataset should contain at least a longitudinal component. Panel data have the advantage of allowing the researcher to control for unobserved heterogeneity (a household may save a lot because it likes saving). However, population composition changes over time, and re-interviewing the same people might reduce the 'representativeness' of the sample, which is an important element for policy research. This tension may be solved by having both a panel and a cross-sectional component.

Even though at the European level some surveys focusing on household income and wealth already exist (ECHP/EU-SILC; LIS/LWS; SHARE), the data infrastructure falls short at present, in a number of respects. SHARE covers only a limited set of European countries, and samples are rather small; since other datasets have either been terminated (ECHP) or do not constitute an adequate replacement (EU-SILC), investing in SHARE represents an important opportunity for progress in the study of retirement savings adequacy (among other topics) with a multidisciplinary approach. A recent initiative is underway to respond to the need for comparable European micro-data on household income and wealth. The Household Finance and Consumption Project, promoted by the European Central Bank (ECB), aims at creating a comprehensive survey on household finances and consumption for the Euro area.

As far as the methodological approach is concerned, more effort is indeed required to integrate other disciplines. Economic behavioural assumptions have long been challenged by psychologists but, until recently, to little effect. Things have begun to change, however: results have been obtained through joint efforts among scholars from different disciplines, also studying the neural circuits of the brain, and preliminary testable implications have resulted from more rigorous analyses. Psychological and sociological features (for example, the tendency of people to procrastinate; the status quo bias; the lack of financial literacy; the role of the family), however, have not yet been incorporated into rigorous theoretical models, with clear-cut empirical predictions.

More effort to address these issues could result in better exploitation of the many opportunities available to European research. Pension design and savings adequacy are timely topics in view of current market and demographic circumstances. Furthermore, Europe offers a unique 'laboratory' where different solutions and political styles are exposed to similar external shocks (such as economic- and demographic developments). At the same time, differences between European countries provide opportunities for learning from different strategies and circumstances. All this could also help to fill the gap between research and policy: the *liaison* is still unsatisfactory, and the results of research are regarded – more often than not – as 'irrelevant' or 'too academic' by decision-makers.

5   What (and when) can we deliver on policy questions?

To conclude the chapter, we return to the policy questions stated at the beginning, and provide some provisional answers.

i.   *What consequences will reforms have on the adequacy of retirement provisions?*
There is no necessary trade-off between sustainability and adequacy. Although mainly induced for sustainability reasons, European pension

reforms as such may not represent a direct threat to adequacy, as long as they address past distortions, and provided that measures are taken for enhancing individual saving choices. Reforms are reducing future replacement rates, thus making room for discretionary savings; they are also reducing annuitization, thus leaving individuals more exposed to longevity risks. On the positive side, however, they reduce distortions (particularly towards early retirement), introduce greater freedom of choice, and, by promoting a multiplicity of financial sources, move towards a better-diversified pension design.

ii.   *How will household discretionary savings respond to changes in pension provisions?* According to the theoretical predictions made by conventional models, workers should work longer and save more in response to a reduction in mandatory saving, thus 'spontaneously' filling the gap created by reforms. There is not enough evidence to support this conclusion, however, and we do not know whether the policy will have (negative) unintended consequences (for example, by increasing the number of poor or, conversely, by increasing the future fiscal cost of supporting elderly people). Reforms affect the younger cohorts more than the older ones, but the young also have to face more uncertainty and show greater flexibility in the labour market, and the joint effects of all these changes weaken simple tests of pension-reform effects on household savings. By transferring risks to the individuals, reforms are likely to affect the poor more than the rich, who are better placed to face risks; the homothetic preferences typically adopted by the LCM, however, neutralise the distinction. The same is true for the effects of the greater freedom of choice that will characterise future pensions: it is likely to increase the variability of responses and diminish the predictive power of the model.

iii.   *Are conventional models really able to capture individual behaviour?* The answer is mixed. The life-cycle model is rigorous, rich and flexible – and has proved itself capable of accommodating important facts of life, such as the relevance of precautionary saving not only in youth but also in old age. Its limitations, however, increasingly suggest alternative paths, such as recourse to experimental economics. Also because of these limitations, we cannot simply transfer to Europe the (mostly American) evidence suggesting that the large majority of households behave according to the model and that only a (limited) fraction of households fails to plan adequately for retirement. The existence of datasets specifically focused on European countries would help provide an answer to the question of the extent and distribution of a saving gap.

iv.   *What can policy do to improve retirement saving choices?* Lack of empirical knowledge creates uncertainty for policies directed at inducing workers to participate in (and adequately contribute to) complementary pension plans. Should these policies be general, or should they mainly be targeted to specific groups at risk? While the former suggestion seems supported by evidence of widespread financial illiteracy and inability to plan, the LCM would suggest more selective policies. Financial education programmes should be adopted, and carefully designed default options should always be present in complex choice situations, along with safety nets for the less fortunate. Further, policy measures should vary according to age, given the different phases of the life cycle faced by the young and the old. In building retirement savings, the young face liquidity constraints and, because of pension reforms, a transitory normative framework. Therefore they should be provided with the instruments to make the best possible saving decisions. First, they should be aware of the opportunities – as well as risks – offered by financial markets, and should receive the education and guidance to be able to make informed judgments. Second, to encourage longevity insurance, annuities should be as cheap, easy, and safe as possible – and the government could try to solve several of the problems that make individual annuities expensive, complex and risky (Munnell and Sundén, 2004). The elderly, on the other hand, face a different kind of liquidity constraint: they normally own the house in which they live; it is their main, but scarcely liquid, asset. Instruments to make housing equity more liquid should become more appealing, also through cost reduction. This is a matter for policy-making (particularly at a local level), as well as for markets.

In conclusion, more than turning down the life-cycle theory, the growing empirical documentation that people do not conform to the model could help transform it into a normative benchmark (Deaton, 2005): a theory providing not a description of how people behave, but of outcomes to which they aspire. The main role of policy would then be to help people, through education and incentives, to realise their ambitions – and not to substitute for them by means of more paternalistic programmes.

References

Agarwal, S., J.C. Driscoll, X. Gabaix and D. Laibson (2009) 'The Age of Reason: Financial Decisions over the Lifecycle', *Brookings Papers on Economic Activity* (2), 51-117.

Aguiar, M. and E. Hurst (2005) 'Consumption versus Expenditure', *Journal of Political Economy* 113(5), 919-948.

Alessie, R., A. Lusardi and A. Kapteyn (1995) 'Saving and Wealth Holdings of the Elderly', *Research in Economics* 49, September, 293-315.

Ball, L. and N.G. Mankiw (2007) 'Intergenerational Risk Sharing in the Spirit of Arrow, Debreu and Rawls, with Applications to Social Security Design', *Journal of Political Economy* 115(4), 523-547.

Banks, J., R. Blundell and S. Tanner (1998) 'Is There a Retirement-Savings Puzzle?' *American Economic Review* 88(4), 769-788.

Banks, J. and Z. Oldfield (2006) 'Understanding Pensions: Cognitive Function, Numerical Ability and Retirement Saving', IFS working paper 06/05.

Barr, N. and P. Diamond (2008) 'Reforming Pensions', Center for Retirement Research, Boston College working paper 2008-26.

Bernheim, D. (1995) 'Do Households Appreciate Their Financial Vulnerabilities? An Analysis of Actions, Perceptions and Public Policy', *Tax Policy and Economic Growth*, Washington, DC: American Council for Capital Formation, 1-30.

Bernheim, D. (1998) 'Financial Illiteracy, Education and Retirement Saving' in O.S. Mitchell and S. Schieber (eds), *Living with Defined Contribution Pensions* (Philadelphia: University of Pennsylvania Press), 38-68.

Bernheim, D. and J. K. Scholz (1992) 'Private Saving and Public Policy', NBER working paper 4215.

Bernheim, D., J. Skinner and S. Weinberg (2001) 'What Accounts for the Variation in Retirement Wealth Among US Households?', *American Economic Review* 91(4), 832-857.

Binswanger, J. and D. Schunk (2009) 'What is an Adequate Standard of Living during Retirement?' CentER discussion paper 2008-82, Tilburg University.

Börsch-Supan, A. and A. Brugiavini, (2001) 'Savings: The Policy Debate in Europe', *Oxford Review of Economic Policy* 17(1), 116-143.

Börsch-Supan, A. and A. Lusardi (2003) 'Saving: Cross-National Perspective' in A. Börsch-Supan (ed.), *Life-Cycle Savings and Public Policy: A Cross-National Study in Six Countries* (New York: Academic Press).

Brown, J. (2009) 'Understanding the Role of Annuities in Retirement Planning' in A. Lusardi (ed.), *Overcoming the Saving Slump: How to Increase the Effectiveness of Financial Education and Saving Programs* (Chicago: University of Chicago Press).

Brown, J. and J. Poterba (2000) 'Joint Life Annuities and Annuity Demand by Married Couples', *Journal of Risk and Insurance* 67, 527-554.

Bucher-Koenen, T. and A. Lusardi (2009). 'Financial Literacy and Retirement Planning in Germany', paper presented at the Netspar International Pension Workshop, Zurich, June 2010.

Calcagno, R., E. Fornero and M. Rossi (2009) 'The Effect of House Prices on Household Saving', *Journal of Real Estate Finance and Economics* 39 (3), 284-300.

Campbell, J. (2006) 'Household Finance', *Journal of Finance* 61(4), 1553-1604.

Castellino, O. and E. Fornero (2006) 'Private Pension in Europe and in the United States: A Comparison of Recent and Likely Future Developments', *ICFAI Journal of Employment Law* iv(4), 8-34.

Castellino, O. and E. Fornero (2007) 'Pension Reform in Europe. Some Observations on Franco Modigliani's Contribution' in A. Muralidhar and S. Allegrezza (eds) *Reforming European Pension Systems* (Amsterdam: Dutch University Press).

Choi, J., D. Laibson, B. Madrian and A. Metrick (2004) 'For Better or For Worse: Default Effects and 401(k) Savings Behavior' in D.A. Wise (ed.) *Perspectives in the Economics of Aging* (Chicago: University of Chicago Press).

Choi, J., D. Laibson, B. Madrian and A. Metrick (2006) 'Saving for Retirement on the Path of Least Resistance' in E. McCaffery and Joel Slemrod (eds), *Behavioral Public Finance: Toward a New Agenda* (New York: Russell Sage Foundation), 304-351.

Clery, E., S. McKay, M. Phillips and C. Robinson (2007) *Attitudes to Pensions: The 2006 Survey*, Department for Work and Pensions research report 434.

Davidoff, T., J. Brown and P. Diamond (2003) 'Annuities and Individual Welfare', MIT working paper 03-15.

Deaton, A. (2005) 'Franco Modigliani and the Life Cycle Theory of Consumption', Princeton University mimeo.

Dushi, I. and A. Webb (2004) 'Annuitization: Keeping Your Options Open', CRR working paper 2004-04.

Engen, E., W. Gale and C. Uccello (1999) 'The Adequacy of Household Saving', *Brookings Papers on Economic Activity* 2, 65-187.

Gomes, F. and A. Michaelides (2003) 'Aggregate Implications of Social Security Reform', prepared for the Fifth Annual Joint Conference of the Retirement Research Consortium 'Securing Retirement Income for Tomorrow's Retirees' May 2003, Washington, DC.

Gustman, A. and T. Steinmeier (1989) 'An Analysis of Pension Benefit Formulas, Pension Wealth and Incentives from Pensions', *Research In Labor Economics* 10, 53-106.

Gustman, A. and T. Steinmeier (1999) 'Effects of Pensions on Savings: Analysis with Data from the Health and Retirement Study', *Carnegie-Rochester Conference Series on Public Policy* 50, 271-324.

Gustman, A. and T. Steinmeier (2004) 'What People Don't Know about their Pensions and Social Security', in W. Gale, J. Shoven and M. Warshawsky (eds), *Private Pensions and Public Policies*, Washington, DC: Brookings Institution, 57-125.

Gustman, A., T. Steinmeier and N. Tabatabai (2009) 'Do Workers Know about Their Pensions? Comparing Workers' and Employers' Pension Information,' in A. Lusardi (ed.), *Overcoming the Saving Slump: How to Increase the Effectiveness of Financial Education and Saving Programs* (Chicago: University of Chicago Press).

Haider, S. and M. Stephens, (2007) 'Is There a Retirement-Consumption Puzzle? Evidence Using Subjective Retirement Expectations', *Review of Economics and Statistics* 89(2), 247-264.

Haveman, R., K. Holden, B. Wolfe and S.M. Sherlund (2006) 'Do Newly Retired Workers in the United States Have Sufficient Resources to Maintain Well-Being?', *Economic Inquiry* 44(2), 249-264.

Hedges, A. (1998) *Pensions and Retirement Planning*, Department of Social Security research report 83.

Hilgert, M., J. Hogarth and S. Beverly (2003) 'Household Financial Management: The Connection between Knowledge and Behavior', *Federal Reserve Bulletin*, 309-322.

Holzmann, R. and M. Pallarès-Miralles (2005) 'The Role, Limits of, and Alternatives to Financial Education in Support of Retirement Saving in the OECD, Eastern Europe and beyond', The World Bank, prepared for presentation at the Sixth Annual Conference 'Individual Behavior with Respect to Retirement Saving' 14 October 2005 at CeRP, Turin, Italy.

Hurd, M. and S. Rohwedder (2006) 'Some Answers to the Retirement-Consumption Puzzle' RAND working paper WR-342.

Hurd, M. and S. Rohwedder (2008) 'Adequacy of Resources in Retirement', prepared for the 10th Annual Joint Conference of the Retirement Research Consortium 'Creating a Secure Retirement System', August 7-8, 2008, Washington, DC.

Kifmann, M. (2008) 'The Design of Pension Pay-Out Options When the Health Status During Retirement is Uncertain', CESifo working paper 2211.

Kimball, M. and T. Shumway (2006) 'Investor Sophistication, and the Participation, Home Bias, Diversification and Employer Stock Puzzles', University of Michigan mimeo.

Kotlikoff, L., A. Spivak and L. Summers (1982) 'The Adequacy of Savings', *American Economic Review* 72(5), 1056-1069.

Laibson, D. (1998) 'Life-cycle Consumption and Hyperbolic Discount Functions', *European Economic Review* 42, 861-871.

Lindbeck, A. and M. Persson (2003) 'The Gains from Pension Reform', *Journal of Economic Literature* 41(1), 74-112.

Love, D., P. Smith and L. McNair (2008) 'A New Look at the Wealth Adequacy of Older US Households', Finance and Economics discussion series 2008-20, Federal Reserve Board Divisions of Research & Statistics and Monetary Affairs, Washington, DC.

Lusardi, A. (1999) 'Information, Expectations, and Savings for Retirement' in H. Aaron (ed.) *Behavioral Dimensions of Retirement Economics* (Washington, DC: Brookings Institution and Russell Sage Foundation), 81-115.

Lusardi, A. (2008) 'Household Saving Behavior: The Role of Financial Literacy, Information and Financial Education Programs', NBER working paper 1382.

Lusardi, A. and O.S. Mitchell (2006) 'Financial Literacy and Planning: Implications for Retirement Wellbeing', Pension Research Council, Wharton School, University of Pennsylvania working paper.

Lusardi, A. and O.S. Mitchell (2007a) 'Baby Boomer Retirement Security: The Role of Planning, Financial Literacy, and Housing Wealth', *Journal of Monetary Economics* 54, 205-224.

Lusardi, A. and O. Mitchell (2007b) 'Financial Literacy and Retirement Preparedness. Evidence and Implications for Financial Education,' *Business Economics*, 35-44.

Lusardi, A. and P. Tufano (2008) 'Debt Literacy, Financial Experiences and Overindebtedness', Dartmouth College working paper.

Lusardi, A., P.A. Keller and A.M. Keller (2009) 'New Ways to Make People Save: A Social Marketing Approach' in A. Lusardi (ed.) *Overcoming the Saving Slump: How to Increase the Effectiveness of Financial Education and Saving Programs* (Chicago: University of Chicago Press).

Madrian, B. and D. Shea (2001) 'The Power of Suggestion: Inertia in 401(k) Participation and Savings Behavior,' *Quarterly Journal of Economics* 116, 1149-1525.

Miniaci, R., C. Monfardini and G. Weber (2003) 'Is There a Retirement Consumption Puzzle in Italy?', IFS working paper 03/14.

Mitchell, O. (1988) 'Worker Knowledge of Pensions Provisions,' *Journal of Labor Economics* 6, 21-39.

Monticone, C. (2010). 'How Much Does Wealth Matter in the Acquisition of Financial Literacy?' *Journal of Consumer Affairs* 44(2) (forthcoming Summer 2010).

Moore, J. and O. Mitchell (1997) 'Projected Retirement Wealth and Savings Adequacy In the Health and Retirement Study', NBER working paper 6240.

Munnell, A. and A. Sunden (2004) *Coming Up Short: The Challenge of 401(k) Plans* (Washington, DC: Brookings Institution Press).

Munnell, A., A. Webb and L. Delorme (2006) 'A New National Retirement Risk Index,' An Issue in Brief, Center for Retirement Research at Boston College, June, 8.

OECD (2008) *Improving Financial Education and Awareness on Insurance and Private Pensions* (Paris: OECD).

Pratt, J. (1995) 'Forward' in L. Eeckhoudt and C. Gollier, *Risk: Evaluation, Management and Sharing* (New York: Harvester Wheatsheaf).

Scholz, J.K. and A. Seshadri (2007) 'Children and Household Wealth', mimeo.

Scholz, J.K. and A. Seshadri (2008) 'Are All Americans Saving 'Optimally' for Retirement?', prepared for the 10th Annual Joint Conference of the Retirement Research Consortium 'Creating a Secure Retirement System' August 7-8, 2008, Washington, DC.

Scholz, J.K., A. Seshadri and S. Khitatrakun, (2006) 'Are Americans Saving 'Optimally' for Retirement?', *Journal of Political Economy* 14(4), 607-643.

Shiller, R. (1998) 'Social Security and Institutions for Intergenerational, Intragenerational and International Risk Sharing', paper prepared for the Carnegie–Rochester Public Policy Conference, April 24-25, 1998.

Sinclair, S. and K. Smetters (2004) 'Health Shocks and the Demand for Annuities', Congressional Budget Office technical paper 2004-09.

Skinner, J. (2007), 'Are You Sure You Are Saving Enough for Retirement?' *Journal of Economic Perspectives* 21(3), 59-80.

Smith, S. (2006) 'The Retirement-consumption Puzzle and Involuntary Retirement: Evidence from the British Household Panel Survey', *Economic Journal* 116.

Stango, V. and J. Zinman (2007) 'Fuzzy Math and Red Ink: When the Opportunity Cost of Consumption is Not What It Seems', Dartmouth College working paper.

Stiglitz, J. (2008) 'Uncertainties in the Life Cycle and How They Should Be Addressed', *Trends in Savings and Wealth*, http://www.savingsandwealth.com/features/uncertainties_in_the_ life_cycle.jhtml

Thaler, R. (1994) 'Psychology and Savings Policies', *American Economic Review* 84(2), 186-192.

Thaler, R. and S. Benartzi (2004) 'Save More Tomorrow: Using Behavioral Economics to Increase Employee Savings', *Journal of Political Economy* 112(1), Part 2: S164-87.

Turra, C. and O. Mitchell (2004) 'The Impact of Health Status and Out-of-Pocket Medical Expenditures on Annuity Valuation', Pension Research Council working paper 2004-2, Philadelphia, PA.

van Rooij, M., A. Lusardi and R. Alessie (2007) 'Financial Literacy and Stock Market Participation', NBER working paper 13565.

Yaari, M. (1965) 'Uncertain Lifetime, Life Insurance, and the Theory of the Consumer', *Review of Economic Studies* 32(2), 137-150.

Yakoboski, P. and J. Dickemper (1997) 'Increased Saving but Little Planning: Results of the 1997 Retirement Confidence Survey', EBRI, Issue Brief 191, November.

Adequacy of Saving for Old Age

Comments by Arie Kapteyn / Maarten van Rooij

Are savings for old age adequate? Elsa Fornero, Annamaria Lusardi and Chiara Monticone provide an excellent overview and demonstrate unequivocally that what might look like an easy question is, in fact, a topic with many different facets. For one thing, there is the micro-macro dimension: thus, regardless of the question whether macro-savings are adequate or inadequate, individual households could either save too little or too much, which is very difficult to gauge given the vast – mostly unobserved – heterogeneity in individual expectations, preferences and intentions. In addition, there is a sometimes subtle interconnection between the institutional design of pension systems with a variety of mandatory elements and incentives and its effect on the accumulation of retirement savings, the date of retirement, payout decisions, and perhaps on the investment in human capital or the acquisition of financial knowledge. The contribution by Fornero, Lusardi and Monticone demonstrates that there is considerable disagreement in opinion – even on very basic questions such as whether households save enough for retirement and how serious the problem of under-saving is. This has important consequences for the trade-off between investments required and benefits expected of research efforts and policy initiatives directed at improving retirement saving decisions.

Although the elderly in Europe might seem better prepared for retirement than their counterparts in the US are, several European countries, unfortunately, have unsustainable old-age pension systems. Recent pension reforms in those countries are characterised by a reduction in future replacement rates and by a shifting of the responsibility for additional discretionary retirement savings to the individual. Households subsequently run higher risks of under-saving and under-annuitization, which increases their exposure to health- and income shocks, as well as to longevity risk. As we have learned from numer-

38

ous examples, individuals are often financially illiterate and have a hard time making long-term financial decisions. This suggests that unless we succeed in developing financial education programmes that effectively boost household financial skills, there is a pressing need for intelligent engineering of choice options. These also happen to be two of the most challenging research questions we face. Can educational initiatives be developed that will substantially improve the capability of citizens to take complex financial decisions, and how can we design an institutional set-up that balances freedom of choice with characteristics that offer help to those consumers who need it most? In addition, if these programmes prove less than sufficiently effective, what are the societal trade-offs – in terms of welfare consequences of restricting those who are quite capable of taking care of adequate retirement savings decisions themselves versus minimising the risk for households who are not able to do so of ending up in an unfortunate situation (which may bring about additional costs or undesirable side-effects for society as a whole)? While these decisions ultimately require a choice between the right of all citizens to make their own choices versus a more paternalistic approach in the interest of society as a whole, the research community may deliver information and tools to guide these decisions.

These basic research questions are the umbrella under which a large number of more specific issues, as identified by Fornero, Lusardi and Monticone, need to be investigated. Indeed, knowledge on age-varying needs has thus far been largely ignored, and definitely requires more research. Workers who nowadays retire at age 65 often have the prospect of living in good health for at least another decade – and still have the flexibility of using their earning capacity to address a mismatch between retirement income and desired expense patterns, if necessary. There are huge differences between these individuals and the older elderly, who face different needs and restrictions – especially if they are faced with deteriorating physical or mental health. In addition, more research effort could be directed at the different cohorts within the population, as this is important for the future development of savings adequacy. Each cohort has a different background, with different experiences and preferences, and is faced with a changing institutional environment – and hence will make different choices. In that sense, neither the situation nor the decision behaviour of current elderly generations who enjoy retirement (or are on the verge of retirement) need be representative of the behaviour of younger cohorts. However, to differentiate effectively between these different cohorts, there is a need for large datasets.

Another important phenomenon is indeed the lack of voluntary annuitization – and while it is 'solved' in many countries by making it mandatory, this issue is likely to become increasingly relevant with the diminishing relative importance of the mandatory part of pension savings. The question why people do not liquidate housing wealth is to some extent the flipside of the lack of annuitization. Other questions that perhaps remain somewhat underexposed, but certainly need to be addressed, are the following: what is the role of the financial services industry, the provision of information, the incentives to be transparent on risk and return of financial products, and the sensitivity of consumers for advertising and marketing? What is the role of regulation in helping consumers to make financial choices leading to adequate retirement savings? How do consumers deal with inflation risk? While a great deal of research focuses on the capability of consumers to assess the risks associated with different portfolio investment strategies in the accumulation phase of pension wealth, understanding the inflation risk is especially important for decisions that are related to the decumulation phase.

Much of what we have learned so far on consumer behaviour and financial decisions comes from the US. This is fine, since we can also learn from mistakes that others make as well as from the degree of success of policy initiatives in other countries. The emphasis on the US – even among many European researchers – is partly due to a longer history in creating and maintaining data collection efforts. However, as mentioned by the authors, a number of valuable European data initiatives have been taken, and effective and high-quality networks with an increasing interest in ageing and pensions have been initiated (SHARE, ELSA, European Wealth Data). However, the data infrastructure is yet incomplete. SHARE does not cover all EU countries, samples are relatively small and it is under-funded. Other data are also not adequate (thus, ECHP was terminated when it finally seemed to get somewhere, and EU-SILC does not seem to provide a reasonable replacement). More general data collected by official agencies are not always adequate for research purposes, or even for answering policy questions. There is a widespread lack of innovation (with exceptions such as ELSA), and legal obstacles do little to improve the situation. Overall, there is an insufficient connection in Europe between research and policy.

Nevertheless, Europe also provides unique opportunities for research. Europe is an ideal laboratory in which to study all aspects of ageing and then to export that knowledge to other countries (in Asia, for example). More than just studying, Europe can also experiment – in some regions of the EU, or at least can exploit natural experiments. Europe can attract more researchers from the US market and strengthen the interactions between academics with a different background (economics, sociology and psychology), policymakers, and financial industry experts. An example could be the creation of a European Behavioural Finance Forum comparable to its forerunner in the US (www.befi-forum.org/members/index.php).

All in all, there are ample opportunities to strengthen the European infrastructure and to foster research and discussion focused on the specific European context of an ageing society. The lack of central funding for data collection, however, endangers continuity. To be able to use Europe as a laboratory, data have to be collected across countries in a highly coordinated way; decentralised funding will undermine this. Data quality requires substantial investment, and standards suffer if there is an unwillingness to invest properly. Budgets are under pressure due to the financial crisis. Yet, costs of policy changes based on an inadequate understanding of economic behaviour are likely to be much higher.

3

Risk and Portfolio Choices in Individual and Collective Pension Plans

Christian Gollier

1 Key policy questions

The current financial crisis is expected to have a considerable impact on the pension benefits to be paid in the future in most countries in which funded schemes play an important role in the social security system. Typically, pension funds invest more than 50% of their reserves in equity – with the remaining being invested in bonds, bills, real estate or private equity. Because of the crash in equity markets, and the bad health of other asset markets, Argentina had to nationalise its insolvent pension funds, whereas many defined-benefit (DB) pension funds in both the US and the UK are expected to rely on national solidarity to be able to pay the promised pension benefits to their members. At the same time, many workers with an individual pension plan (such as the 401k in the US) are considering the possibility of compensating their own huge losses of pension wealth by increasing their contributions, or by delaying their retirement. Does this mean that funded pension systems and individual workers around the world have relied too much on risky assets to improve their expected pension benefits? More generally, what are the determinants of the optimal allocation of pension wealth, for individual and collective plans? Given the difficulty of most households to plan for the long term, what default options should defined-contribution (DC) retirement plans propose to their members (Cui, 2008)?

Because of the exponential nature of compounded interest, asset allocation plays a crucial role in the expected pension wealth at retirement age. As an illustration, consider the case of US financial markets, in which the mean real rates of return of bonds and stocks have been respectively equal to 1.5% and 6.9% over the 20th century.[1] Consider a funded DC pension plan with a yearly real contribution of €1000 over 40 years. If the reserves of the plan are fully invested in bonds, the members of the plan should then end up with pension wealth equaling €54,268 on average. This mean pension wealth rises to €194,562 if the fund would have been invested in stocks. The investment duration of 40 years magnifies the well-known 'equity premium puzzle' (Mehra and Prescott, 1985), which recognises the triumph of the risk-takers over the last one hundred years. Markets have been compensating for risk on such a large scale that it is hard to explain why most individual investors invested mostly in risk-free assets. We believe that the recent drop in equity return in 2008 is not likely to change the conclusion from this long-term perspective – but this remains to be formally proved.

In addition to this question associated with the overall lifetime exposure to portfolio risk, this chapter also addresses the following policy questions:

Should investors implement market-timing strategies?

Collectively funded pension schemes usually implement a stable asset alloca-tion, whereas individual investors often rebalance their portfolios frequently based on their fluctuating beliefs about the market conditions. These fluctuating beliefs may be irrational (thereby potentially explaining bubbles on financial markets), or they may have a more scientific basis. This discrepancy comes from a different perception of the world – in which most individual investors believe that future asset returns are partially predictable (from estimated predictors such as the price-earning ratio or the past equity return), whereas most insti-tutional investors believe that asset returns are unpredictable. For example, it is suggested that the current historically low price-earnings ratio – a situation that is often claimed to be usually followed by historically larger excess stock returns – provides a motive to rebalance the portfolio of pension funds towards equity. Is this right? More generally, can investors beat the market by using predictors to 'time' the markets? If this is the case, how can we help individual investors to time the market more effectively?

This is an especially important policy question these days, as investors are often encouraged to behave pro-cyclically (that is, to sell assets under bear-market conditions, and to buy them when markets are bullish). If we believe that stock markets mean-revert, then long-term investors (such as pension funds

---

1   See Dimson, Marsh and Staunton (2002).

and younger workers with an individual pension account) should adopt instead counter-cyclical investment behaviour, thereby stabilising markets.

Should individuals reduce portfolio risk at older age?

It is often recommended that pension portfolios be gradually rebalanced from stocks to bonds when the investment horizon shrinks. This recommendation is often transformed into legal rules. For example, in the French *Plan d'Epargne Retraite Populaire (PERP)*, created in 2005, individual pension accounts must be fully invested in bonds and bills when their owners are older than 60 years of age. The idea is to 'lock in' past returns and to protect old employees against a last-minute crash on financial markets. Young employees, in contrast, should invest most of their pension wealth in equity, in order to take advantage of the large equity premium. Common wisdom suggests that young investors can recoup their transient losses by future gains. Are these financial guidelines and formal portfolio constraints socially efficient? If yes, what is the optimal age profile of the optimal asset allocation?

Should collective pension schemes be promoted in preference to individual ones?

Countries with funded pension systems have organised different degrees of solidarity among generations of households. In an individual DC system, each employee saves for him- or herself, and bears the entire portfolio risk at retirement age. In a collective DC system, each employee contributes to the accumulation of a global fund, which is redistributed to retirees. Contributions and benefits are contingent on the financial health of the fund, which allows for intergenerational risk sharing. In countries without a funded system (like France, Spain or Italy), life insurance companies have played a similar role, but under more stringent solvency regulations. What are the costs and benefits of a more collective system? How should one shape the rules of the collective system in terms of fixing contributions and accruing benefits to each generation? These rules, which should be made transparent, determine the degree of intergenerational risk sharing of the scheme. Given the observed heterogeneity of risk attitudes in the population, how should one shape the menus of risk-sharing contracts within each cohort of employees? For example, should one provide full portfolio insurance for the most fragile segment of the population?

The long-term nature of the liabilities of collective funds that is generated by this intergenerational solidarity makes them typically long-term institutional investors. How does it affect the structure of their portfolios? Because collective schemes allow for a better time diversification of financial risks, should this induce them to take more portfolio risk?

If independent collective funds compete to attract the savings of individuals – as is the case in many countries with a funded system or with life insurance – how should the capital requirements of these funds be regulated in order to preserve their solvency? If asset returns are predictable with mean-reversion, for example, should the capital adequacy rules be adjusted according to the position of the economy in the financial cycle?[2] What is the optimal level of protection of contributors against insolvent funds? Should there be an insurer of last resort?

How is the optimal asset allocation affected by various institutional contexts, and by other economic reforms?

The optimal portfolio allocation of a funded pension system cannot be determined in a vacuum. It is a function of other economic institutions that characterise the country, and it will be affected by other reforms in the economy. Let us consider various illustrations of this fact, and the associated policy questions:

- *Flexibility of the labour market*: Should an employee take more portfolio risk if he or she can compensate capital losses by working longer or harder after an adverse financial shock? It is often assumed that employees incurring big losses on their pension wealth will compensate for these losses by retiring later. Does flexible labour supply justify taking more portfolio risk along the employee's career? In countries where the retirement age is compulsory, in contrast, should more conservative investment strategies be recommended for individual pension accounts? Is there a symmetric answer for collective schemes?

- *Flexible contributions and benefits*: Most pension schemes impose constraints on the level of contributions and benefits – such as minimum contribution level, minimum guaranteed benefits and prohibition of early liquidation of the individual account. How do these constraints affect the optimal portfolio allocation?

- *Liquidity of other assets*: The risk tolerance of households is enhanced by the ability of these households to adapt their lifestyle to their actual wealth (including pension wealth). Introducing more liquidity on the real estate

---

2   In the case of life insurance, this question refers to whether a 'dampener' should be implemented to smooth shocks on stock returns, as currently discussed in the Solvency 2 reform.

markets will induce households to move to a smaller house after incurring losses on their financial wealth. In contrast, if their real estate assets are illiquid, they will have a harder time compensating losses efficiently. Should pension funds be advised to take more portfolio risk in countries with a more flexible real estate market?

- *Insurability of longevity risk*: One of the most important uncertainties in life is longevity risk. In Pay-As-You-Go (PAYG) systems, this longevity risk is covered by the annuitization of benefits. In some funded systems, the longevity risk is efficiently shared through compulsory annuitization of the pension wealth. However, some funded systems (such as the 401k programme in the US) allow employees to exit without annuitising the pension wealth. Given the important inefficiencies on annuity markets due to asymmetric information, how do these exit options affect portfolio management during the lifetime?

What is the relevant rate of return of the funded system compared to the return of the PAYG system in different countries?

The Pay-As-You-Go system has a return approximately equal to the growth rate of the economy. The relative performance of a more funded system depends upon the investment strategy of the fund. Proponents of funded systems typically compare the relatively low financial performance of the PAYG system to the high expected return on the equity market. This comparison is misleading, for at least three reasons. First, equity is risky, which implies that account should be taken of the cost of risk when estimating the welfare surplus of the reform. Second, a 100% investment in equity is usually not optimal along the contribution period. Policymakers considering a switch from one system to the other must determine the optimal dynamic investment strategy of the fund and estimate the welfare impact of the riskier options. Finally, this reasoning does not take into account transitional issues, in which the newly created funded system will have to service the PAYG debt.

2   Major progress in understanding key economic concepts

2.1   Historical benchmark results

In a static framework, the portfolio choice problem involves finding a best compromise between risk and performance. It is classically formalised by introducing a concave utility function on final consumption. Under the standard axioms on decision-making under uncertainty (von Neumann-Morgenstern, 1944), households will select the portfolio that maximises the expected utility of their final consumption. The degree of concavity of the utility function, as defined by Arrow (1971) and Pratt (1965), characterises their degree of absolute risk aversion. This can be measured in various ways: revealed preferences, questionnaires on risk choices,[3] or tests in experimental labs. This information, combined with the specific distribution of returns to financial assets, makes it possible to compute the optimal portfolio. An increase in absolute risk aversion reduces the demand for risky assets. In the same spirit, an increase in wealth boosts the demand for risky assets if absolute risk aversion is decreasing. This mechanism is well understood and is cited in most textbooks in market finance.

Important progress was made in the late 1960s by introducing time into portfolio strategies. Usually, households have long-term objectives – such as financing consumption after retirement – when they invest in financial markets. How does the length of the time horizon affect the optimal structure of the portfolio? In other words, how does the option to purchase stocks tomorrow modify the attitude towards portfolio risk today? The pioneering papers of Mossin (1968), Merton (1969) and Samuelson (1969) answered this question in a straightforward but intriguing way: the intrinsically dynamic structure of portfolio management has no effect on the solution. Under their assumptions, the sequence of portfolio structures that is statically optimal is also dynamically optimal. Expressed differently, households should act as if each period of investment were the last before retirement. Myopia is optimal. But it is often forgotten that this result holds only under very restrictive conditions on the utility function. That is, myopia is optimal only if the utility function exhibits constant relative risk aversion, if the asset returns are serially uncorrelated, if background risk is absent (Guiso, Jappelli and Terlizzese, 1996), and if investors cannot compensate their transient capital losses by saving more. The role of flexibility in future decision-making has been shown to be crucial in the optimal age-profile of the asset allocation (Bodie, Merton and Samuelson, 1992; Gollier, 2002a, 2002b; Gomes, Kotlikoff and Viceira, 2008). The role of rigidities in the real estate market has also been examined.[4]

2.2   *Time diversification*

In the benchmark model presented above, the only decision to be made at each instant in time involves the asset allocation. In the real world, households also

---

3   See, for example, Barsky, Juster, Kimball and Shapiro (1997).

4   Cocco (2005) and de Jong, Driessen and van Hemert (2007) are good examples.

control their labour supply or their saving effort. Wealth is a multi-task buffer stock. When households have been lucky on their previous investments, they can reduce their saving rate and cash-in the benefits of the larger-than-expected portfolio returns immediately. They could also increase their saving rate after an adverse shock on portfolio returns. This means that households do not bear the accumulated lifetime portfolio risk at the time of retirement. Rather, they can spread this risk over their lifetime consumption pattern. This flexible approach to risk management allows for some time diversification through consumption smoothing. Gollier (2001) showed that this time diversification effect has a deep impact on the relationship between the investor's time horizon and the share of wealth invested in risky assets. In a fully flexible framework in which contributions and benefits to the pension plan can be adapted in a continuous way to the financial situation, the share of wealth that is invested in stocks should be proportional to the remaining investment horizon.

However, time diversification is limited by various constraints, such as the no-borrowing constraint. If households are unable to use their future labour incomes – or their pension wealth – as collateral for borrowing in bad times, this will preclude the transfer of capital risks over time through consumption smoothing. It will induce more conservative portfolio management for those households that are most likely to face a binding liquidity constraint. Since Deaton (1991), who first recognised liquid wealth as a buffer stock, an important literature has emerged that examines the optimal wealth accumulation and portfolio decisions over the household's lifetime.[5]

More recently, different authors have applied the same ideas to collective pension plans. When participation is compulsory, future contributions can be securitised, thereby improving the intergenerational risk-sharing efficiency of the system. This implies allowing plans to smooth financial shocks across generations, or enlarging the investment horizons of plans beyond each generation's lifetime. Poterba, Rauh, Venti and Wise (2005), Cui, de Jong and Ponds (2006), Teulings and de Vries (2006), Krueger and Kubler (2006) and Gollier (2008) estimated the welfare surplus generated by this better allocation of risk and its impact on the optimal asset demand. However, many constraints limit our ability to extract this social surplus. For example, the flexibility of pension benefits is limited by various guarantees offered by the plans, such as a minimum return or automatic indexation. Moreover, solvency constraints imposed on pension plans limit their ability to transfer adverse financial shocks across

---

5   See, for example, Carrol (1992, 1994, 1997), Gourinchas and Parker (2002), Cocco, Gomes and Maenhout (2005), Gomes and Michaelides (2005) and Agarwal, Driscoll, Gabaix and Laibson (2007).

generations. The plans are thus compelled to adopt more conservative investment strategies, thereby limiting the welfare surplus of the scheme.

2.3   Predictability of asset returns

The accepted wisdom in the modern theory of finance has long been that asset returns are unpredictable. But the most striking recent finding of financial econometrics is that the investment opportunity set available to investors is not constant over time. Moreover, it fluctuates stochastically, and these fluctuations are serially correlated with economic and financial variables that are observable (that is, they are predictable). For example, there is evidence of mean-reversion in stock returns, as first shown by Poterba and Summers (1988) for the US markets. This implies that the future equity premium is negatively correlated to the past returns of equity. The negative correlation of stock returns implies, then, that stocks are relatively safer for long-term investors than for short-term ones. In the US, the implied standard deviation of ten-year returns is 23.7% – much smaller than the 45.2% value implied by the standard deviation of monthly returns. Similarly, Fama and French (1988) showed that the price-earnings ratio is negatively correlated with future equity returns. Moreover, French, Schwert and Stambaugh (1987) have shown that the volatility of stock returns is predictable, because it is negatively correlated with past stock returns.

The existence of these various forms of predictability in asset returns has generated a revival of interest in Merton's notion of intertemporal hedging demand (see Merton, 1969, 1971; Mossin, 1968). Campbell and Viceira (2002) provide an overview of this recent literature, and they quantify the importance of the impact of predictability on the optimal portfolio structure. The predictability of future asset returns has two important consequences for the optimal dynamic portfolio strategy. First, a change in the instantaneous interest rate, in the equity premium or in the volatility of asset returns modifies the optimal structure of the portfolio. For example, an increase in the volatility of the stock returns should typically reduce the demand for stocks by all risk-averse investors. This 'market timing' of the optimal portfolio would be the typical reaction of myopic investors. But, when determining their optimal portfolio today, rational investors should also take account of the change in the set of future investment opportunities. This yields a second term to the asset demand, which is usually referred to as the 'hedging demand'. Quantifying the hedging demand for bonds and equity is a numerical challenge.

The easy – but unrealistic – case is when the utility function is logarithmic, which implies that the hedging demand for the risky asset is zero. Myopia remains optimal in this case, in spite of the predictability of asset returns. On the contrary, when relative risk aversion is larger than unity,[6] one obtains the intuitive sign for the hedging demand. For example, Brennan, Schwartz and Lagnado (1997) showed that these preferences imply that the rational investor should invest more in long bonds than the myopic agent would, in order to take account of the negative correlation between the instantaneous (risky) return of this asset and the future interest rate. Under the same assumption on the investor's preferences, Kim and Omberg (1996) showed that the existence of mean-reversion in stock returns yields a positive hedging demand for stocks. These two studies illustrate the idea that the risky asset is safer in the long run, due to the negative serial correlation of returns.

These developments in academia have had little influence on the practice of financial advisors and of individual- and institutional investors. The main reason for this is that dynamic portfolio optimisation is technically very complex, once predictability is added to the model. Analytical solutions do not exist, except in very special cases. Moreover, numerical methods are still unreliable when the underlying model is aimed at describing a realistic environment (as when a vector autoregression [VAR] for asset returns is considered). Campbell and Viceira (1999, 2002), Campbell, Chan and Viceira (2003) and Diris, Palm and Schotman (2008) have recently developed an approximate solution method when the dynamics of a set of asset returns and predictive variables is described by a vector autoregression.

3   Gaps in the literature

3.1   *What are the practical consequences of the predictability of asset returns for European long-term investors?*

The existing literature on the predictability of asset returns provides scant information on how pension funds should be strategically managed in Europe. There are two reasons for that. First, the optimal rebalancing of strategic portfolios is seldom quantified explicitly in the published papers. When authors have been able to solve the complex dynamic optimisation problem, one may believe that much of these results are protected by intellectual property rights. It implies that most practical questions about the impact of predictability on

---

6   Gollier (2007) provides intuition for why the sign of the hedging demand is reversed when relative risk aversion is less than unity.

the asset allocation remain unanswered. What is the intensity of the portfolio rebalancing that occurred after the autumn 2008 crash? What is the relationship between the time horizon of the investor and the share of his or her pension wealth invested in stocks, bonds and bills? What is the Value-at-Risk (VAR) of stocks and bonds along the financial cycle (that is, as a function of the level of the predictive variables)? These questions must be answered in order to be able to provide operational rules to manage pension funds efficiently.

But another challenging problem limits the value of the existing literature. Most, if not all, published papers on predictability rely on US data. To the best of our knowledge, no realistic test of predictability has been performed based on European financial markets data.[7] Even European teams that have worked in this field used US data. The reason for that is well-known: Relevant European financial datasets with long historical series – if they exist – are scarce, unreliable, and offer only a limited degree of comparability across different countries. They usually start in the early 1970s, which limits the quarterly data to no more than 150 observations – which is quite limited when testing VAR models with six or seven variables and one or two lags. The bottom line is that we don't really know whether the existence of predictability is a purely American-specific phenomenon, or whether European markets share this statistical trait with the US.

Another problem that needs to be fixed is related to the statistical significance of the predictability effects. In an oft-cited paper by Lettau and Ludvigson (2001)[8], '*it is now widely accepted that excess returns are predictable by variables such as dividend-price ratios, earnings-price ratios, dividend-earnings ratios, and an assortment of other financial indicators*'. But in a recent paper, Welch and Goyal (2008) claim that no predictor variable can forecast future excess stock returns better than the historical average excess stock return itself! They claim that while predictive regressions performed well 'in-sample', they usually performed poorly 'out-of-sample'. In other words, the coefficients of the predictive equation seem to be unstable. This was particularly the case when the dividend-price ratio was used; the ratio was very low in the second half of 1990s, therefore predicting extraordinarily low returns – a prediction that had long been contradicted by the facts (until the internet crash of 2000). Thus, the existence of predictability on financial markets is a contestable recent finding. It is thus crucial to investigate the ambiguity of the intensity of predictability on the optimal portfolio strategy, as has been initiated by Diris, Palm and Schotman (2008).

---

7   An exception is Bec and Gollier (2008), who test a VAR model for stocks, bonds and bills using quarterly French data.

8   Cited by Welch and Goyal (2007).

The role of uncertainty and predictability of investment in real estate, which usually represents a large share of the portfolio of households, should also be addressed. The lack of data on both sides of the Atlantic has clearly been a factor behind the limited progress in our understanding of this crucial aspect of household investment behaviour.

3.2 The role of model uncertainty on portfolio choices

The estimation of the distribution of various assets' excess returns is subject to imprecision, and the parameters are unstable. Traditional decision models take no account of this model uncertainty. More generally, most of the existing literature is based on the assumption that the stochastic process describing the future evolution of asset prices is unambiguous. In the real world, it is difficult to assess the precise probability distribution to describe the uncertainty faced by investors. The problem of imprecise probability is particularly crucial for very unlikely events (for which the speed at which one can learn from the empirical frequency is low).

The ambiguous probability distribution of asset returns raises two questions that are relevant for both researchers and portfolio managers. First, one should anticipate that households will update their beliefs in the future. Should investors wait to get this better information before taking more portfolio risk? It is important to better understand the relationship between risk-taking, learning and the investor's time horizon.[9] Second, one must recognise that the ambiguity affecting the probability distribution of returns modifies the investor's attitude towards uncertainty. Most experimental tests indicate that betting on an event that has an objective probability of p is not the same thing as betting on an event with the same pay-off, but with an ambiguous probability whose expectation is p. After Ellsberg (1961), many experimental studies have confirmed this characteristic of human preferences.[10] We say that one is ambiguity-averse if one prefers betting on the unambiguous event instead of on the ambiguous one. One important difficulty is that while the existing literature tests whether people are ambiguity-averse, we know very little about their degree of ambiguity aversion. Various approaches based on surveys, experiments and neuro-economics could be used to fill this gap.

Future research should explore the consequences of ambiguity aversion on the optimal dynamic portfolio strategy. The general idea is that ambiguity aversion reinforces risk aversion to make people more reluctant to undergo

9 A first step in that direction is provided by Diris, Palm, and Schotman (2008).
10 See for example Ahn, Choï, Gale, and Kariv (2007) and Camerer, Bhatt, and Hsu (2007).

ambiguous risky investments. But we still have limited knowledge about the role of ambiguity aversion on the optimal market timing, or on the age profile of the investors' optimal asset allocation.[11]

3.3   Fat tails, extreme events, and insurer of last resort

Robert Barro (2006, 2009) recently claimed that the existence of 'fat tails' (that is, small probability catastrophic events) plays a key role in risk evaluation, and that the often-assumed (log-) normality of the probability distribution is misleading. Martin Weitzman (2007) shows that fat tails of the subjective beliefs occur, in particular, in a Bayesian framework with an uncertain variance. Following this new trend of the literature, we should examine the optimal dynamic risk management of a fat-tailed risk, when beliefs about the probability of disasters are updated through time by observing the frequency of disasters.

The existence of extreme financial events (such as those we have been facing since the summer of 2007) is also likely to change our view about how pension funds should be managed. One should in particular address the complex question of the role of an insurer of last resort for highly distressed funds hit by a liquidity/solvency crisis. How should we shape the principles governing such institutions? What will their impact be on intergenerational welfare and on the funds' portfolio strategy?

3.4   The intergenerational commitment problem, and its impact on the portfolio of collective funds

Intergenerational risk sharing is possible in a collectively funded pension system only if one can commit future generations to participation in the pension scheme. This may be problematic after a financial crash, when the financial reserve of the pension fund needs to be replenished with the contributions of the new generations. These households may prefer to reform the system in favour of individual accounts, thereby raising doubt about our ability to share risk efficiently across generations. Our modeling of collective pension systems must recognise that it is hard to believe that one can commit future generations to participate. One should characterise the impact of the absence of commitment on the degree of risk aversion of the collective fund. The absence of commitment is expected to go against the natural tendency to smooth benefits and contribu-

---

11   See Gollier (2007b) and Chen, Ju and Miao (2009) for a first analysis of the effect of ambiguity aversion on the optimal asset allocation.

tions over time. This will, in turn, affect the marginal value of wealth, and thus the societal tolerance to risk. The general methodology to solve this problem is to impose a participation constraint in the fund's optimisation programme (as was done, for example, by Kocherlakota [1996] in another context).

3.5   *Taking account of household psychology*

The normative theory of strategic portfolio management is based on standard assumptions: expected utility (possibly generalised to ambiguity aversion), rational expectations, and exponential discounting of future felicity. Although this theory has a strong normative basis (the independence axiom, for example), it has relatively weak positive power to predict real behaviour. Recent developments at the frontier between psychology and economics have revealed that people often evaluate, take and judge decisions under uncertainty in a different way than predicted by the standard theory. New models need to be developed and tested in order to enhance the understanding of individual behaviour in the face of long-term risks. The role of the different aspects of prospect theory (Kahneman and Tversky, 1979; Tversky and Kahneman, 1992), such as loss aversion (von Gaudecker, van Soest and Wengström, 2008) or the frequency of resetting the reference point in the determination of the optimal portfolio allocation, needs to be explored in greater detail.

An important aspect of the political economy of pension fund management is related to the way in which people evaluate and judge collective decisions after new information is obtained about the risk. From a normative point of view, information that is not available at the time of the decision should be irrelevant in evaluating the optimality of the initial decision in the face of uncertainty. In reality, people have some difficulty in behaving in this way. They may, for example, feel regret (Bell, 1982). Regret is a psychological reaction to making a wrong decision, where the quality of decision is predicated on the basis of actual outcomes rather than on the information available at the time of the decision. In the context of pension fund management, people may feel two different types of regret. The first type is felt *ex post*, when investing in bonds before a bull stock market. The second is felt when investing in stocks before a stock market crash. How do investors integrate this risk of regret when making their portfolio allocation? How does the aversion to regret affect the optimal portfolio risk of individual and collective pension systems? How do public decision-makers respond to the expectation that they will be judged by regret-sensitive agents?

3.6   The interaction between flexibility and optimal financial risk exposure

The previous section explained that the flexibility of contributions and benefits plays a crucial role in the individual and collective attitude towards portfolio risk. More generally, does flexibility always enhance risk tolerance? This seems to be confirmed by Bodie, Merton and Samuelson (1992), who showed that the flexibility of the household's labour supply raises the household's tolerance to risk on pension wealth, when the utility function is Cobb-Douglas. However, Machina (1982) used an exponential quadratic utility function to show numerically that it is not true in general that flexibility induces more risk-prone behaviour. It may thus be possible that the loss of flexibility in housing- and labour decisions raises the demand for risky assets. This ambiguous theoretical result implies that more effort must now be made to examine realistic calibrations of portfolio choices in order to predict the effect of a more flexible environment in relation to intra-household solidarity, housing, human capital, or other financial assets.

There are many reasons why the level of workers' contributions and pension benefits needs to be re-evaluated periodically. In order to make the scheme sustainable, shocks on demographic variables, on unemployment, on growth or on the financial performance of the pension fund need to be compensated by changes in the flows of revenues and expenses of the fund. Following Gabaix and Laibson (2002), an examination should be made of the effect of modifying the frequency at which the pension contributions are re-evaluated in the pension scheme, given the new information that became available since the previous reset date. The system is more flexible if the frequency of the reset dates is higher.

3.7   What is known about the intensity of intergenerational solidarity within collective pension systems and life insurance?

The main social role of life insurers and collective pension funds is to organise the sharing of risk across different generations of long-term saving contracts. The latter are assumed to offer a better return than the market for unlucky generations that faced low market returns during their lifetime, whereas lucky generations get a return on their saving that is smaller than what financial markets delivered during their own lifetime. As explained earlier, this social contract is constrained by the opportunistic behaviour of future generations, and by regulatory constraints that prevent the occurrence of such an event.

We propose collecting data to estimate the extent to which these institutions have been able to share risk efficiently over time. This would require collecting information about the flow of contributions and benefits that different generations of households have faced over their lifetime, about the asset allocation of the collective saving institutions, and about the return of these assets. The main objective of this international project is to compare different national institutional frameworks for organising intergenerational risk sharing and to come up with policy recommendations for reforming the inefficient schemes.

4　Current situation of European research infrastructures

Because national contexts of the long-term saving industry are very heterogeneous in the European Union, the emergence of an international club of European researchers specialised in the management of financial risks of pension funds has been relatively slow. This is in striking contrast with what has occurred in the US, where several preeminent researchers emerged as soon as the country switched to a more funded pension system. One could name Peter Diamond, Jim Poterba, Larry Summers, or Larry Kotlikoff, for example, who animated a healthy scientific debate on social security reforms in the country. They put forward their research agenda on this topic in the most prestigious American universities, such as Harvard and MIT. They have also actively participated in the public debate. It is fair to say that their scientific research shaped most of the social security reforms in their country.

It is far from my intention to suggest that European research institutions do not have the human capital to play a similar role on this side of the Atlantic – quite the contrary, in fact. Excellent economists scattered across the continent have been working on and publishing on these important challenges, in prominent European universities such as Tilburg, Amsterdam, Mannheim (MEA), Frankfurt, University College London (IFS), London Business School, Stockholm, Florence, Naples, Torino, Paris and Toulouse.

Two recent steps have improved the overall picture of research in pension economics. First, various European networks have emerged to structure the research field. This was initiated in the 1990s by the European Training Network, and followed more recently by the cross-national initiative of the Survey of Health Aging and Retirement in Europe (SHARE). The second instrumental step was an increase in funding of academic researchers in the field. The most obvious example is the Netspar programme in the Netherlands. Other national initiatives have been undertaken that are focused more on financial markets than on pension finance. One can mention the Lausanne-based Swiss Finance Institute, or the Paris-based Europlace Institute of Finance as illustrations of this move.

In the spring of 2008, the European Savings Institute (OEE) published a call of tenders with five topics. One of these concerned the 'Estimation of Assets Returns Predictability in Europe and its Implications on Long-term Individual Assets Management'. Five European teams submitted a project, but only one could be financed. These projects were submitted, respectively, by Michael Schröder (ZEW, Mannheim), Mikael Petitjean (CeSAM, UCL), Carlo A. Favero (IGIER/Bocconi), Michael Rockinger (SFI, Lausanne) and Patrice Fontaine (Grenoble). This suggests that a lively research network could be structured around this crucial challenge.

5   Recommendation for a more innovative research programme in the field

Some remaining gaps and problems in knowledge were identified in section 3, the most important of which include the following: the absence of consensus about the existence and importance of the predictability of asset returns, the existence of ambiguity regarding the distribution of returns, and the modeling of fat tails and extreme events. We have shown that it remains to be clarified what the consequences of these facts are for the optimal portfolio strategy, in relation to the households' psychology, and to their institutional environment (flexibility, commitment).

The limited progress that has been made on these policy questions often originates from the low quality of historical financial data and from disagreements about investors' objectives and psychology. The most important necessary condition for the mobilisation of prominent economists and researchers in finance in Europe is thus the access to these data. The United States has a competitive advantage in research in finance because extremely useful datasets are available for research purposes. These datasets provide very detailed information about financial and macroeconomic variables over long periods. They have been screened and tested by generations of researchers in finance. Moreover, they go back quite far in the past, a crucial characteristic if one wants to test for the presence of mean-reversion in stock returns, for example.

By comparison, the available datasets (on asset returns, on fund returns, on macro variables, and so forth) in Europe are usually very dispersed among various public and private providers (national statistics institutes, Datastream, Morningstar, Bloomberg…). Most often, they are not harmonised and feature heterogeneous methodologies and norms. They are usually quite expensive. The bottom line is that many European researchers use US data to calibrate or test their models. Unfortunately, therefore, public decision-makers cannot rely on good economic theory tested in their national context. It would be nice to establish a new institute in charge of collecting country-specific data that can be made available for researchers in a more standardised form.

Non-coordinated efforts have been made to explore the characteristics of household preferences towards portfolio risks, as described in section 3. Various methodologies have been used to test preference hypotheses, from experiments in the lab to surveys of consumers, or even to neuro-scientific studies. There is no doubt that a great deal of progress can be made for a better understanding of investor psychology. Important differences, both cultural and institutional, exist across European countries, which should facilitate measurement of the role of these characteristics on actual financial decisions. This could be useful in fine-tuning policy recommendations in favour of a more responsible management of long-term savings – at both the individual and collective levels.

Beyond dealing with this problem, another recommendation is to increase the frequency and quality of the interaction of the meetings (workshops, conferences, summer schools) organised in relation to portfolio management and regulation in the long-term saving industry. This could also attract more good researchers into the field of long-term finance. The action with the largest potential impact for pension research would, in my opinion, be to provide explicit financial incentives for doing research in the field. Such incentives could be used to induce the most talented researchers to modify their research agenda. This could take different forms. For example, an annual prize (with a sizeable amount of money attached) could be offered for the best paper published on European pension economics. A good example of this is the Paul Samuelson Award in the United States.

Another action that might be useful in some European countries is to finance more policy-oriented papers. Many research results remain within the boundaries of academic circles, and are not made available to public and private decision-makers. Again, the Netherlands seems to be an exception, thanks to the existence of Netspar, which organises the transfer of knowledge from academia to practitioners. In some other countries, this transfer relies on fragile personal relationships between some researchers and their relationship with the media or within the public sphere (pension institutions, insurers and politicians). In addition to financing fundamental research, one should also implement a strategy aiming at a better transfer of knowledge. Commissioning policy papers in the field is one strategy going in that direction.

6   Expected outcomes of the research agenda

The current financial crisis clearly shows that there is no consensus – in the public sector or among experts – about how to manage the financial risks associated with funded pension systems. The main aim of a cross-national research programme is to provide a strong scientific basis for portfolio management of long-term investors.

The emergence of a lively network of prominent researchers in this field would be beneficial to the long-term saving industry (pension funds, life insurers, sovereign funds), and eventually to the present and future generations of pensioners. This research agenda is not likely to eliminate the occurrence of financial crises of funded systems, such as the one that we currently witness in most countries that have implemented such schemes. Rather, the practical outcomes will be related to a better understanding of the importance of intergenerational risk sharing, flexible pension rules and market timing. The provision of operational rules to manage the risks of funded schemes will allow us to improve their social return. A side benefit of this research effort will be to provide a better way to compare the social advantages of funded- and PAYG pension systems. In these days of upheaval on financial markets, this will not be the least benefit of this programme.

References

Agarwal, S., J. Driscoll, X. Gabaix and D. Laibson (2007) 'The Age of Reason: Financial Decisions over the Lifecycle,' NBER working paper 13191.

Ahn, D., S. Choï, D. Gale and S. Kariv (2008) 'Estimating Ambiguity Aversion in a Portfolio Choice Experiment', Department of Economics, UCLA.

Arrow, K.J. (1965) 'The Theory of Risk Aversion', lecture 2 in Arrow (1971).

Arrow, K.J. (1971) *Essays in the Theory of Risk Bearing* (Chicago: Markham Publishing Co).

Barro, R. (2006) 'Rare Disasters and Asset Markets in the Twentieth Century,' *Quarterly Journal of Economics* 121, 823-866.

Barro, R. (2009) 'Rare Disasters, Asset Prices and Welfare Costs, *American Economic Review*, forthcoming.

Barsky, R.B., F.T. Juster, M.S. Kimball and M.D. Shapiro (1997) 'Preference Parameters and Behavioural Heterogeneity: An Experimental Approach in the Health and Retirement Study,' *Quarterly Journal of Economics* 112, 537-579.

Bec, F. and C. Gollier (2008) 'Assets Returns Volatility and Investment Horizon: The French Case,' Toulouse School of Economics.

Bell, D.E. (1982) 'Regret in Decision Making under Uncertainty,' *Operations Research* 30, 961-981.

Bodie, Z., R.C. Merton and W.F. Samuelson (1992) 'Labor Supply Flexibility and Portfolio Choice in a Life-Cycle Model,' *Journal of Economic Dynamics and Control* 16, 427-449.

Brennan, M.J., E.S. Schwartz and R. Lagnado (1997) 'Strategic Asset Allocation,' *Journal of Economic Dynamics and Control* 21, 1377-1403.

Camerer, C., M. Bhatt and M. Hsu (2007) 'Neuroeconomics: Illustrated by fMRI and Lesion-Patient Evidence of Ambiguity-Aversion,' in B. Frey and A. Stutzer (eds) *Economics and Psychology: a Promising New Cross-Disciplinary Field* (Cambridge: MIT Press).

Campbell, J.Y. and L.M. Viceira (1999) 'Consumption and Portfolio Decisions when Expected Returns are Time Varying,' *Quarterly Journal of Economics* 114, 433-495.

Campbell, J.Y. and L.M. Viceira (2002) *Strategic Asset Allocation* (Oxford: Oxford University Press).

Campbell, J.Y., Y.L. Chan and L.M. Viceira (2003) 'A Multivariate Model of Strategic Asset Allocation,' *Journal of Financial Economics* 67, 41-80. Winner of Fama/DFA Second Prize, 2003.

Carroll, C.D. (1992) 'The Buffer Stock Theory of Saving: Some Macroeconomic Evidence,' *Brookings Papers on Economic Activity* 2, 61-135.

Carroll, C.D. (1994) 'How Does Future Income Affect Current Consumption?' *Quarterly Journal of Economics* 109, 111-147.

Carroll, C.D. (1997) 'The Buffer Stock Saving and the Life Cycle/Permanent Income Hypothesis,' *Quarterly Journal of Economics* 107, 1-56.

Chen, H., N. Ju and J. Miao (2009) 'Dynamic Asset Allocation with Ambiguous Return Predictability,' MIT Sloan School.

Cocco, J.F. (2005) 'Portfolio Choice in the Presence of Housing,' *Review of Financial Studies* 18(2) 535-567.

Cocco, J.F., F.J. Gomes and P.J. Maenhout (2005) 'Consumption and Portfolio Choice over the Life Cycle,' *Review of Financial Studies* 18, 492-533.

Cui, J. (2008) 'DC Pension Plan Defaults and Individual Welfare,' Netspar discussion paper 09/2008 – 034.

Cui, J., F. de Jong and E. Ponds (2006) 'The Value of Intergenerational Transfers within Funded Pension Schemes,' Netspar working paper.

De Jong, F., J. Driessen and O. van Hemert (2007) 'Hedging House Price Risk: Portfolio Choice with Housing Futures,' Netspar discussion paper 2007 – 022.

Deaton , A. (1991) 'Saving and liquidity constraints,' *Econometrica* 59, 1221-48.

Dimson, E., P. Marsh and M. Staunton (2002) *Triumph of the Optimists: 101 Years of Global Investment Returns* (Princeton University Press: Princeton).

Diris, B., F. Palm and P. Schotman (2008) 'Long-term Strategic Asset Allocation: An Out-of-sample Evaluation,' Maastricht University mimeo.

Ellsberg, D. (1961) 'Risk, Ambiguity, and the Savage Axioms,' *Quarterly Journal of Economics* 75, 643-669.

Epstein, L.G. and S. Zin (1991) 'Substitution, Risk Aversion and the Temporal Behaviour of Consumption and Asset Returns: An Empirical Framework,' *Journal of Political Economy* 99, 263-286.

Fama, E. and K. French (1988) 'Dividend Yields and Expected Stock Returns,' *Journal of Financial Economics* 22, 3-27.

Gollier, C. (2001) *The Economics of Risk and Time* (Cambridge: MIT Press).

Gollier, C. (2002a) 'What Does the Classical Theory Have to Say about Household Portfolios?,' in L. Guiso, M. Haliassos and T. Jappelli (eds) *Household portfolios* (Cambridge: MIT Press), 27-54.

Gollier, C. (2002b) 'Time Diversification, Liquidity Constraints, and Decreasing Aversion to Risk on Wealth,' *Journal of Monetary Economics* 49, 1439-1459.

Gollier, C. (2004) 'Optimal Portfolio Risk with First-Order and Second-Order Predictability,' *Contributions to Theoretical Economics*, http://www.bepress.com/ bejte/contributions/vol4/iss1/art4.

Gollier, C. (2007) 'Saving and Portfolio Choices with Predictable Changes in Asset Returns,' *Journal of Mathematical Economics* 44, 445-458.

Gollier, C. (2007b) 'Does Ambiguity Aversion Reinforce Risk Aversion? Applications to Portfolio Choices and Asset Pricing,' Toulouse School of Economics.

Gollier, C. (2008) 'Intergenerational Risk-sharing and Risk-taking of a Pension Fund,' *Journal of Public Economics* 92(5-6), 1463-1485.

Gollier, C. and R.J. Zeckhauser (2002) 'Time Horizon and Portfolio Risk,' *Journal of Risk and Uncertainty* 24(3) 195-212.

Gomes, F.J. and A. Michaelides (2005) 'Optimal Life-Cycle Asset Allocation: Understanding the Empirical Evidence,' *Journal of Finance* 60(2) 869-904.

Gomes, F.J., L.J. Kotlikoff and L.M. Viceira (2008) 'Optimal Life-Cycle Investing with Flexible Labor Supply: A Welfare Analysis of Life-Cycle Funds,' NBER working paper W13966.

Gourinchas, P., and J.A. Parker (2002) 'Consumption over the Life Cycle,' *Econometrica* 71(1), 47-89.

Guiso, L., T. Jappelli and D. Terlizzese (1996) 'Income Risk, Borrowing Constraints, and Portfolio Choice,' *American Economic Review* 86, 158-172.

Kahneman, D.V. and A.V. Tversky (1979) 'Prospect Theory: An Analysis of Decision Under Risk,' *Econometrica* 47, 263-291.

Kim, T.S. and E. Omberg (1996) 'Dynamic Nonmyopic Portfolio Behaviour,' *Review of Financial Studies* 9, 141-61.

Kocherlakota, N.R. (1996) 'Implications of Efficient Risk Sharing without Commitment,' *Review of Economic Studies* 63, 595-609.

Krueger, D. and F. Kubler (2006) 'Pareto-improving Social Security Reform when Financial Markets are Incomplete,' *American Economic Review* 96, 737-755.

Lettau, M. and S. Ludvigson (2001) 'Consumption, Aggregate Wealth, and Expected Stock Returns,' *Journal of Finance* 56(3) 815-49.

Machina, M.J. (1982) 'Flexibility and the Demand for Risky Assets,' *Economics Letters* 10, 71-76.

Mehra, R. and E. Prescott (1985) 'The Equity Premium: A Puzzle,' *Journal of Monetary Economics* 10, 335-339.

Merton, R.C. (1969) 'Lifetime Portfolio Selection under Uncertainty: The Continuous-time Case,' *Review of Economics and Statistics* 51, 247-257.

Merton, R.C. (1971) 'Optimum Consumption and Portfolio Rules in a Continuous-time Model,' *Journal of Economic Theory* 3, 373-413.

Mossin, J. (1968) 'Optimal Multiperiod Portfolio Policies,' *Journal of Business*, 215-229.

Poterba, J. and L.H. Summers (1988) 'Mean Reversion in Stock Returns: Evidence and Implications,' *Journal of Financial Economics* 22, 27-60.

Poterba, J., J. Rauh, S. Venti and D. Wise (2005) 'Utility Evaluation of Risk in Retirement Saving Accounts,' in D. Wise (ed.) *Analyses in the Economics of Aging* (Chicago: University of Chicago Press), 13-52.

Pratt, J. (1964) 'Risk Aversion in the Small and in the Large', *Econometrica* 32, 122-136.

Samuelson, P.A. (1969) 'Lifetime Portfolio Selection by Dynamic Stochastic Programming,' *Review of Economics and Statistics* 51, 239-246.

Teulings, C.N. and C.G. de Vries (2006) 'General Accounting, Solidarity and Pension Losses. *De Economist* 154(1) 63-83.

Tversky, A.V. and D.V. Kahneman (1992) 'Advances in Prospect Theory: Cumulative Representation of Uncertainty,' *Journal of Risk and Uncertainty* 5(4) 297-323.

von Gaudecker, H.M., A. van Soest and E. Wengström (2008) 'Risk Preferences in the Small for a Large Population,' Netspar discussion paper 2008 – 011.

von Neumann, J. and O. Morgenstern (1944) *Theory of Games and Economic Behaviour* (Princeton: Princeton University Press).

Weitzman, M.L. (2007) 'Subjective Expectations and Asset-return Puzzles,' *American Economic Review* 97, 1102-1130.

Welch, I. and A. Goyal (2008) 'A Comprehensive Look at the Empirical Performance of Equity Premium Prediction,' *Review of Financial Studies* 21, 1455-1508.

Risk and Portfolio Choices in Individual and Collective Pension Plans

Comments by Frank de Jong / Antoon Pelsser

The chapter by Christian Gollier provides a nice and useful review of the current research on asset allocation for pension funds. The chapter identifies several key policy questions that should be addressed in future research. These topics will be discussed below, along with some suggestions for a few additional research topics that we believe are also important.

Should investors 'time' the market?

This is quite a controversial issue, and a great deal of work has already been done in this area. If parameter uncertainty is taken into account, then the value of timing is far from clear. Peter Schotman and his co-authors have written several papers on this (see, for example, Hoevenaars et al., 2008; and Diris et al., 2009). We therefore doubt whether this issue is the most important issue to address.

Should individuals reduce portfolio risk at older age?

This is a much more established research question, as the effects of human capital are well understood. An open issue is how much correlation there is between labour income and stock returns. Such correlations will be different for people working in different sectors. A young employee with a (stream of future) labour income that is highly correlated with stock returns – for example, someone working in the financial sector – should be careful not to become over-exposed to financial risk by having both his / her pension and future labour income invested in financial markets. Furthermore, we should be careful to

64

distinguish between short- and long-run correlations. The correlations between stock returns and wages are typically small in the short run, but may be much larger in the long run (see Benzoni et al., 2007). A further under-explored issue is the effect of flexible labour supply. Employees have control over the number of hours they work, and can also control their retirement date. The effect and impact on portfolio choice of optimally controlling these decision variables still has to be investigated.

Should collective pension schemes be promoted in preference to individual ones?

This is a difficult question, as it involves a trade-off between many issues. Arguments in favour of a collective solution include the following: the collective offers economies of scale (see Bikker and De Dreu, 2007); mandatory participation in the collective scheme makes it possible to avoid problems of adverse selection; and the collective solution allows us to reap the benefits of intergenerational risk sharing. Arguments in favour of individual schemes include the following: the one-size-fits-all solution of a collective pension scheme is too restrictive; individual schemes can provide tailor-made solutions that pay attention to the background risk and other assets of the participant; and it is possible in individual schemes to include age-dependent risk allocation and portfolio choices. A problem with individual schemes has its root in the behavioural problems of individuals: it is difficult for individuals to make good decisions about benefits that are far away in the future. Evidence from the US (and other countries) suggests that individuals tend to delay these decisions – and typically invest too little money in their individual pension scheme. A great deal of work remains to be done in this area.

Asset allocation: institutional context and economic reforms

The issues raised in this part are not entirely clear to us. We would encourage the author to further sharpen the issues and questions.

Relevant rates of return in funded systems and PAYG systems

We are not sure that the question that Gollier asks is framed in the right way. Not only do funded systems have a return on assets (and not just a return equal to the population growth rate), but also the risk profile differs greatly from the notional returns of a Pay-As-You-Go system. Some interesting observations can be found in Lucas (2009). Furthermore, the fact that the markets for wage indexation are incomplete should be taken into consideration. Making a fair comparison requires further modeling for the pricing of such risks.

Other important research topics

We conclude our discussion by suggesting a few other relevant research questions:

- Annuitization: recent research has indicated that pensioners should invest their (lump-sum) pension benefits into annuities. However, very few individuals that have a free choice actually do so. Therefore, should annuitization be made mandatory? A related question is the importance of variable- and indexed annuities.
- Private retirement provision: should we promote and/or create a market for retirement products? Is it possible to design cost-effective and yet flexible products?

References

Benzoni, L., P. Collin-Dufresne and R.S. Goldstein (2007) 'Portfolio Choice over the Life-Cycle when the Stock and Labor Markets are Cointegrated,' *Journal of Finance*, 62, 2123-2167.

Bikker, J.A. and J. De Dreu (2007) 'Operating Costs of Pension Funds: The Impact of Scale, Governance, and Plan Design,' *Journal of Pension Economics and Finance*, 8(1), 63-89.

Diris, B.F., F.C. Palm and P.C. Schotman (2009) 'Long-term Strategic Asset Allocation: An Out-of-sample Evaluation', Maastricht University working paper.

Hoevenaars, R.P.M.M., R.D.J. Molenaar, P.C. Schotman and T.B.M. Steenkamp (2008) 'Strategic Asset Allocation with Liabilities: Beyond Stocks and Bonds', *Journal of Economic Dynamics and Control*, 32(9), 2939-2970.

Lucas, D. (2009) 'Issues in DB Pension Plan Valuation and Asset Allocation, Part I', notes for Netspar Workshop, January 28, 2009.

4

Innovative Institutions and Products for Retirement Provision in Europe

Lans Bovenberg / Theo Nijman[1]

1　Introduction

The institutional settings related to pension provision differ widely within Europe. In some countries, income transfers within families are still very important. In other countries, major parts of pension provision are delegated to large financial conglomerates. Some countries rely almost exclusively on public pay-as-you-go (PAYG) systems. In others, private funded pension systems are important. A wide diversity of funded systems exists in Europe. Some countries feature DB (defined benefit) plans in which participation is mandatory, the sponsor absorbs most risks, pension entitlements are paid out as annuities and uniform pension products are offered to all participants. Funded systems in other countries are of the individual DC (defined contribution) type in which participants themselves take all savings and investment decisions and usually face substantial investment, inflation and conversion risk.

The pension systems in various European countries not only differ widely but also are developing rapidly. More funded components are being added in countries relying mostly on family transfers and public PAYG-based systems. In countries with large funded systems, the recent turbulence in financial markets is stimulating reforms aimed at better risk management. European legislation

---

1　Scientific Directors Netspar, Tilburg University; see www.netspar.nl. This chapter builds on Bovenberg et al. (2007) and Bovenberg and Nijman (2008). It is a revised draft from the draft presented at the first Forward Looks workshop in Turin in September 2008.

is also an increasingly important force in creating a single European market for pension-related services. Together with the rapid ageing of the population and the increasingly competitive and dynamic world economy, this trend changes the environment in which national pension systems develop.

This chapter focuses on the analysis of optimal institutions and contracts for efficient risk sharing in funded pension systems. The outline of the rest of this chapter is as follows. Section 2 outlines the main policy questions related to pension system design. Section 3 presents the benchmark model for thinking about optimal life-cycle financial planning. Section 4 investigates how the recent academic literature has extended this model and explored how people do actually behave in saving, investing and insuring longevity risk. This section also considers the remaining research gaps and the associated needs for additional research infrastructures. Section 5 reviews the current European research infrastructures and networks that contribute to the research surveyed in this chapter. On the basis of the analysis in section 4, section 6 systematically investigates the infrastructure that would be required to address the research challenges most efficiently. Section 7 concludes by formulating what academic research in this field can contribute to addressing major policy questions.

2   The main policy questions

2.1   *Supplementing PAYG pensions by funded pensions*

Retirement schemes can be either funded or PAYG. PAYG schemes pay the retirement benefits of the older generations by levying contributions on the younger, working generations. The retirement promise is thus not backed by financial assets, but rather by the power of the government to force the younger generations to transfer resources to the elderly. In the larger continental European countries (including France, Germany, Italy and Spain), the pension system is based almost exclusively on PAYG financing. This makes these countries especially vulnerable to lower fertility, because PAYG schemes rely on human capital of the young to finance the pensions of older generations. The large continental European countries that depend almost exclusively on PAYG financing for the provision of retirement income have integrated the two main functions of pensions – poverty alleviation and old-age insurance – into a single comprehensive public pension system. Several of these countries now consider reducing the generosity of these schemes by no longer fully indexing them to the standard of living. This would yield a better-balanced portfolio between funded- and PAYG schemes, as workers with middle- and higher incomes substitute private, funded pensions for public PAYG benefits. Also Eastern European countries and several emerging economies are developing a funded pillar to supplement public PAYG systems.

2.2 Various funded pension schemes

Broadly speaking, two kinds of funded pension systems can be distinguished. In traditional defined-benefit (DB) plans, companies guarantee fixed pension benefits by absorbing financial market and demographic risks. In these plans, years of service and a reference wage typically determine benefit entitlements, while participation in the scheme is generally mandatory for all employees. In defined-contribution (DC) plans, in contrast, individuals themselves are responsible for planning how much to save for retirement, how to invest their savings in the capital market, and how to insure individual longevity risk by converting pension capital into annuity income. A thorough understanding of the advantages and disadvantages of the various types of funded systems, as well as suggestions for innovative hybrid systems that combine the strengths of DB and DC systems, will help countries develop reliable funded pension systems.

2.3 Collective versus individual decision-making

Pension saving requires a wide range of decisions. During the working life, one needs to determine how much to contribute to the pension scheme and in which assets the pension wealth is to be invested. At retirement, one has to decide on how one wants to draw down wealth, and whether one wants to buy an annuity insuring longevity risk (that is, the risk that one outlives one's financial wealth). As regards insurance, also decisions involving other individual- and family-specific risks are to be made during the working life – for example, disability insurance or provisions for partners and children in case of premature death of the main breadwinner.

These decisions can be taken either by individuals themselves (as is usually the case in individual schemes, which are often the DC type) or collectively and then be imposed on individuals (as is traditionally the case in DB schemes). It is well-known by now that a disproportionately large fraction of individuals suffer from financial illiteracy and that many actual decisions involving saving and investment are hard to rationalise. Indeed, many individuals indicate that they require help in taking pension-related decisions. In view of this, economists draw more and more on insights from psychology and sociology. Moreover, financial illiteracy implies that households have to delegate these complex decisions to professionals and financial institutions, which do not necessarily act in the interests of their clients.

2.4   Risk management

Another important policy question facing funded schemes is risk management. Pension provision requires long-term saving and investment strategies that are affected by many risk factors – including investment risk, inflation risk, longevity risk and interest rate risk. These risks are to be managed through adequate savings as well as asset-allocation decisions. In DC schemes, pension capital can be heavily affected by investment risks during the accumulation phase. Contribution levels as well as asset allocations are to be adjusted accordingly. If at some stage the pension capital is converted into pension income in the form of annuities, the individual involved faces sizable conversion risk. In particular, the conversion factor employed by an insurance company providing the annuities strongly depends on the interest rate and life expectancy at the time the accumulated capital is converted into an annuity. During the period when the pension income is paid out, the beneficiary typically faces inflation risk because most annuities are fixed in nominal rather than real terms. Rather than macro risk at the time of retirement, individual (micro) longevity risk becomes a dominant risk factor if the individual does not convert the pension capital into an annuity at retirement, but gradually draws down the pension capital when retired.

In DC schemes, the various macro risks are taken individually. The central policy question is what the optimal strategy for saving and risk-taking is for each individual, and how this strategy depends on personal characteristics such as age or wealth. Moreover, the welfare losses of suboptimal strategies are to be determined. A closely related question is to what extent policy interventions that allow for trade of additional risk factors (such as macro longevity risk) are welfare-improving.

Sectoral collective schemes can share risks collectively by trading these risks not on the capital market but among the various stakeholders of the schemes. In particular, these schemes can arrange internal trade of risk factors that cannot be traded on financial markets, including macro longevity risk, conversion risk and specific inflation risks. Risks in corporate plans are also often shared with the sponsoring company. For various reasons, however, many companies are no longer willing to underwrite the risks of their pension funds (see Bovenberg and Nijman, 2009). Collective schemes also trade risk factors on financial markets. Due to their scale and professional expertise, they can often do this in a more cost-effective way than individual investors, who suffer from financial illiteracy and do not benefit from scale economies. The relevant policy questions related to collective schemes is to what extent they are welfare-improving and to what extent hybrid institutions can be developed that combine the advantages of DC and DB schemes – and avoid their disadvantages.

3   Optimal saving, investment and insurance over the life cycle

The benchmark model to analyse optimal saving- and portfolio decisions over the life cycle is the model put forward by Merton (1971) and Merton and Samuelson (1974). This model is based on the following core assumptions on financial markets, labour markets and preferences:

Financial markets:
- Equity-market risk is the only aggregate risk factor, which is traded through equity.
- A risk-free asset (a bond) is available.
- The interest rate, inflation, the volatility of equity, and the equity risk premium are exogenous and constant over time.
- Log stock returns are identically and independently distributed according to a normal distribution.
- Financial markets are dynamically complete, in that stocks and bonds can be traded without constraints.

Labour markets:
- The after-tax wage during the working career is constant and risk-free.
- Labour supply is fixed, and the retirement age is exogenously fixed.
- Wages are exogenous, and pension premia are paid by workers. Pension premia thus reduce disposable incomes one-for-one.
- Death is predictable, or perfect insurance of individual longevity risk is available. Aggregate longevity risk is absent.

Preferences

- Individuals aim to maximise lifetime utility, which is the weighted sum over time of expected utility at each point in time.
- Utility at a given point in time depends only on consumption at that time. Habit formation is thus absent.
- Preferences feature positive and constant relative risk aversion.
- Bequest motives are absent.

3.1   *Optimal saving*

Optimal saving (that is, optimal consumption smoothing over time) and optimal risk-taking (that is, optimal consumption smoothing over various contingencies implemented through portfolio decisions) in this model is well understood. The rate at which overall wealth is consumed depends on age. Since older people feature a shorter planning horizon, they consume a larger share of their overall wealth. If both a young and an old person obtain one additional euro, the old person consumes the euro more rapidly. However, if both agents obtain x% more wealth, both agents increase their consumption by x% during the rest of their lives.

3.2   *Optimal risk-taking*

The portfolio decision implements optimal risk-taking. Total wealth consists of human wealth and financial wealth, which can be invested in either stocks or bonds. The share of total wealth invested in the risky asset (that is, equity) does not depend on the investment horizon (that is, age) but only on the risk premium, the volatility of the risky asset and the agent's risk aversion. Human capital is not tradable. Accordingly, financial rather than human wealth should be allocated optimally to achieve the optimal exposure to risk. In particular, riskless labour income implies that human capital is equivalent to the risk-free asset, so that all exposure to risk should come from the financial wealth.

Young agents should thus invest a larger fraction of their *financial* wealth in risky assets than old agents – for two reasons. First, the absolute amount of wealth invested in equity tends to fall with time as an individual consumes part of overall wealth (recall that this includes human wealth) during the working life. Second, the stock of financial wealth tends to increase during the working life as the individual transforms part of his human capital into financial capital by engaging in saving. The economic intuition why the young should hold a larger share of their financial wealth in risky stocks is that the young are less dependent on financial wealth for their consumption because they have an alternative riskless income source in the form of labour income. They thus can afford to take more risk with their financial wealth than elderly agents can (as the latter depend almost entirely on financial wealth for their livelihood).

This model's recommendation that the young should invest a larger fraction of their financial wealth in equities is widely recognised in financial planning advice. In fact, the rule of thumb that one should invest a financial-wealth share of 100 minus one's age in equities is often found in the popular press. Life-cycle funds and target-date funds, which gradually reduce the equity exposure as one ages, are popular in a number of countries, and have been recently added as approved defaults in American 401(k) plans. The Dutch financial supervisor recommends this type of investment strategy for DC schemes. The next section addresses the robustness of the recommendations of the benchmark model.

3.3 Optimal longevity insurance

As regards longevity insurance, full annuitization of wealth is the benchmark setting. The benchmark models in the literature are Yaari (1965) and Davidoff, Brown and Diamond (2005). The core assumption in the Davidoff, Brown and Diamond (2005) model, which generalises the Yaari (1965) model, is the presence of a complete set of annuity markets; every traded asset can be traded also in annuitised form, so that survivors in the insurance pool benefit from a mortality credit if others in the insurance pool pass away. Furthermore, their model assumes the absence of non-traded or insurable exogenous shocks (such as health costs), the absence of bequest motives and (sufficiently) fair annuity markets.

4   Recent developments and remaining gaps in the literature

This section explores a number of recent developments in the literature, and identifies remaining gaps in the literature that call for further analysis and extensions. Sub-section 4.1 investigates the robustness of the benchmark model of saving, investing and insuring over the life cycle. Sub-section 4.2 addresses recent developments in the literature on individual behaviour and how this can be used to optimally structure the choice architecture of life-cycle financial planning. Sub-section 4.3 analyses the governance and solvency of the institutions involved.

4.1 Optimal saving and investment over the life cycle

A rapidly developing literature analyses to what extent the core conclusions of the benchmark model discussed in the previous section are robust to alternative assumptions with respect to human capital, to the properties of financial markets and to the specification of individual preferences.

4.1.1 Human capital and labour markets

Wage risk and risk-taking

Many papers analyse the case where human capital is risky because of uncertainties in an individual's career path. Often these papers also include borrowing constraints, as young workers often cannot borrow against future labour income. With borrowing constraints, the individual is not able to optimally diversify labour income risks over time. As a direct consequence of the reduced ability to smooth risks intertemporally, the individual will take on less risk in his or her investment portfolio.

Cocco, Gomes and Maenhout (2005) find that young investors choose portfolios that are less tilted towards equity than the portfolios of middle-aged investors. The desire to hold a safe, liquid stock of precautionary savings has become one of the explanations for the so-called equity premium puzzle (that is, the large rewards for risk-taking that are hard to explain on the basis of the benchmark model presented in sub-section 3.1) and is the reason why young households do not participate much in the stock market (see Constantinides and Duffie, 1996).

Other issues arise if the aggregate component to wage risk is correlated with equity risk. With positive correlation between these two risks, equity and human capital become substitutes. This reduces the equity holdings of young agents with ample human capital. A recent paper by Benzoni et al. (2007) strengthens this conclusion. This study indicates that labour income is cointegrated with equity prices. This makes human capital a closer equity substitute for younger workers with a longer investment horizon than for older workers and retirees with a shorter investment horizon. Consequently, young workers invest less in stocks than older agents do.

WAGE RISKS AND SAVING

Liquidity constraints not only make investment behaviour more conserva-tive but also strengthen precautionary saving motives. Indeed, by investing conservatively and setting aside resources through saving, individuals ensure the presence of a financial buffer that helps them to optimally time-diversify temporary risks. Precautionary motives rather than saving for retirement tend to be the main reason why young households save (see Cocco, Gomes and Maenhout, 2005). As a direct consequence of the additional saving, financial wealth will be higher later on in life, which will tend to increase equity expo-sure when old.

ENDOGENOUS LABOUR SUPPLY

Another crucial assumption on human capital in the benchmark model is that labour supply is fixed. In reality, people often face the opportunity to adjust their supply of labour – by working additional or fewer hours per week, for example, or by adjusting the retirement date. Bodie et al. (1992) show that endogenous labour supply enhances the capacity to bear risk. Shocks can be absorbed by adjusting consumption of not only produced commodities, but also leisure. This increases demand for risk-taking.

If labour supply is flexible, the share of financial wealth invested in risk-bearing assets varies even more with age (see, for instance, Gomes, Kotlikoff and Viceira, 2008). This changes if the labour supply is especially flexible around retirement, because agents can adjust the time and speed with which they retire. In that case, also older workers can afford to invest in risk-bearing assets. This points to the importance of a flexible labour market for older workers to sustain the supply of risk-bearing capital in an ageing economy. Farhi and Panageas (2007) provide a formal analysis of the impact of flexibility of the retirement date on the optimal risk-taking, and find that this increases the fraction of wealth that is invested in stocks.

LONGEVITY AND HEALTH RISKS

The key result from the benchmark model – that full annuitization should be optimal – is challenging because empirical evidence suggests that many people prefer lump sums and avoid annuities when they are given the choice to do so (see Brown, 2007). This raises the policy question whether compulsory an-nuitization should be stimulated.

Two types of longevity risk have been distinguished in the literature. Macro longevity risk refers to uncertain developments in future mortality rates due to improved healthcare, changes in nutrition and habits, etc. Macro longevity affects the average mortality of a pool of individuals (for example, in a pension fund or insurance pool). Micro longevity risk refers to the uncertainty for an individual regarding whether or not he or she will survive, irrespective of changes in mortality rates.

Micro longevity risk cannot easily be traded on financial markets because of adverse selection. Older individuals typically have superior information on their survival probabilities than the issuer of the annuity. Since the asymmetry in information becomes more important for older agents with poor health conditions, annuities are irreversible. This gives rise to a fundamental trade-off between liquidity and longevity insurance. Adverse selection forces the issuer of voluntary annuities to offer prices that are not actuarially fair for the population at large. Mandatory participation and compulsory annuitization in a collective pension scheme alleviate these adverse-selection problems.

The benchmark view on the optimality of annuitization is based on the absence of background risks. However, non-insured medical expenses can be an important source of idiosyncratic background risk, especially for older households. Hence, older households face idiosyncratic risk with implications for saving behaviour (that is, precautionary saving) and investment behaviour (and also the willingness to take out annuities). In particular, these households will save for precautionary reasons, invest more conservatively and are discouraged from taking out illiquid annuities.

Remaining research questions

DOCUMENT IDIOSYNCRATIC HUMAN CAPITAL RISK

An important task for future research is to estimate idiosyncratic human capital risk in various European countries and investigate how it has developed over time. Both unemployment risk and wage risk are to be considered. On the one hand, reduced coverage of social insurance, and increased competition (and the associated creative destruction) may have increased idiosyncratic human capital risk. On the other hand, a more flexible labour market may reduce the length of unemployment spells and increase the opportunities for outsiders to find work, thereby reducing persistent labour-market risks. Within idiosyncratic risks, one can distinguish between temporary and permanent wage shocks. Abowd and Card (1989) and Topel and Ward (1992) showed that log wages in the United States are close to a random walk, where the standard deviation of innovations is large, namely about 10% per year. Long data series on wages are required to investigate the long-run correlation between labour-market risks and financial-market risks.

This research agenda calls for close cooperation between finance, labour economics and health economics. The issues investigated by the finance literature exploring the importance of labour-market flexibility are in fact close to the micro-econometric literature on retirement behaviour (see the contribution of van Soest to this book). To understand the nature and origin of various labour-market risks, we need a good understanding of the labour market (both supply- and demand factors) and the social insurance institutions in the various European countries. Moreover, health economics and medical sciences can shed light on the health and morbidity risks facing older workers. For example, the relationship between increased longevity and morbidity is crucial for a possible link between longevity and the effective retirement age.

HEALTH SHOCKS AND LONGEVITY INSURANCE

An important topic for future research is to distinguish between liquid precautionary saving and illiquid retirement saving. Most of the literature does not distinguish between these two categories, and assumes that all saving is liquid. In the same vein, more attention to the interaction between health insurance and longevity insurance is called for. Linking reverse life insurance through annuities to healthcare insurance can combat selection. Bad risks for an annuity company tend to be good risks for health insurers – and the other way around. Moreover, by providing health insurance, an insurance company reduces the need for liquidity, thereby making annuities more attractive. Moreover, moral hazard may be important in health insurance, especially for relatively small risks such as personal services required around the home. For these risks, precautionary saving may thus be appropriate. This implies that annuities should be complemented with liquid private saving.

This research agenda links the finance literature to that in health economics. It necessitates also an understanding of not only pension institutions but also those in health insurance, in general, and old-age care, in particular.

COMPULSORY PARTICIPATION, LABOUR MARKETS AND RISK SHARING

The benchmark model discussed in section 3 implies that one should borrow at the beginning of one's career and invest the proceeds in the stock market to acquire sufficient exposure to the equity market. Unfortunately, adverse selection and moral hazard typically preclude borrowing against future labour income. Compulsory participation in collective pension schemes can alleviate adverse selection and moral hazard by securing the human capital of the younger generations as collateral. However, relieving the borrowing constraints through compulsory participation is not costless if, in contrast to our benchmark model, labour supply in the sector to which the compulsory participation applies is endogenous. In that case, mandatory participation gives rise to labour-market distortions. In particular, agents reduce their labour supply to the sector if adverse developments in equity markets saddle them with negative pension wealth (that is, pension debt) that they have to finance with recovery premia on the basis of their labour supply. Sectoral labour supply can be reduced by moving to other sectors or countries, by becoming self-employed, by working part-time or by retiring early.

The analysis of the trade-offs associated with compulsory participation in pension schemes calls for close cooperation between finance and labour economics and a good understanding of the local labour-market and pension institutions in various European countries.

4.1.2 Financial markets

Apart from human capital risk, the main risk factors that individuals face over the life cycle are interest-rate risk, inflation risk, equity risk and longevity risk. In the base-case model, longevity risk, interest rate risk and inflation risk are ignored and stock returns are assumed to be non-predictable. This sub-section explores how different models of financial market risks affect the results from the benchmark model.

INTEREST-RATE RISK

If (real) interest rates fluctuate over time, the investor is faced with an additional risk factor. Fluctuations in interest rates affect the value of the human capital as well as the value of the bond portfolio. Early in the life cycle, bond portfolios are to be held with long durations to hedge the impact of interest-rate risk on consumption during retirement. If bonds or interest-rate derivatives would be available with very long maturities, then the part of financial wealth that is invested in fixed income can be invested in these assets to obtain the optimal duration for total wealth. In the more realistic case in which the maturity of traded bonds and derivatives is limited, the fraction of financial wealth invested in bonds is increased at the expense of equity. In this way, the duration of the assets is more closely aligned with the duration of the consumption pattern, while the remaining interest-rate risk is balanced with equity risk. Campbell and Viceira (2001), Brennan and Xia (2000) and Munk and Sorensen (2005) have considered the implications of time-varying interest rates on the optimal demand for long-term bonds, and analyse the welfare gains of hedging variations in real rates.

A number of recent studies report that the expected bond returns are clearly time-varying; that is, bonds are more attractive in some periods than in others (see, for example, Cochrane and Piazessi, 2005). In particular, the real (as well as the nominal) bond-risk premium is increasing in the real rate. Sangvinatsos and Wachter (2005) show that unconstrained long-term investors can realise large gains by exploiting these time variations in bond premia. Koijen, Nijman and Werker (2009) consider the case of borrowing-constrained life-cycle investors. They find that short-sell constraints reduce the potential gains associated with bond/timing strategies because these gains can be obtained only at the expense of reduced equity exposure – unless very long maturity bonds would be traded. The economic gains realised by bond timing strategies peak around the age of 50 (when agents have built up substantial financial wealth and are free of borrowing constraints), and are hump-shaped over the life cycle.

INFLATION RISK

Section 3 assumed that inflation risk is negligible or, equivalently, that all assets that have been considered are inflation-indexed. In reality, inflation-linked assets are rarely traded, and inflation is an important risk factor for long-term financial planning. Investors who do not have access to indexed bonds face the question to what extent indexed bonds can be replicated by nominal bonds. If investors are forced to bear inflation risk, they will shorten the maturity of the bond portfolio (see Campbell and Viceira, 2005). This implies that the optimal duration of the fixed-income portfolio is smaller than that of the consumption profile. Indeed, in the absence of inflation-linked bonds, investors face a trade-off between hedging real interest-rate risk and hedging inflation risk. Lengthening the duration of the nominal bond portfolio reduces real interest-rate risk, but exposes the investor to inflation risk.

EQUITY RISK

There is a lively debate in the literature as to whether the expected stock returns are time-varying. Many authors claim that the expected return on equity depends on variables such as the dividend yield or the term spread, so that equity returns are to some extent predictable. The typical finding in the literature is that – at the low frequencies that are relevant for financial planning over the life cycle – stock returns are negatively correlated, which implies that the stock market is mean-reverting. If stock returns are mean-reverting, stocks are less risky for long-term investors than for short-term investors, so that young investors should invest an even larger part of their financial wealth in equities. Campbell, Chan and Viceira (2003) and Campbell and Viceira (2005) show that the annualised standard deviation of real returns of about 17% of a diversified stock portfolio is reduced to an annualised standard deviation of only 8% on a 20-year horizon, which is in fact comparable to the annualised variance of the real returns of nominal bonds with the same maturity that are held to maturity. Much longer data series for more countries are essential in order to obtain robust evidence on mean-reversion.

A second assumption on the stock-market process adopted in the benchmark model is that the volatility of stock returns is time-invariant. The literature contains abundant evidence that this is not the case at shorter horizons, so that time-invariant volatility can be an approximation at lower frequencies, at best. Chacko and Viceira (2005) and Liu (2007) consider an incomplete market setting with time-varying volatility of equity returns. These papers show that the impact of time variation in volatility on the optimal asset allocation is small – particularly in the empirically relevant case of volatility being negatively correlated with stock returns (so that stocks become less risky if returns are high).

Remaining research questions

TIMING OF ANNUITIZATION

Although the risk of outliving one's assets is one of the main risks in retirement provision, individuals are often reluctant to invest a large fraction of their wealth to purchase annuity products in order to avoid this risk (see also section 4.2.1). Recent research (see, for example, Scott, Watson and Wu, 2006; and Gong and Webb, 2007) has indicated that a substantial part of the risk can be hedged by investing only a small fraction of wealth in deferred annuities around the retirement date. A deferred annuity generates income at very old age, and thereby significantly reduces the risk of outliving one's assets. More research is needed to see how robust this result is – for example, to inclusion of additional risk factors, such as health costs.

OWNER-OCCUPIED HOUSING AND RETIREMENT

With human capital and retirement saving, owner-occupied housing belongs to the most important assets in the portfolio of individuals. Owner-occupied housing is quite illiquid, and raises all kinds of issues in connection to retirement saving. In particular, can owner-occupied housing help hedge inflation and interest-rate risks? Can innovative financial products (for example, various types of reversed mortgages) help individuals to access the wealth accumulated in their own housing? To what extent can owner-occupied housing hedge health risks (because an adverse health shock may induce people to move house and free the equity in their home)? Can building societies take on part of the house price risks so that various hybrid types of ownership become possible? To illustrate, individuals may first buy a larger part of their home in middle age, but later on sell their home back to a building society. As regards risk-taking by younger individuals, investment in housing financed by mortgages can contribute to efficient risk bearing at younger ages if housing risk is correlated with equity risk. The current state of knowledge is well summarised in Michelangeli (2008) and Munnel et al. (2007).

Bonds linked to inflation and growth

Inflation-linked bonds have been analysed widely in the literature. The bonds that are actually traded are related to specific inflation indices (for example, European price inflation). Further analysis is needed to analyse how useful these bonds are in managing the risks of insurance companies and pension funds that provide pension products. Likewise, more analysis is required on the advantages and disadvantages for governments to issue such bonds linked to their local inflation or economic growth. Better access of the institutional arrangements is crucial here.

Longevity bonds

To enable insurance companies and pension funds to provide optimal retirement products by managing the accumulated (macro) longevity risk of the provider, the introduction of longevity-related financial instruments (such as longevity bonds) is potentially also of significant importance. A new literature is developing here (see, for example, Blake, Cairns and Dowd, 2008). Interaction with industry and policymakers is essential here to develop a market for longevity risk.

Information on institutional arrangements

The financial instruments referred to above (inflation-linked bonds, longevity bonds, deferred annuities, reverse mortgages) are traded in some European countries. More accessible information on institutional arrangements in the various countries would be very useful to analyse the added value of introducing them also in other countries. Moreover, as regards retail products such as deferred annuities and reverse mortgages, facilities for experimental research are essential to analyse how the impact of inflation and longevity risk, for instance, is to be made transparent for individuals.

4.1.3 Preferences

The results of the base-case model are dependent on the assumption of constant risk aversion. Two lines of research analyse the implications of alternative individual preferences.

HABIT FORMATION

The first stream of papers explores the implications of habit formation, so that well-being depends on current consumption relative to recent consumption. Habit formation is a popular way of explaining a number of asset-pricing phenomena – including a large equity premium that varies counter-cyclically. In particular, investors demand a high return on stocks because stocks perform poorly in recessions, when it is difficult to keep consumption in line with habits that were formed in the past.

Habit formation also has important implications for life-cycle saving and investment. As regards consumption smoothing over time, habit formation may explain why consumption levels typically rise over the life cycle. In particular, if young agents start with low habit levels, they optimally should save more than without habit formation. The reason is that with habit formation high initial consumption levels raise consumption needs later in life. As people get older, they raise consumption levels, as the horizons become shorter over which the negative future external effects of current consumption levels materialise.

Another implication for saving of habit formation is that, compared to the case without habit formation, the immediate adjustment of consumption to a given wealth shock lessens, but adjustment later is larger (as habits have had time to adjust to the changed consumption level). Habit formation thus explains why pension schemes adjust their premia levels only gradually following an unexpected shock, and why consumption is less volatile than predicted by the benchmark model of section 3.

As regards risk-taking over the life course, habit formation implies that the ability to absorb unexpected shocks increases if the time horizon increases. Accordingly, younger agents with longer horizons can adjust more easily to shocks, and thus should invest a larger share of their wealth in risk-bearing assets. Intuitively, they have more time to adjust their habits to unexpected shocks. Habit formation is thus another reason why young agents should optimally invest a larger share of their portfolio in equity than older agents do, and why pension contracts may optimally take on more of a defined-benefit nature (with more guarantees) when people grow older.

LOSS AVERSION

Recent experimental work suggests that human behaviour is not always well described by constant relative risk aversion. Tversky and Kahneman (1992) therefore developed prospect theory to better describe human behaviour with respect to risk. In particular, behaviour seems better captured by the type of utility function with a kink at what is called the reference point: a gain from this reference point yields only a small increase in utility, while a loss yields a large decrease. On the basis of experimental work, Tversky and Kahneman (1992) find that losses tend to weigh about 2 1/4 times more heavily than gains. The kink implies that relative risk aversion is not constant, but depends on the actual consumption level and the particular losses and gains considered. Risk aversion is thus especially large at consumption levels close to the reference point. A second important element of prospect theory is that agents are risk-averse with respect to gains (twice-as-large gains yield an increase in utility that is less than twice as much), and risk-loving with respect to losses (twice-as-large losses yield a decrease in utility that is less than twice as much).

Prospect theory (and the kink at the reference point, in particular) has important implications for risk-taking over time. With kinked utility, the marginal benefit of risk-taking (that is, a higher expected return) increases more rapidly with the investment horizon than the associated marginal cost (that is, more risk). As a direct consequence, optimal risk-taking increases with the time horizon. This yields another reason why young agents should bear more risk than old agents: loss aversion as described by prospect theory implies that young agents with long horizons should invest more in equity than old agents.

Benartzi and Thaler (1995) argue that people are reluctant to invest in stocks because of a combination of loss aversion and a short planning horizon. People evaluate stocks over periods that are shorter than their true long-run horizon. This myopic loss aversion explains why agents require a high risk premium on equity. Pension funds can correct this distortion by investing more in equity than individuals would do. Myopic loss aversion also suggests that pension schemes should not report their wealth levels too frequently to their members, because this would stimulate the tendency of the members to evaluate their wealth on an excessively short time horizon.

The literature has explored what happens if the reference point adjusts over time in response to current consumption levels and rational investors anticipate this. Berkelaar, Kouwenberg and Post (2004) find that updating the reference point implies more risk-taking. Intuitively, preferences are more flexible: with adverse shocks the point from which gains and losses are being evaluated, and to which one wants to be close, is adjusted downwards.

Barberis, Huang and Santos (2001) explore a setting in which investors derive direct utility from not only consumption but also fluctuations in their financial wealth. After a run-up in stock prices, the agent gets further away from this reference point and thus becomes less risk-averse. History thus determines risk aversion. Just as in Campbell and Cochrane (2000), this variation in risk aversion allows the equilibrium returns to be much more volatile than the underlying fundamentals.

Remaining research questions

Elicit preferences
Recent theoretical advances have explored how robust results on optimal saving and investment are with respect to alternative specifications of preferences. Economists are cooperating more with psychologists to understand the formation of preferences. Various methodological tools are available to determine the actual preferences of people. Laboratory experiments are increasingly being used to elicit preferences. The main advantages of this type of experiment are observation and control. The experimenter can perfectly observe and document behaviour. Moreover, the environment in which subjects make their decisions can be controlled, which allows for a careful analysis of causality.

The main drawback of laboratory experiments is external validity: does the human behaviour in an artificial laboratory setting really reflect behaviour of actual people in the real world, with often bigger stakes? Laboratory experiments should therefore ideally be supplemented by survey experiments and field experiments. Survey experiments can be conducted on the basis of panels of more representative populations than the usual participants in laboratory experiments (that is, students). Field experiments are conducted by systematically varying treatments of individuals when they make real economic decisions. In field experiments, the stakes are real and the subjects make decisions in a natural environment. Moreover, sample sizes typically exceed those in survey experiments. The costs of field experiments, however, are generally quite large.

4.2 Choice architecture and governance

A substantial literature on how individuals actually decide on saving and investment over their lifetime has developed recently. This subsection provides a brief overview of this literature.

4.2.1 Actual behaviour

ACTUAL RETIREMENT PLANNING

Individuals spend little time and effort on planning for retirement. Lusardi (1999) reports that even in the US, where individuals have many more pension-related decisions to make, one-third of the workers had 'hardly thought about retirement' only ten years before retirement. Ameriks, Caplin and Leahy (2003) report, also for the US, that more than 50% of a sample of highly educated wealthy individuals had *'not spent a great deal of time developing a financial plan'*. Lack of planning has serious consequences: those who do not plan have lower wealth holdings and are less likely to report that they experience a satisfying retirement. Many recent papers analyse the impact of improved information (for example, through benefit information fairs) and financial education (see, for example, Venti, 2006; and Clark et al., 2006). At best, the results indicate modest improvements in the effort spent on pension-related decisions and on the quality of these decisions. Choi et al. (2005), for example, report that tax incentives for retirement saving were not utilised – even after they had been explicitly pointed out in financial training courses.

ACTUAL SAVING BEHAVIOUR

Individuals selecting their pension contract face two decisions: how much to save and how to invest the financial wealth that is accumulated. The existing evidence, which is primarily on the US, indicates that in American DC schemes individuals do not save enough, and that they experience unanticipated drops in consumption at retirement. Moreover, a number of surveys have found that the vast majority of individuals think that they should be saving more for retirement (see, for example, Choi et al., 2002). One explanation that is well documented in psychology is that people lack the self-control that is required to implement a savings plan. A convenient way to model this behaviour is hyperbolic discounting. This model assumes that nearby discount rates are much larger than long-term discount rates. As a consequence, consumers exhibit time-inconsistent behaviour, while actual behaviour diverges from planned behaviour.

ACTUAL RISK-TAKING

With respect to asset allocation, an even more extensive literature indicates that individuals seem to make significant mistakes. Many households take decisions that are difficult to reconcile with the advice given to them by financial planners or other experts. A well-established stylised fact in the literature is that only small parts of the population in countries like Italy (Guiso and Japelli, 2005) or the US (Haliassos and Bertaut, 1995) hold stocks, directly or indirectly. Participants in retirement saving plans rarely rebalance their portfolio or alter the allocation of their contributions over the life cycle. Many households do not diversify and hold only a few stocks, often with a local bias – and in 401(k) schemes often even with a bias to the stock of their own employer (see, for example, Munnel and Sunden, 2004). Participants in 401(k) plans display a tendency to split their contributions evenly among investment options, irrespective of the type of options that is offered (Benartzi and Thaler, 2001). Such a '1/n' rule-of -thumb is clearly at odds with optimal diversification and adequate risk-taking.

ACTUAL ANNUITY CHOICE

In many countries agents can choose whether or not to convert their pension wealth into annuities. As discussed in section 3, the benchmark model suggests that annuities are very attractive. Brown (2007) surveys the extensive literature documenting the fact that individuals hardly ever voluntarily buy annuities, and analyses to what extent these findings can be reconciled with economic theory. Brown argues that the main ingredient for a good grasp of actual behaviour is a better understanding of the psychological biases involved. This requires improved interaction between economists and psychologists (see also section 5 on the required research infrastructure).

THE IMPORTANCE OF DEFAULTS

Strong evidence has been presented that individual behaviour in retirement plans can be strongly influenced by the specification of the default option (see Benartzi and Thaler, 2007). In case of automatic enrolment in pension schemes – where the default is that people will participate in the scheme but can opt-out – participation is far larger than would be the case if people have to actively opt-in to participate in the scheme. Likewise, the specification of the default is important in the selection of the contribution rate, the asset allocation or the choice between a lump sum and an annuity.

Remaining research questions

The core challenges in understanding actual decision-making include a better understanding of not only how the various decisions depend on background variables, but also how they interact and how the decisions of others (for example, of other members in the same household) affect decision-making. This clearly requires substantial efforts in data collection.

4.2.2 Optimal choice architecture

TAILORING PENSION SCHEMES TO INDIVIDUAL CHARACTERISTICS

A better understanding of optimal saving and risk-taking and its dependence on individual characteristics under different assumptions is crucial for efficient pension provision. Bovenberg et al. (2007) indicated that pension products that are not tailor-made to individual preferences and circumstances can imply substantial welfare losses. More tailor-made financial-planning solutions tend to become increasingly important because individual life cycles have become more heterogeneous, and ICT (Information and Communication Technology) allows for more tailor-made products. Another reason for paying more attention to individual solutions is the increasing number of self-employed individuals who are not served by sectoral- and company pension funds, and have to make active decisions themselves on retirement insurance.

FREE CHOICE

Pension contracts can be tailored to the characteristics and preferences of an individual in a variety of ways. A first possibility could be to offer a wide range of choices, and leave it to the individual to select the most adequate strategy. Subsection 4.1, however, discussed evidence indicating that people struggle to make adequate choices on life-cycle saving, investing and insuring.

COMPULSION

One option, then, is to compel mandatory participation in a collective scheme in which decisions are taken by professionals. One of the main drawbacks is that collective schemes typically do not take into account individual characteristics and preferences. At the same time, the costs of more tailor-made features in collective schemes should be traded off against the associated additional transaction costs and the potential for adverse selection.

As a third alternative, Thaler and Sunstein (2008) argue in favour of libertarian paternalism to help people to make adequate choices. The term *libertarian paternalism* indicates that the freedom of choice is not restricted (that is, libertarian) but that paternalistic 'choice architects' (that is, the designers of a pension scheme) help agents to make adequate decisions. Important instruments in this regard include adequate defaults with respect to savings rates and asset allocation. Another well-known example is the 'Save More Tomorrow' plan proposed by Thaler and Benartzi (2004). The commitment mechanism allows individuals to commit to increasing their savings rate at some later date. Yet another example is the utilisation of peer effects through the organisation of benefit-information fairs. Duflo and Saez (2003) show that this impacts the behaviour of colleagues that had not been invited to attend.

Thaler and Sunstein (2008) maintain that it is impossible to avoid influencing people's choices. The designers of choice schemes should be aware, for example, that the order in which alternatives are presented will affect decisions. Once the power of defaults and other 'nudges' is appreciated, a host of policy-related issues arise (see Kooreman and Prast, 2007). What behaviour is preferred? How should various nudges be used, and how should they be combined with other instruments such as improved information and/or financial education?

Remaining research questions

To make further progress on these issues, one needs micro-econometric data to document the heterogeneity across individuals. Administrative data on income sources, health, assets and pension entitlements should be combined with survey information aimed at eliciting preferences. Various types of experiments could be used to study the behavioural impact of various types of nudges (for example, defaults, information provision, framing of choices) and how these nudges can be exploited to help people make better choices. Experiments can also shed light on the inherent social dimension of pensions and insurance. This includes not only peer-group effects, but also the extent to which people feel a sense of solidarity and perceive pension schemes to be fair and equitable. Both experiments and large databases that document individual behaviour and heterogeneity in individual circumstances typically bring together researchers from different disciplines studying human behaviour (for example, psychologists, sociologists, economists). These methodological tools are thus an important stimulus for cooperation across various disciplines.

4.3　Governance and solvency of pension schemes

GOVERNANCE ISSUES

Pension schemes raise important governance issues. First of all, once one recognises the power of the choice architecture, in general, and defaults, in particular, the question arises how the governance of the 'choice architect' be arranged so that the architect acts in the interests of the individuals involved. More generally, governance arrangements should thus address principal-agent issues that arise if unsophisticated consumers delegate complex financial decisions to professionals.

The incomplete character of pension contracts in collective pension schemes also raises important political economy- and governance issues. In particular, the trustees of a pension scheme typically have the discretion to pay out the surplus as dividends to current stakeholders (in the form of additional pension benefits or additional future benefit claims that are not fully covered by additional premia) or to save the surplus for future participants. This discretionary power of the trustees has the advantage that all of the possible contingencies do not have to be thought through *ex ante*, thereby reducing transaction costs. The disadvantage is that the collective decision-making within the board of the scheme may give rise to political risks.

SOLVENCY REQUIREMENTS AND INTERGENERATIONAL RISK SHARING

Another important governance issue involves intergenerational risk sharing. It is theoretically optimal to diversify risk over as many generations as possible – including generations that are not yet active or even not yet born. Shocks are thus smoothed out as widely as possible so that each separate individual in each period is affected as little as possible by the shocks.

In practice, risk sharing among non-overlapping generations is limited. The largest pool for enforcing this risk sharing is the national state. In particular, in combination with public debt, the government can employ its tax power to commit future generations to share risks with current generations. In a democracy, however, the state can commit only imperfectly. Young generations can always exert political pressure to change laws that force them to pick up the tab of risks that the older generations have shifted onto them. Similarly, current generations, who control the votes, may be tempted to consume the buffers that were initially assigned to the future generations. This raises various political economy issues.

Competition constrains the intergenerational risk sharing conducted by private pension schemes. If young workers face substantial negative buffers on entry into a pension scheme, they may vote not only with their voice but also with their feet by seeking employment elsewhere or reducing labour supply. This is why regulators adopt what is termed the *discontinuity principle* in setting solvency requirements: a pension fund should be able to comply with its obligations also if, starting today, no new generations would be willing to enter the scheme. Recognising that pension schemes cannot easily secure the human capital of future generations, this discontinuity principle limits the scope for risk sharing among non-overlapping generations.

The solvency requirements have consequences for optimal saving and investment. As regards saving, the solvency constraints strengthen precautionary saving – especially if the fund is close to the solvency constraints. Indeed, the fund then sets aside additional funds to escape the limit on intergenerational risk sharing; the financial reserves of the fund act, in fact, as a buffer stock. As regards investment, the fund invests mainly in bonds if it has a small amount of assets, and is thus close to the solvency constraint. Intuitively, the pension fund constructs a put option by trading dynamically and exchanging its equity for bonds if equity declines in value in line with a dynamic strategy that replicates the put option. The fund thus behaves as if it exhibits decreasing absolute risk aversion. Gollier (2006) shows that the welfare gains from intergenerational risk sharing are substantially reduced when solvency constraints are binding.

Remaining research questions

In some European countries pension schemes issue nominal guarantees to their participants. The solvency requirements for these guarantees imposed by the regulator increasingly lead them to reduce the mismatch risk between their assets and these nominal guarantees by extending the maturity of their assets to that of their nominal liabilities. If individuals care about their purchasing power rather than the nominal value of their pension, then even the most risk-averse individuals would prefer a less-than-full reduction of the duration mismatch between the assets and the liabilities. The trade-offs between managing nominal guarantees and real ambitions provide an important topic for future research.

Analysis of governance issues for pension plans can benefit from input from psychologists, sociologists as well as economists studying collective decision-making. More empirical evidence on the competences and behaviour of trustees, in particular, would be welcome (see, for example, Clark et al., 2006). A better understanding of the exact institutional arrangements is also crucial here.

5   European research infrastructures and networks

European research centres

European research on the many challenging questions outlined in the previous section does not have to start from scratch. A number of research centres in Europe have groups or prominent individuals that focus their research efforts on issues related to optimal pension design for an ageing population. Examples of universities that are active in this field are the universities of Toulouse, Paris, Turin, Frankfurt, Mannheim, Stockholm, Copenhagen, London School of Economics, London Business School, City University London, and a number of Dutch universities that participate in Netspar (Network for Studies on Pensions Aging and Retirement).

Networking

The more senior researchers are well-connected and meet at all kinds of conferences. Typically, they are also well-connected to the US scene. Networks like CES-Ifo, CEPR and Netspar play a crucial role in this interaction by organising conferences, publishing papers and providing expert reports on preliminary drafts of discussion papers. The Survey of Health Aging and Retirement in Europe (SHARE) plays a vital role in providing consistent European data for analysis of individual behaviour. Potential reforms in pension systems are sometimes motivated by reforms in other European countries. Cross-country contacts between researchers are therefore of central importance.

Interaction academia and private and public pension sectors

Analysis of European pension systems requires interaction not only between academics in different countries, but also between academia and the pension and insurance sector. The European financial industry has recently undertaken a number of successful attempts to improve the interaction with academia. Leading examples are the French Europlace Institute, the Swiss Finance Institute and the Dutch Duisenberg School. These initiatives have a wider scope than a focus only on pensions. Initiatives like the Mannheim Research Institute for the Economics of Aging, the Pensions Institute (London) and Netspar (Netherlands) focus specifically on pension-related issues and benefit from excellent contacts with the local pension industry.

6   Required research infrastructure

This section discusses the following aspects of a European research infrastructure:
- Exchange of scholars and networks
- Access to relevant data sources
- Facilities for experimental research
- Accessible information on European institutional arrangements
- Networks for interaction with industry and policymakers

Exchange of scholars and networks

While the contacts between senior researchers in small groups are worthwhile, more effort should be placed on exchange of PhD students and junior researchers. Also, participation in networks of a broader range of universities and European countries would be welcome.

Another limitation of the existing networks is that the existing contacts tend to be limited by disciplinary or even subdisciplinary boundaries. Micro-economists meet with micro-economists, financial economists with financial economists, psychologists with psychologists, and so forth. Academic knowledge – and especially the contribution of this knowledge to society – can be enhanced by combining expertise from different disciplines on a single topic. Section 4 identified a number of areas (for example, setting of defaults, intergenerational risk sharing, integration of health- and pension insurance with owner-occupied housing, and so forth) in which cooperation between various disciplines (for example, finance, labour economics, medical sciences, psychology, sociology, law) is crucial. Researchers from different disciplines studying human behaviour are often brought together around methodological tools, such as large databases documenting individual behaviour and various types of experiments. We now turn to these tools.

Access to relevant data sources

Access to rich data sources on individual behaviour is of vital importance for exploring how individuals behave and how the availability of additional pension products and / or changes in institutional settings would affect behaviour. Ideally, such data sources should combine information on past behaviour, current choice options and possible scenarios for the future. Data on labour-market history and opportunities are as relevant as data on consumption patterns, health and wealth. Europe needs to develop a data infrastructure that documents how human, financial and social capital are developing over the life cycle in the various regions, and how different national institutions affect these capital resources. This data infrastructure not only facilitates country-by-country comparisons, but also stimulates the interdisciplinary research (social sciences, psychology, medicine) that is needed to address the challenges raised by aging.

The availability of SHARE data has improved access to relevant micro-data for addressing the many research questions that are generated by ageing. However, SHARE contains hardly any data on the specifications of an individual's pension contract, which means that information is lacking on the incentives for additional saving, labour supply, adjustment of the asset allocation or the choice of a lump sum rather than annuities. Another limitation of these data is that they are not checked or enriched with more reliable administrative records of governments, employers or pension providers. Indeed, data from administrative records (tax returns, wealth) must be combined with data from pension providers (pension entitlements, provision for early or partial retirement) and data from questionnaires on personal views, characteristics and preferences (life expectancy, perceived health, preference for specific consumption goals, household composition, social networks). In the Netherlands, Netspar is cooperating with Statistics Netherlands and pension providers to merge administrative datasets with data from questionnaires, and to make 'anonimised' micro data available for academic research. Similar initiatives on a European scale would be extremely useful.

Facilities for experimental research

Section 4.2 discussed the existing literature on the impact of the choice architecture and governance. In order to offer successful innovative products, to define optimal institutional rules, to specify the right defaults and nudges, and to elicit preferences, it is important to experiment before certain product specifications or institutional arrangements are offered on a wide scale. Experiments can also shed light on the inherent social dimension of pensions and insurance.

Small-scale facilities for laboratory experiments are available at a number of universities, and a number of questionnaires and panels offer useful opportunities to run experiments. These facilities should be extended and made more generally available. In Europe there are very few examples of field experiments in which employers, pension providers, government and knowledge institutions collaborate to improve understanding of individual behaviour related to pensions and retirement. Such field experiments are quite relevant from an academic perspective, and would reduce the societal costs of too-frequent changes in institutions and product specifications on a country-wide scale.

Accessible information on European institutional arrangements

One of the crucial aspects of pension provision in Europe is the wide variety of institutions and legal structures across and within countries. This is a rich potential source of mutual learning. Research on the most policy-relevant questions can be improved significantly by enhanced access and analysis of the specific institutional arrangements in each country. Improved information and comparison of institutions in different European countries will help also to improve pension schemes by learning from other countries. At the same time, European legislation is becoming more important also in the pension and retirement domain. In order to be able to set adequate European legislation, policymakers need a thorough understanding of the existing systems and their advantages and disadvantages.

Accessible information and cross-country comparisons of the institutional arrangements for retirement provision are important goals of international organisations such as the OECD. Too many researchers are hardly connected to information on institutional arrangements in other European countries, and do not contribute to cross-country analyses. The knowledge institutions could collaborate with the OECD to improve cooperation. This should enhance not only research quality, but also the policy relevance of the research output and the relevance of the data collected by the OECD. In particular, information on pensions and other institutions in the various countries could be combined with micro-econometric panel data in order to come up with better information on the incentives faced by the various individuals in these panels.

Networks for interaction with industry and policymakers

Interaction with the private pension industry (insurance companies, pension funds, banks) and policymakers in governments is of significant importance for academic research to be able to contribute to the most urgent policy questions and to have access to relevant expertise and data. Section 5 indicated that some research centres maintain excellent links with the local pension industry and the local government.

A European network for research on ageing and retirement provision that is set up and run jointly by knowledge institutions, national and European government institutions and the pension industry could be mutually advantageous. Stimulated by the recent legislation on Institutions for Occupational Retirement Provision (IORPs) and increased labour mobility within Europe, pension provision is becoming an international (rather than only a local) matter. Moreover, the European diversity in pension institutions allows for mutual learning. Europe can become a laboratory for social innovation. A European network could fund research projects that combine pure academic research with contributions to policy-relevant questions. Availability of such funding stimulates researchers to work on innovative and relevant questions and products and improves the information flows on the specification of the most pressing research questions. The pension industry may well be willing to cover part of the costs of such research projects if they have a say in the research topics covered.

7   Relevance for policy questions

Section 2 classified the main policy questions on innovative institutions and products for retirement provision in Europe into two groups: risk management and individual versus collective decision-making. The research questions and research infrastructure outlined in this chapter can help enlighten risk management by stimulating research and knowledge exchange on various pension products and by providing a thorough and independent analysis of the advantages and disadvantages of different institutions and products. Among the research topics are the proper balance between first-, second- and third-pillar arrangements, the choice between defined-contribution and defined-benefit schemes (or one of many of hybrid forms between these extremes), the recommended asset allocation and possible integration of retirement provision with health insurance and housing.

The intended research programme will be equally important to provide answers to policy questions on optimal life-cycle decision-making. Analysis of the behavioural impact of the rich European diversity in institutions and products for retirement provision will provide answers on how people actually behave and on how pensions and other institutions can help individuals improve life-cycle planning in financial- and labour markets.

References

Abowd, J.M. and D. Card (1989) 'On the Covariance Structure of Earnings and Hours Changes', *Econometrica*, 57, 411-445.

Ameriks, J., A. Caplin and J. Leahy (2003) 'Wealth Accumulation and Propensity to Plan', Quarterly *Journal of Economics*, 1007-1047.

Barberis, N., M. Huang and T. Santos (2001) 'Prospect Theory and Asset Prices', *Quarterly Journal of Economics*, 116, 1-53.

Benartzi, S. and R. Thaler (1995) 'Myopic Loss Aversion and the Equity Premium Puzzle', *Quarterly Journal of Economics*, 110, 73-92.

Benartzi, S. and R. Thaler (2001) 'Naive Diversification Strategies in Defined Contribution Savings Plans', *American Economic Review*, 91, 79-98.

Benartzi, S. and R. Thaler (2007) 'Heuristics and Biases in Retirement Savings Behaviour', *Journal of Economic Perspectives*, 21(3), 81-104.

Benzoni, L., P. Collin-Dufresne and R.S. Goldstein (2007) 'Portfolio Choice over the Life-Cycle when the Stock and Labor Markets are Cointegrated', *Journal of Finance*, 62, 2123-2167.

Berkelaar, A., R. Kouwenberg and T. Post (2004) 'Optimal Portfolio Choice under Loss Aversion', *Review of Economics and Statistics*, 86, 973-986.

Blake, D., A. Cairns and K. Dowd (2008) 'Modelling and Management of Mortality Risk: A Review', *Scandinavian Actuarial Journal*, 2008(2&3), 79-113.

Bodie, Z., R.C. Merton and W.F. Samuelson (1992) 'Labor Supply Flexibility and Portfolio Choice in a Life-Cycle Model', *Journal of Economic Dynamics and Control*, 16, 427-449.

Bovenberg, L., R. Koijen, T. Nijman and C. Teulings (2007) 'Saving and Investing over the Life Cycle and the Role of Collective Pension Funds', Netspar panel paper nr. 1.

Bovenberg, L. and T. Nijman (2009) 'Dutch Stand-alone Collective Pension Schemes: The Best of Both Worlds?', *International Tax and Public Finance*, 16, 443- 467.

Brennan, M. and Y. Xia (2000) 'Stochastic Interest Rates and the Bond-Stock Mix', *European Finance Review*, 4, 197-210.

Brown, J. (2007) 'Rational and Behavioral Perspectives on the Role of Annuities in Retirement Planning', NBER working paper 13357.

Campbell, J. and J. Cochrane (2000) 'Explaining the poor Performance of Consumption based Asset Pricing Models', *Journal of Finance*, 55, 2863-2887.

Campbell, J. and L. Viceira (2001) 'Who should buy Long Term Bonds?', *American Economic Review*, 91, 99-127.

Campbell, J. and L. Viceira (2005) 'The Term Structure of the Risk-Return Trade-off', *Financial Analysts Journal*, 61(1), 34-44.

Campbell, J., Y.L. Chan and L.M. Viceira (2003) 'A Multivariate Model of Strategic Asset Allocation', *Journal of Financial Economics*, 67, 41-80.

Chacko, G. and L. Viceira (2005) 'Dynamic Consumption and Portfolio Choice with Stochastic Volatility in Incomplete Markets', *Review of Financial Studies*, 8(4), 1369-1402.

Choi, J., B. Madrian, A. Metrick and D. Laibson (2002) 'DC Pensions: Plan Rules, Participant Decisions, and the Path of Least Resistance', *Tax Policy and the Economy*, 16, 67-114.

Choi, J., D. Laibson and B. Madrian (2005) '$100 Bills on the Sidewalk: Suboptimal Saving in 401(k) Plans', NBER working paper.

Clark, G., E. Caerlewy and J. Marshall (2006) 'Pension Fund Trustee Competence: Decision Making in Problems Relevant to Practice', *Journal of Pension Economics and Finance*, 5(1), 91-110.

Cocco, J.F., F. J. Gomes and P.J. Maenhout (2005) 'Consumption and Portfolio Choice over the Life Cycle', *Review of Financial Studies*, 18, 492-533.

Cochrane, J. and M. Piazessi (2005) 'Bond Risk Premia', *American Economic Review*, 95, 138-160.

Constantinides, G.M., and D. Duffie (1996) 'Asset Pricing with Heterogeneous Consumers', *Journal of Political Economy*, 105, 219-240.

Davidoff, T., J.A. Brown and P. Diamond, 2005, 'Annuities and Individual Welfare', *American Economic Review*, 95, 1573-1590.

Duflo, E. and E. Saez (2003) 'The Role Of Information And Social Interactions in Retirement Plan Decisions: Evidence from a Randomized Experiment', *Quarterly Journal of Economics*, MIT Press, 118(3), 815-842.

Farhi, E. and S. Panageas (2007) 'Saving and Investing for Early Retirement: A Theoretical Analysis', *Journal of Financial Economics*, 83, 87-121.

Gollier, C. (2006) 'Intergenerational Risk Sharing and Risk Taking in a Pension Fund', University of Toulouse working paper.

Gomes, F.J., L.J. Kotlikoff and L.M. Viceira (2008) 'Optimal Life-cycle Investing with Flexible Labor Supply: A Welfare Analysis of Life-Cycle Funds', *American Economic Review*, 98(2), 0. 297-303.

Gong. G., and A. Webb (2007) 'Evaluating the Advanced Life Deferred Annuity – An Annuity People Might Actually Buy', Center for Retirement Research working paper 2007-15.

Guiso, L. and T. Japelli (2005) 'Awareness and Stock Market Participation', *Review of Finance*, 9, 537-567.

Haliassos, M. and C. Bertaut (1995) 'Why do so Few Hold Stocks?', *Economic Journal*, 105, 1110-1129.

Koijen, R., T. Nijman and B. Werker (2009) 'When can Lifecycle Investors Benefit from Time-varying Bond Risk Premia?', Netspar discussion paper.

Kooreman, P. and H. Prast (2007) 'What does Behavioral Economics Mean for Policy: Challenges to Savings and Health Policies in the Netherlands', Netspar panel paper.

Liu, J. (2007) Portfolio Selection in Stochastic Environments, *Review of Financial Studies*, 20(1), 39.

Lusardi, A. (1999) 'Information, Expectations and Savings for Retirement' in H. Aron (ed.) *Behavioral Dimensions of Retirement Economics* (Washington DC: Brookings Institution), 81-115.

Merton, R.C. (1971) 'Optimum Consumption and Portfolio Rules in a Continuous Time Model', *Journal of Economic Theory 3*, 373-413.

Michelangeli, V. (2008) 'Does it Pay to get a Reverse Mortgage?', Boston University working paper.

Munk, C. and C. Sorensen (2010) 'Dynamic Asset Allocation with Stochastic Income Interest Rates', *Journal of Financial Economics*, 96, 433-462.

Munnel, A. and A. Sunden (2004) 'Coming up Short: The Challenge of 401(k) Plans', Brookings Institution, Washington DC.

Munnel, A.H., M. Soto and J.P. Aubrey (2007) 'Do People Plan to Tap their Home Equity in Retirement?', Report 7-7, Center for Retirement Management, Boston College.

Sangvinatsos, A. and J. Wachter (2005) 'Does the Failure of the Expectations Hypothesis Matter for Long Term Investors?', *Journal of Finance*, 60, 179-230.

Scott, J., J. Watson and W.Y. Hu (2006) 'Efficient Annuitization with Delayed Pay-out Annuities', SSRN, working paper.

Soest, van, A. (2009) 'Labour Supply and Employment of Older Workers,' Forward Looks paper.

Thaler, R. and S. Benartzi (2004) 'Save More Tomorrow', *Journal of Political Economy*, 112(1), 164-187.

Thaler, R.H. and C.R Sunstein (2008) *Nudge: Improving Decisions about Health, Wealth and Happiness* (Yale University Press).

Topel, R.H. and M.P. Ward (1992) 'Job Mobility and the Careers of Young Men', *Quarterly Journal of Economics*, 107, 439-479.

Tversky, A. and D. Kahnemann (1992) 'Advances in Prospect Theory: Cumulative Representation of Uncertainty', *Journal of Risk and Uncertainty*, 5, 297-323.

Venti, S. (2006) 'Choice, Behaviour and Retirement Savings', in G. Clark, A. Munnel and M. Orszag (eds) *The Oxford Handbook of Pensions and Retirement Income* (Oxford University Press).

Yaari, M.E. (1965) 'Uncertain Life Time, Life Insurance and the Theory of the Consumer', *Review of Economic Studies*, 32, 137.

Innovative Institutions and Products for Retirement Provision in Europe

Comments by Eric French

This chapter focuses on the key issues that must be answered before we can consider what optimal pension schemes look like. The chapter considers how pension schemes might affect the adequacy of savings and the optimality of risk sharing.

The chapter uses a good benchmark model for thinking about optimal savings and risk sharing: the Merton-Samuelson model. It is a simple model with clear welfare implications and empirical predictions. The Merton-Samuelson model gives a framework for thinking about two key questions. First, what is the optimal level of savings? The model gives us a framework for thinking about whether people are adequately prepared for retirement. Second, what is the optimal risk-sharing arrangement? This question has several parts, as people face many different risks. For example, most people should probably have some exposure to the stock market, as the stock market tends to provide consistently higher rates of returns over long time spans. But as we have recently learned, stock prices can go down. Thus, the optimal portfolio problem trades off the higher returns from stocks against their higher risk. The Merton-Samuelson model can also be augmented to consider the risk of outliving one's financial resources, and how best to insure against this risk.

The Merton-Samuelson model imposes very strong assumptions, which Bovenberg and Nijman describe in greater detail. The model assumes that individuals feature constant relative risk aversion, have rational expectations, enjoy a certain lifespan, and can borrow and lend without constraints. The original model is often augmented to have zero initial financial capital, so that initial capital is riskless and human capital exogenous. Furthermore, individuals have two financial instruments: riskless bonds and risky stocks. Individuals have two choices to make at any point in time. The first choice is how much to consume and how much to save. The second choice is over the share of lifetime wealth, how much should be invested in stocks versus bonds.

The optimal decision rule in this problem is that the share of remaining lifetime wealth (both financial and human) in equities is a constant. In other words, the share of lifetime wealth invested in equities does not depend on age and does not depend on wealth. Everyone should partake in some stock-market risk in order to get some of the upside risk of the market. But this does not necessarily mean that their current financial wealth will be split evenly between stocks and bonds, because – especially when young – most of their lifetime wealth is in the form of human capital. If human capital is riskless, then most of their wealth is tied up in a riskless asset. If this is true, then young people should buy stocks on margin to enjoy some of the upside of the market.

There are two key predictions of the Merton-Samuelson model that are at odds with the evidence. First, most people hold few financial assets in the form of equity. This goes against the notion that everyone should have some exposure to the stock market. Second, young people invest in stocks less than middle-aged people. The chapter describes research on two potential explanations for why the model is at odds with the empirical evidence. First, it is possible that people are rational forward-looking agents. Second, it is possible that the basic model ignores key risks and budget constraints. Personally, I agree more strongly with the second explanation.

One potential reason why the Merton-Samuelson model fails to accurately predict human behaviour is that it does not capture the full richness of people's financial lives. Consider the following depiction of the financial lives of most people. When people are in their twenties, their income is low. Thus, individuals should optimally set consumption close to income (but save a little for precautionary reasons, such as job loss). Once in their thirties, people are making more money, often buy a house, and have children. When in their forties and fifties, they pay off the mortgage and start sending the children off to college. After retirement, income falls. However, public and private pensions often replace more than 70% of pre-retirement income. Furthermore, there are no work-related expenses and the mortgage on the house is largely paid off. Lastly, the kids are out of the house and are no longer an expense. Thus, the pension is largely adequate to finance retirement. We still do not have models that capture the level of richness described above. The models that come close show that it is an optimal rule not to save very much.

For example, Scholz, Seshadri and Kitatrikun (2006) formulate a model of the savings behaviour of US households. The model features a realistic pension system, idiosyncratic earnings and medical expense risk, and expenses associated with children. The authors show that this model predicts the distribution of wealth at retirement extremely well. This shows that the low levels of wealth of many people need not be the result of any rejection of optimising behaviour. Instead, these levels are merely indicative of households behaving optimally, given their incentives. Likewise, DeNardi et al. (2009) show that when adding sufficient realism to a model of savings behaviour after retirement, savings behaviour predicted by the model lines up very closely with the data. What comes out of many of these models is that certain risks in models reinforce one another. For example, DeNardi et al. show that an important determinant of saving is the risk of outliving one's financial resources, much more than facing catastrophic expenses. Considering one aspect of a model (just uncertain lifetimes or uncertain medical expenses) in isolation is misleading.

One drawback of the optimal portfolio literature is that it is still somewhat crude. While several models have some realistic aspects, no single model has all the sources of realism. Thus, while financial planning mistakes have been well documented, it is still not totally clear to what extent the observed portfolio holdings of people are the result of optimal behaviour or the result of mistakes – and until we know that, we cannot be sure of how to modify institutions in order to provide for more savings adequacy and better risk sharing for the elderly.

References

De Nardi, M., E. French, and J.B. Jones (2009) 'Life Expectancy and Old-Age Savings,' *American Economic Review: Papers and Proceedings*, 99(2):110–115.

Scholz, J.K., A. Seshadri and S. Khitatrakun (2006) 'Are Americans Saving Optimally for Retirement?', *Journal of Political Economy*, 114(4): 607-643.

Innovative Institutions and Products for Retirement Provision in Europe

Comments by Henriëtte Prast

Systems for providing income to the elderly differ widely from country to country. In most cases there is a first layer provided by government or quasi-governmental organisations that is intended to have broad coverage of the population and eliminate poverty among the elderly. First-layer benefits may or may not be related to one's history of contributions. In Italy, for example, Social Security benefits depend on one's labour history, whereas in the Netherlands benefits are the same amount for every citizen – provided he or she has lived for a long enough time in the Netherlands. Many countries also have a second layer of pension income that is intended to enable people to replace their wages or salary after they retire from the labour force. It is usually provided through an employer or trade union, and hence referred to as an occupational pension. Occupational plans may or may not be mandatory, and they may be funded DB plans, pay-as-you-go DB plans, collective or individual DC plans, or a mixture. There is often a third layer of voluntary tax-advantaged personal saving, intended to enable the elderly to maintain their standard of living in old age.

Not only do pension systems differ across Europe, there are also huge cross-country differences in retirement behaviour. Even with a standard retirement age of 65 in most European countries, actual behaviour differs considerably (Kapteyn and Andreyeva, 2008). Despite these differences, there is a common characteristic: the systems are under pressure. The ageing of the population, the new international accounting standards, new views on optimal life-cycle planning – making use of new-generation financial products and techniques –, labour-market developments, internationalisation, a general policy trend of risk-shifting toward (financial) consumers, and, last but not least, behavioural economics evidence call for a rethinking of existing pension plans. Moreover, some countries are setting up pension systems from scratch and need guidance. In the US there is widespread dissatisfaction with 401k-type individual retirement accounts, and in Europe there is a growing recognition that traditional collective plans need to be modified or even completely redesigned.

This offers huge challenges, and the chapter by Bovenberg and Nijman offers an extremely valuable overview of the work that still has to be done. In this discussion I would like to focus on useful advances in the behavioural economics field and their implications for the research and policy agenda.

Over the past decades the policy trend of deregulation and liberalisation has gone hand-in-hand with increased attempts by governments to affect individual behaviour without the use of explicit distortionary instruments. In a world in which choice and risk are shifted to individuals, the focus is on making markets work (hence the establishment of antitrust authorities and market conduct watchdogs), and on ensuring that individuals are well-informed in a world in which risk is shifted towards them (hence, the programmes for educating financial consumers, and the new rules for transparency with regard to the quality of services and products).

The assumption underlying a policy aimed at helping individuals make good choices by informing and persuading them – where persuasion often uses facts to try to convince people – is that well-informed individuals make choices that are consistent with their preferences. While the trend in policymaking has been to help people make calculated decisions by making markets work, creating a level playing field, and providing individuals with knowledge, behavioural economics research has shown that decision-making errors are much more prevalent and systematic than has been assumed in economics and among policymakers.

Core findings in psychology and economics are that people deviate from the rational-choice model in all of the following respects: they have nonstandard time-, risk- and social preferences; they exhibit systematic biases in the gathering and processing of information; and they exhibit systematic and predictable biases in decision-making (DellaVigna, 2009).

These findings have several implications for policymakers. First, they expand the scope of paternalistic regulation. Second, they call into question the effectiveness of policies that rely on information and education: if people do not gather and process information rationally, and systematically make choices that are inconsistent with their preferences (even if they are well informed), then a policy based on information and education will not yield the results predicted by rational-choice theory. In fact, we now know that whereas information and education can alter intention, they do not alter action. The latter requires the design of helpful access, clever decision-aids and insightful contexts. The issue is even more relevant as third parties may benefit from decision errors and intentionally frame decisions accordingly. Moreover, in some areas there is little scope to learn from mistakes, as decisions are taken infrequently or even 'once in a lifetime' – retirement saving being a case in point (Gabaix and Laibson, 2006). The powerful effect of choice architecture requires that this architecture be designed carefully. As Bovenberg and Nijman correctly point out in this chapter, further progress in this area will occur when more research is done – for example, by using micro-econometric data to document heterogeneity across individuals. We now know that saving for retirement is an area in which the factors that contribute to differences between revealed- and normative preferences abound (Beshears et al., 2010). Research using US data also shows that private parties exploit cognitive errors by people over 60 (Agarwal et al., 2010). Here, also, there is plenty of scope for more European research.

References

Agarwal, S., J.C. Driscoll, X. Gabaix and D. Laibson (2010) 'The Age of Reason: Financial Decisions over the Life Cycle and Implications for Regulation', forthcoming Brookings Papers on Economic Activity.

Kapteyn, A. and T. Andreyeva (2008) 'Retirement Patterns in Europe and the US', Netspar panel paper (6).

Beshears, J., J.J. Choi, D. Laibson and B.C. Madrian, (2010) 'How are Preferences Revealed?', forthcoming in Journal of Public Economics.

Della Vigna, S. (2009) 'Psychology and Economics: Evidence from the Field,' *Journal of Economic Literature*, XLVIL (2).

Gabaix, X. and D. Laibson (2006) 'Shrouded Attributes, Consumer Myopia, and Information Suppression in Competitive Markets,' *Quarterly Journal of Economics*, 121(2): 505-540.

Part II:
Well-being of the Elderly

5

Socioeconomic and Psychosocial Determinants of Well-being in Early Old Age

Johannes Siegrist / Morten Wahrendorf

1 Introduction and policy questions

Many actions and decisions taken by welfare institutions and healthcare organisations in modern societies are contingent on a medically defined condition of disease or disability. Medical knowledge on the development and course of diseases and disabilities, and their diagnosis, treatment and prevention, serves to legitimise access to scarce resources. This access may take the form of a disability pension or some form of early retirement, use of hospitals or rehabilitation centres, or allocation of specific benefits provided by health insurance organisations (disease management programmes, for instance). While convincing reasons back this approach, positive notions of health and well-being as phenomena that deserve an analysis on their own are rare – perhaps with the exception of the International Classification of Function (WHO, 2001). Rather, health and well-being are defined as states that are characterised by the absence of signs of disease or disability. More elaborate positive notions of health and well-being might be instrumental, however, in supporting attempts of welfare- and healthcare organisations to act pro-actively rather than reactively to needs (for example, to develop incentives to maintain or promote good health and well-being over the life course). Developing incentives towards maintaining and promoting good health seems particularly relevant in view of rapidly ageing populations in modern societies, including most European countries. Given the economic, social and healthcare burden of ageing societies, such attempts to promote 'healthy ageing' (Rowe and Kahn, 1997) must be considered a policy challenge of high priority.

Several decades ago, an attempt was made to forge a more elaborate definition of health by René Dubos, who maintained that health is best defined as one's ability to achieve self-defined or other-defined goals by own agency on the basis of relatively stable physiological functions (Dubos, 1969). This definition gives key significance to goal-oriented agency. Hence, welfare institutions and (more generally) societal opportunity structures that support and strengthen the capability of individuals to achieve goals by means of autonomous activity might contribute to healthy ageing at a population level, rather than merely the individual level. Examples of such institutional arrangements include specific policy programmes that favour participation in productive activities after labour-market exit, such as insurance protection for volunteers or tax-relief measures for this group of people.

This chapter, using a comparison of different European countries, considers the potential of societal opportunity structures, including welfare institutions, to provide options of goal-oriented agency to early old-age populations. An analysis is also made of how different types of goal-oriented agency of older people affect their health status and well-being (use is made of the most recent data from the Survey of Health, Aging and Retirement in Europe [SHARE] (Börsch-Supan et al., 2005) and comparable investigations, in addition to a review of currently available evidence; see section 2). This analysis makes it possible to identify gaps of knowledge, and to develop suggestions for future research (section 3). Finally, the chapter discusses the extent to which the currently available scientific evidence can provide answers to relevant policy questions – and what is needed to reduce the gap between science and policy (section 4).

This contribution addresses the following two sets of policy questions:

(1)   What needs to be done to retain as many older people as possible in work – either employed or self-employed? What might be done to reduce the proportion of employees with early exit from the labour market? What measures might be taken to better protect and improve the health and well-being of middle-aged to early old-aged working men and women?

(2)   What could be done to enable retired people to continue or to initiate socially productive activities (such as volunteering, being engaged in informal help or caring for a sick or disabled person)? How might the proportion of socially productive early old-aged people be augmented? What are the costs and benefits of extending respective opportunities and incentives?

As can be seen, a distinction has been made of two types of goal-oriented agency with potentially beneficial effects on health and well-being: paid work and voluntary or informal work. In either case, an analysis must be made of the conditions under which such potentially beneficial effects may actually occur. While this is the main task of the following section, it might be helpful to first explain why early old age – as a distinct phase in the life course – deserves special attention with regard to these two types of goal-oriented agency, and to what extent well-being in this phase of the life course is socially patterned.

2   Understanding the links between socially productive activities and well-being: A novel approach and a review of the evidence

2.1   Well-being in early old age

While life expectancy has increased among most age groups in Europe during the past decades, this increase has been particularly pronounced in middle-aged populations. In England, for instance, life expectancy at middle age has increased since 1970 by more than it did during the period from 1900 to 1970 combined (Blane et al., 2007). As a result of demographic ageing in combination with substantial improvements of population health, a new phase of the life course has evolved: early old age, or the 'third age' (Laslett, 1989). While this phase lacks precise age limits, it typically includes people with an age range from 50 or 55 years to about 75 or 80 years. Most people in early old age are still in good health, and free from physical dependency – but also free from earlier responsibilities for work and family, which opens opportunities for individual freedom, hobbies and other options of self-realisation. At the same time, this phase lacks a clear societal definition in terms of social roles and social status, legitimised expectations, norms and values. While most middle-aged people experience a secure sense of social identity by maintaining core social roles (such as the work role, family roles or civic roles), social identity during the third age becomes more fragmented and insecure – often in combination with a reduced intensity of contacts in social networks and with reduced income opportunities (Banks et al., 2006).

This is exactly the place to consider the role of goal-oriented agency in early old age. As most men and women in this age group are able to continue to engage in some type of activity, the question arises to what extent they are offered options of paid (formal) or unpaid (informal) work, and under what conditions these options can be realised. Despite distinct differences, formal- and informal work have some things in common, as they represent two manifestations of what can be termed a 'socially productive activity'. Socially productive activity has been defined as 'any agreed-upon continued activity that generates goods or services that are socially or economically valued by the recipient(s), whether or not based upon a formal contract' (Siegrist et al., 2004: 4).

This definition, which includes paid work as well as volunteering and other types of informal work, has two relevant implications. First, it emphasises the voluntary nature of a continued transaction – irrespective of its degree of formalisation. Second, it underscores the expression and transmission of value, either monetary or non-monetary, from recipients to providers as part of a reciprocal exchange. Thus, socially productive activities are based on a fundamental principle of interpersonal exchange, the norm of reciprocity. According to this norm, any action or service provided by person A to person B that has some utility to B is expected to be returned by person B to A (Gouldner, 1960). Exchange expectancy does not implicate that the service in return corresponds exactly to the service provided – but is assumed to meet some agreed-upon standard of equivalence.

The norm of reciprocity is thus considered to be a general principle governing voluntary social exchange that includes productive activity. Valuing productive activities through rewards occurs both in formal and informal transactions.

Research has demonstrated that the experience of appropriate rewards that match prolonged effort expended ('recurrent reciprocity') elicits a pronounced activation of the brain reward circuits, and that this activation is associated with strong positive emotions (Schultz et al., 1997). It is likely that the experience of reward evolving from reciprocal exchange reinforces the provider's positive sense of self-esteem and, by doing so, acts as a powerful motivation to continue goal-oriented agency. Given its basic motivational function, this experience is expected to exert powerful effects throughout the life course (that is, in midlife as well as in early old age). A theoretical model (the 'effort-reward imbalance' model) claims that the continued experience of reciprocity between effort expended and rewards received promotes health and well-being, whereas violations of this principle under circumstances where efforts outmatch rewards ('recurrent non-reciprocity') elicit sustained stressful experience with adverse long-term effects on health and well-being (Siegrist, 2005) (see below 2.3 and 2.4).

Experiencing reciprocity in a goal-oriented productive activity has been emphasised as a potentially health-protective psychosocial resource. Experiencing autonomy and control over one's productive activity must be considered an additional health-protective psychosocial resource. Control over one's agency reinforces feelings of self-efficacy and mastery, thereby reducing feelings of uncertainty, threat and anxiety (Haidt and Rodin, 1999). Theoretical concepts of objective and subjective control have been successfully tested with respect to a variety of health outcomes (Skinner, 1996; Karasek and Theorell, 1990; Wahrendorf, 2009; see below 2.3 and 2.4).

Based on these assumptions, the following hypotheses elucidate the links between socially productive activity and well-being:

First, people who are engaged in a socially productive activity experience, in general, better health and well-being compared to those who are not thus engaged. This effect is attributed to the experience of positive emotions of agency and meaningfulness.

Second, people engaged in a productive activity whose efforts are compensated by appropriate rewards ('recurrent reciprocity') experience, in general, better health and well-being compared to those whose efforts are not adequately rewarded ('recurrent non-reciprocity'). This effect is attributed to the experience of strong positive emotions of esteem/recognition/compensation in relation to their commitment.

Third, people engaged in a productive activity that provides control and autonomy experience, in general, better health and well-being compared to those whose activity limits or excludes control and autonomy. This effect is attributed to the experience of strong positive emotions of self-efficacy and mastery.

The socio-emotional consequences of socially productive activities may be particularly relevant in early old age, where options of agency, control and reward resulting from core social roles are becoming less frequent and less pronounced. Therefore, being socially productive in a formal activity (continued paid work, for example) or informal activity (volunteering, for instance) that provides recurrent reciprocity and autonomy is assumed to exert beneficial effects on health and well-being in early old age. Conversely, experiencing recurrent non-reciprocity and lack of autonomy in such an activity reduces the probability of healthy ageing – other things being equal (Siegrist et al., 2004).

Sections 2.3 - 2.5 review the evidence that supports or contradicts these assumptions. In case of robust support, a policy implication would point to the need of maintaining or creating opportunities of socially productive activities in early old age that provide opportunities for reward and control. In this framework, rewards are not confined to the material dimension, but include the socio-emotional dimension of esteem and recognition, given its significant role in triggering positive self-experience, personal need fulfilment and its beneficial health effects.

2.2   The social distribution of well-being in early old age

Having outlined a conceptual framework for the analysis of social productivity and well-being among older people, we must consider the social distribution of well-being in early old age. Previous research conducted in middle-aged populations in a variety of Western countries has demonstrated a social gradient of morbidity and mortality across the whole of a society. Results indicate that with each step one moves up on the social ladder, the better one's health is. Mean differences in life expectancy between those at the top and at the bottom of a society's social structure are substantial, ranging from four to ten years (Marmot and Wilkinson, 2006; Mackenbach and Bakker, 2002; Siegrist and Marmot, 2006). Whether this social gradient persists into older age is less well known, however. Given a high inter-individual variability of healthy ageing, and in view of selective mortality, it is likely that social inequalities in health and well-being will be diminished or even absent in older populations. What evidence is there of a social gradient of health and well-being in early old age?

Answers to this question are not easy to obtain, because the traditional measures of social inequality (that is, social status as defined by occupational position, income and education) need to be supplemented by measures that reflect more accurately the material and non-material conditions of older people (such as housing tenure and wealth, or access to sources of tangible support). Thus far, however, very few studies have used these indicators (Knesebeck et al., 2007). Moreover, the association of social status with health is more likely to be bi-directional in older age, where ill health leads to income loss and deterioration in some other measures of social standing, restricting the evidence to longitudinal investigations.

Several studies so far have explored the associations of socioeconomic position with health and well-being in the elderly. Their results can be summarised as follows.

First, concerning mortality, a social gradient is still observed if measures of previous occupational position or education are used, but the steepness of the gradient clearly declines with advanced age (van Rossum et al., 2000; Huisman et al., 2004). Although the relative inequalities between social groups decrease, it is likely that the absolute differences remain substantial, given the general increase of mortality in older age (Karasek and Theorell, 1990). For example, taking the period 1997-1999 in the United Kingdom, life expectancy at age 65 was 30% longer for men in the highest of the five social strata (17.5 years) compared to men in the lowest social group (12.3 years) (Siegrist and

Marmot, 2006). In addition to total mortality, a social gradient was documented for several leading causes of death (including cardiovascular disease, certain cancers, respiratory disease, liver disease and neurodegenerative disease; see van Rossum et al., 2000).

Second, a recent review documents robust evidence of educational degree being associated with risk of disability at older age, leaving those with lower degrees at higher risk (McMunn et al., 2006). This association was found to be less consistent if occupational position, income or wealth was used as indicator of social standing (McMunn et al., 2006). Yet, a study in Great Britain indicates that occupational grade assessed at midlife strongly predicted level of functioning some 29 years later (Breeze et al., 2001). Moreover, the risk of experiencing a longstanding illness or incidental mobility problems increases with lower income or wealth (McMunn et al., 2006).

Third, several studies have investigated social inequalities of well-being among older people. Two types of measures are used in this regard, self-assessed health and quality of life. Self-reported health was found to be related to educational degree and (independent of education) to financial wealth in several reports (McMunn et al., 2006). Quality of life may represent a more adequate indicator of well-being, as it captures different dimensions of everyday experience. One such measure, the *CASP questionnaire* (Hyde et al., 2003), has proved to be particularly useful in ageing studies. This measure is a summative index consisting of 19 (or, in the short version, 12) Likert-scale items assessing the four domains of Control, Autonomy, Self-realisation and Pleasure (hence, CASP-19; Blane et al., 2007). These domains are thought to describe aspects of well-being that are typical for third-age populations, while not being confounded by potential determinants such as health status or financial circumstances. The association between socioeconomic position and CASP was tested in two large-scale investigations.

The first of these examined this association with multiple social indicators among some 15,000 men and women 50 years or more of age in ten European countries (Knesebeck et al., 2007). The main findings indicate that education, income and assets are related to quality of life – although associations vary by country. Moreover, no indication was found that the socioeconomic differences in quality of life diminished after retirement. The second investigation used data from the *English Longitudinal Study of Ageing,* and applied the newly developed National Statistics Socio-Economic Classification, which defines seven different social strata mainly based on occupational characteristics (Blane et al., 2007). Results show that social position explains quality of life at older ages in a hierarchical manner, with a gradient similar to the one observed in previous morbidity and mortality studies.

In summary, substantial social inequalities in health and well-being are present in third-age populations in modern societies. These inequalities were most consistently reported for a variety of Western European countries as well as for the United States. Given this evidence, the next challenge consists of exploring the role of socially productive activities in the context of unequal health and well-being.

2.3   *Social productivity and well-being in early old age: The role of work and employment*

Early retirement from regular employment provides a major challenge to social and health policy in all European countries, as a shrinking proportion of economically active people will have to support a growing number of economically dependent older people. Across European countries, large variations in workforce participation rates among those aged 50 years and older are observed (Brugiavini et al., 2006). At least three types of determinants of early retirement have been identified: financial incentives, poor health and low socioeconomic status associated with poor quality of work.

Concerning financial incentives to retire early, national policies vary substantially with respect to regulations. The same holds true for pension schemes with extended eligibility and alternative income options (Gruber and Wise, 1999). In combination with economic pressure from employers, these national variations may largely account for substantial differences in workforce participation rates (in the age group 55-59, for example, these rates were recently as low as about 20% in Belgium, Italy and France, about 35 to 40% in Germany and Spain, and much higher in Switzerland and Scandinavian countries; see Brugiavini et al., 2006).

The two remaining determinants of early retirement, poor health and low socioeconomic status related to poor quality of work, are closely intertwined. Obviously, workforce participation beyond age 50 is socially patterned in all modern societies from which respective data are available: working people with low skill levels or low socioeconomic position leave the labour market earlier than do higher skilled and higher status people (Blekesaune and Solem, 2005). Part of this effect is due to a higher proportion of people working in physically demanding or even precarious jobs. The association of employment duration and well-being is bi-directional (Smith, 1999). Poor health and disability are powerful predictors of early exit from the labour market, and leaving the job prematurely contributes to diminished well-being.

The English Longitudinal Study of Ageing (ELSA) provides direct evidence of this bi-directionality (Emmerson and Tetlow, 2006). Men and women in paid employment reported better health than those who left their job (with the exception of a privileged, wealthy high-status group that could afford early retirement). During follow-up, both men and women who reported that their health was fair or poor at study onset were more likely to be retired prematurely two years later, compared with those reporting good health. Among men, the relatively largest differences were observed in the age group 50 to 54 years. The same holds true for those who reported some mobility limitations in the beginning. Conversely, those reporting excellent health during the first wave, while not being employed, had a higher probability of returning to work two years later.

The social patterning of disability retirement further supports the argument that health and well-being are powerful determinants of early labour-market exit. Several large-scale studies from Scandinavian countries have found that low occupational status (Krokstad et al., 2002), low income (Blekesaune and Solem, 2005), or low level of education (Hagen et al., 2000) were associated with elevated risk of disability retirement. A recent study from Germany observed that, among men, low income was a stronger predictor of disability retirement than low educational degree, piecework or physically demanding work, whereas, among women, low educational level was the strongest predictor (Dragano, 2007).

There is more, however: poor quality of work not only is more prevalent among working men and women with low socioeconomic status, but also exerts direct effects on health and well-being. Several studies have identified different dimensions of poor quality of work that have direct relevance for health. First, exposure to noxious physical or chemical stressors, noise and physically strenuous job tasks are still widespread across Europe (Parent-Thirion et al., 2007). All of these conditions were shown to increase the risk of early retirement – particularly the risk of receiving disability pension (Krokstad et al., 2002; Dragano, 2007). Second, shift work in combination with night shifts and extended overtime work adversely affects health – and the same holds true for monotonous, repetitive work, such as piecework or work in low-status service jobs (Blekesaune and Solem, 2005; Krokstad et al., 2002; Hagen et al., 2000; Dragano, 2007).

A third dimension of poor quality of work concerns the psychosocial features that elicit chronically stressful experience among exposed people. It is important to define poor psychosocial quality of work in terms of a theoretical model that allows for an identification of stressful aspects at a general level, thereby enabling its application to a wide range of different occupations. Although several theoretical concepts of stressful work have been developed (Cartwright and Cooper, 2008; Antoniou and Cooper, 2005), two models have received special attention in recent international research: the *demand-control model* (Karasek and Theorell, 1990) and the *effort-reward imbalance model* (Siegrist, 2005).

The former model identifies stressful work by job task profiles that are characterised by high demand in combination with low control (in particular, lack of decision authority over one's tasks). The combination of either component, but also the presence of a low level of control or decision latitude alone, is critical for health (Marmot et al., 2006). The latter model builds on the notion of social reciprocity that lies at the core of the employment (or work) contract. It claims that an imbalance between high effort expended and low rewards received in turn (money, esteem, career prospects including job security) adversely affects health. 'High cost/low gain' conditions at work are frequent among those who have no alternatives in the labour market, or those who work in highly competitive jobs. Moreover, the probability of these conditions occurring is increased by certain motivational patterns, such as work-related over-commitment.

Based on psychometrically validated questionnaires (Karasek et al., 1998; Siegrist et al., 2004), both models were tested in a broad range of epidemiological investigations. Recent reviews summarised the current state of research with respect to increased risk of cardiovascular disease (Kivimäki et al., 2006), mental health (in particular, depression) (Stansfeld and Candy, 2006), and other types of stress-related diseases. Moreover, recent studies include the analysis of psychobiological mechanisms and of behavioural outcomes (Siegrist, 2008; Theorell, 2008). Overall, working in a high-demand/low-control job, or encountering high effort in combination with low reward at work is associated with a twofold elevated relative risk of cardiovascular (and especially coronary heart) disease and a similar increase in the risk of incidental depression. In addition, psychosocial stress at work was found to be associated with an increased risk of metabolic syndrome and type-2 diabetes, alcohol dependence, musculoskeletal pain, and reduced physical and mental functioning (Siegrist, 2008; Theorell, 2008).

An adverse psychosocial work environment in terms of these two models exerts negative effects on health among older workers by increasing their probability of long leaves of absence due to sickness, of reduced performance and productivity, and of forced early retirement due to illness or disability. In fact, substantial evidence indicates that low control at work is an independent risk factor of disability pension (Blekesaune and Solem, 2005; Krokstad et al., 2002; Stattin and Järvholm, 2005), and at least one large-scale study demonstrates a similar effect of high effort in combination with low reward at work (Dragano, 2007). Moreover, the intention to leave one's job prematurely was found to be much more prevalent among older employees suffering from poor quality of work. In an analysis of first-wave data from the Survey of Health, Aging and Retirement (SHARE) study, based on data obtained from 3523 men and 3318 women in ten European countries (Sweden, Denmark, the Netherlands, Germany, France, Switzerland, Austria, Italy, Spain, Greece), significantly-elevated odds ratios (OR) of effort-reward imbalance (OR 1.72 – 1.43-2.08) and of low control at work (OR 1.51 – 1.27-1.80) were independently associated with the intention expressed by respondents to retire from work, after adjustment for well-being (Siegrist et al., 2007). This consistent association across all European countries in the study calls for greater investment in improving the quality of work – in particular, increased control and an appropriate balance between effort expended and rewards received at work (see below section 4).

2.4   Prospective evidence from the SHARE study

Data from the second wave of the SHARE study offer an opportunity to replicate earlier findings and to test the hypothesis that low quality of work (in terms of failed reciprocity between efforts spent and rewards received and in terms of low control) predicts reduced health and well-being two years later. The following discussion summarises the respective evidence of these analyses.

First, for both models we observe an overall lower quality of work in Southern and Eastern European countries, compared to Northern and Western European countries (Börsch-Supan et al., 2008). Western countries included the Netherlands, Germany, Belgium, France, Switzerland and Austria, Northern countries included Sweden and Denmark, Southern countries included Italy, Spain and Greece, and Eastern countries included Poland and Czechia.

Second, in all countries the prevalence of low quality of work is highest in the group of men and women who are located in the lowest tertile of income distribution, and similarly in the lowest tertile of educational attainment. Findings demonstrate a clear social gradient with decreasing prevalence of stressful work according to increasing income and increasing educational attainment. Social gradients are steeper for education than for income, and they are much steeper in Eastern and Southern countries, as compared to their Western and Northern counterparts. For instance, whereas 43% of older employees with low education in Eastern Europe experience lack of reciprocity between efforts and rewards, only 8% of those with highest level of educational attainment experience this lack (Börsch-Supan et al., 2008).

Third, to test the main hypothesis, two indicators of well-being were analysed: depressive symptoms, as measured by the EURO-D scale of depression (Dewey and Prince, 2005), and 'decreased self-rated health', as assessed by a single question comparing one's current health status to previous health. The sample, restricted to people with data from both waves, was composed of 5412 participants. Figure 5.1 demonstrates the prevalence of depressive symptoms (at wave 2) for three of the above-mentioned groups of European countries (Eastern Europe being excluded, as only one wave was completed so far), according to whether low quality of work assessed at wave 1 (scores in upper tertiles) was present or not. Despite variations in the prevalence of depressive symptoms, clear-cut differences are obvious in all three of the country groups for effort-reward imbalance and for low control at work. As mentioned, the same analysis was performed with decreased self-rated health as an indicator of well-being – and again, a similar pattern of results emerged (results not shown).

These findings were further analysed in multivariate logistic regression models that were adjusted for important confounders, including level of depressive symptoms at wave 1 and socioeconomic status (see Table 5.1). A remarkable outcome of these latter analyses indicates that the association of education and income with well-being is considerably attenuated if measures of quality of work are introduced into the model. For instance, a significant odds ratio (OR) of low education (OR 1.30) for depression was diminished (OR 1.09) after introducing the two work-stress summary measures, the effort-reward ratio and low control (respectively, significant ORs were 1.36 and 1.28). Similar findings were observed in analyses with decreased self-rated health as an indicator of well-being (results not shown).

In conclusion, a brief review of recent research testing hypotheses 2 and 3 (stated above in section 2.1) supports these assumptions, in general. These hypotheses claim that individuals engaging in productive activities characterised by recurrent reciprocity between effort and rewards and by a high degree of autonomy at work exhibit better health compared to those who are deprived

Figure 5.1: Prevalence of depressive symptoms in wave II (EURO-D) according to low quality of work in wave I (yes= highest tertile effort-reward ratio or low control; no= lower tertiles)

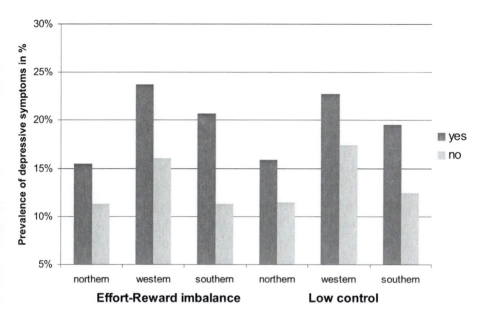

Table 5.1: Quality of work (effort-reward imbalance (ERI); low control) in wave I and depressive symptoms (EURO-D) in wave II, summary of regression models (OR)

		EURO-D	
Variables		Model 1	Model 2
		OR	OR
	Low	1.30**	1.09
Education	Medium	1.02	1.05
	High		
	Low	1.25**	1.05
Income	Medium	1.11	1.02
	High		
ERI	Yes		1.36***
	No		
Low control	Yes		1.28*
	No		

*p < 0,05; ** p< 0,01; *** p < 0,001

Note: All models are adjusted for gender, age, depressive symptoms (wave I) and country affiliation.

of these qualities of work. Our findings demonstrate that poor quality of work (defined in the terms above) is associated with elevated future risks of depressive symptoms and decreased self-rated health – and that these conditions account for part of the social gradient of health among older employees. Clearly, additional prospective evidence will be needed to confirm these findings. Furthermore, additional potential confounders, such as time and intensity of exposure to adverse working conditions, must also be monitored in greater detail.

2.5   *Social productivity and well-being in early old age: The role of informal work*

The definition of social productivity given above includes informal, unpaid work (e.g. in terms of volunteering, informal help, caring) as well as paid work. These activities become more relevant in third-age populations that are retired, given a lack of established societal roles. According to the hypotheses stated above, people who participate in socially productive activities benefit from these actions in terms of their well-being, especially if they provide experiences of reward and control. This section explores the evidence in a comparative European perspective. First, variations in the prevalence of productive activities across countries are described, taking into account age and gender differences. Next, the social gradient of productive activities is demonstrated. Finally, associations of socially productive activity with well-being are analysed, using volunteer work as a prominent example. This latter analysis considers differences in well-being between active versus non-active groups as well as differences within active groups, according to their experience of reciprocity or control.

Concerning variations across countries, baseline data from SHARE with information on people aged 50 and older demonstrate a North-South European gradient of socially productive activities, with more activities in Northern and North-Western countries and fewer activities in the Southern region. Taken together, about 10% of participants are engaged in volunteer work, about 17% in informal help, and about 5% in care for a sick person. Every tenth participant of this study is involved in more than one socially productive activity (Erlinghagen and Hank, 2006). The North-South difference is most pronounced in volunteer work, with 20% of older persons active in the Netherlands, Sweden and Denmark, but only 3% of their counterparts active in Greece and Spain. Possible explanations of these differences have been discussed elsewhere (Wahrendorf, 2009; Erlinghagen and Hank, 2006). In sum, it is assumed that elements of welfare-state regimes are related to patterns of participation in voluntary work – for example, high amounts of social expenditures in combination with a developed social sector offering people opportunities to participate (Salomon and Sokolowski, 2003).

Within each country, age, gender and socioeconomic status are relevant determinants of activities. For instance, informal help and caring is more frequent at a younger age (50 to 64 and 65 to 74), and the same holds true for the prevalence of voluntary work. The frequency of this latter activity sharply declines in the oldest age group (75 and beyond). In most countries, women are more active in caring, and men are more active in volunteering – but in general, gender differences are not very pronounced (Wahrendorf et al., 2008a).

A significant and consistent finding concerns social inequalities in the frequency of productive activities. For instance, those with higher education are three times more likely to volunteer and twice as likely to provide informal help and care compared to those with low education. These differences are consistent across all countries and most pronounced in case of voluntary work (Wahrendorf et al., 2008a).

The next question to be addressed has to do with the association of socially productive activity with well-being. The SHARE study introduced two indicators of well-being: quality of life and depressive symptoms. Quality of life was assessed by the CASP-12, an instrument specifically designed to capture quality of life in early old age. More specifically, quality of life is assessed as the degree to which specific domains of human needs are satisfied – needs that are thought to be particularly relevant in early old age. Referring to the literature of ageing (Laslett, 1996), four domains are used: control (C), autonomy (A), self-realisation (S), and pleasure (P). The degree to which these aspects are perceived as being satisfied is measured with 12 questionnaire items (three for each domain), and a sum score is calculated (range 12-48; with higher scores indicating better quality of life) (Hyde et al., 2003). Depressive symptoms were measured by the EURO-D scale. Figure 5.2 summarises the results from bivariate analyses using volunteering (assessed at wave 1) and quality of life (assessed at wave 2), again stratified according to the three groups of European countries (as above). In this case, participants are categorised into one of the following conditions: (1) those who did not volunteer; (2) those who volunteered and who experienced reciprocity of exchange; (3) those who volunteered, but did not experience reciprocal exchange.

Two striking findings are obvious from Figure 5.2. First, we observe again a North-South gradient in quality of life, irrespective of the participants' activity status. Second, in each group of countries, those who experience their voluntary work as rewarding enjoy significantly better quality of life than those without experienced reciprocity – and inactive participants have the lowest mean quality of life. A similar structure of results is apparent if the prevalence of depressive symptoms is studied.

Figure 5.2: Quality of life in wave II (mean scores of CASP-12 [range 12-48] and standard errors) according to voluntary work (3 categories) in wave I

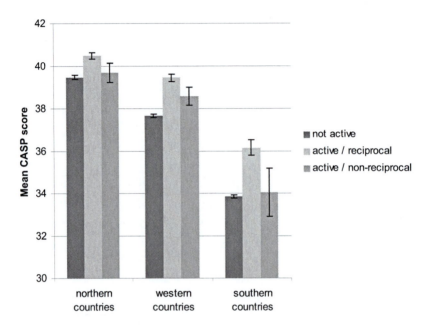

Again, these findings were further analysed in multivariate regression models that were adjusted for socio-demographic characteristics, including social position (model 1 in Table 5.2). In addition, a possible selection bias (people participate because they are in better health) was considered, taking quality of life at study onset into account (model 2). This latter model makes it possible to explore how the explanatory variables are associated with changes in quality of life between wave 1 and wave 2. Analyses were restricted to retired older people, as our particular interest is in the relevance of activities for the post-employment period of life.

Table 5.2: Voluntary work in wave I and quality of life (CASP) in wave II, summary of regression models (unstand. regression coefficient)

Variables		CASP	
		Model 1	Model 2
		Coef.	Coef.
	Low	-1.57***	-0.69***
Education	Medium	-0.65**	-0.11
	High		
	Low	-1.39***	-0.78***
Income	Medium	-0.46**	-0.23
	High		
	None		
Voluntary work	Reciprocal activity	1.52***	0.86***
	Non-reciprocal activity	0.86	0.78
CASP wave 1		0.49***	

*p < 0,05; ** p< 0,01; *** p < 0,001

Note: All models are adjusted for gender, age and country affiliation.

Results confirm the descriptive findings: We observe that both indicators of socioeconomic position (education and income) and participation in voluntary work are associated with higher prospective quality of life (model 1). Yet, with respect to volunteering, a statistically significant association is restricted to the group of people who experience reciprocity in exchange. The regression coefficients are generally reduced after introducing quality of life at the baseline (model 2), but remain almost statistically significant. This finding provides further evidence that higher income and higher education positively affect changes in quality of life. The same holds true for participation in rewarding work as a volunteer. Respective results were found for depressive symptoms. Taken together, results suggest that higher socioeconomic status and participation in rewarding voluntary work beneficially affect well-being in early old age. Additional findings obtained from a French study on an elderly population indicate that well-being (in terms of a low level of depressive symptoms) was relatively highest among those participants who were active in a socially productive activity in which a high level of control and autonomy was perceived (Wahrendorf et al., 2008b).

In conclusion, these results lend some support to the three hypotheses stated above. Being socially productive in informal work in early old age, in a situation in which reward and control are experienced, is associated with better well-being in terms of quality of life (Figure 5.2 and Table 5.2). The findings on informal work to some extent parallel those obtained from working populations, where a high quality of work (in terms of reward and control) was associated with lower level of depression and better self-rated health. Although these findings do not allow disentangling the pathways between socioeconomic position, productive activity and well-being, the prospective study design with additional waves of data collection will offer an opportunity to further test their interrelation in more detail.

3   Future directions of research

In policy terms, the results presented and discussed in section 2 are directed mainly towards the meso- and micro level of sociological analysis, as they involve the study of associations between the activity of older people in core social roles (formal and informal work roles) and the well-being of this group. Yet, the larger sociopolitical contexts that must be addressed at the policy level in order to develop significant and sustainable changes were not included in this analysis.

A significant next step of this research programme addresses the role of national welfare regimes and macroeconomic, labour-market-oriented indicators in modifying the meso-level associations demonstrated so far. For instance, welfare regulations and labour-market policies that explicitly favour employment in older age as well as informal social activities beyond employment are likely to exert beneficial effects on the quality of formal and informal work. An empirical test of this assumption requires identification of relevant macro indicators, such as 'employment rate of older people' (50-64 years), 'long-term unemployment rate of older people' (50-64 years) or 'extent of continued education among adults'. Some of these indicators are available in official administrative datasets at the European Union level (e.g. EUROSTAT, OECD), but additional indicators need to be identified.

Moreover, national policies regarding employment and pension regulations vary substantially, and it is important to search for underlying commonalities and differences. To this end, distinct typologies of welfare have been developed, most importantly the influential typology of European (or Western) welfare systems formulated by Esping-Andersen (Esping-Andersen, 1990). This typology distinguishes between three worlds of welfare-state capitalism: the liberal, the conservative and the social-democratic. These three types differ with regard to the ways in which the two main components of welfare and social policy are handled at national levels: the degree of welfare transfer to people who are not working, and the degree of welfare service provided to the public (healthcare and social care). This typology is more strongly based on the first component than on the second, and as such it has evoked critical discussion (Jensen, 2008). Nevertheless, empirical evidence indicates that differences between the three welfare regimes are most pronounced with respect to the transfers provided to population groups outside of the labour market (in particular, retired, unemployed and permanently sick persons). In this regard, the social-democratic welfare regime ranks highest, the liberal regime ranks lowest, and the conservative regime lies between these two (Dahl et al., 2006).

As pointed out, this new direction of research calls for substantial expansion at the conceptual, methodological and empirical levels. Conceptually, macroeconomic, labour-market oriented indicators should correspond to theories of economic and social development – and the same holds true for welfare typologies. Methodologically, the application of advanced statistical techniques (such as multilevel analysis) will be required to disentangle variations at each level of analysis (between and within countries). Empirically, the number of countries needs to be extended beyond the North-, West- and South-European countries included thus far in the SHARE study. This extension would offer opportunities for more rigorous testing of the above-mentioned hypotheses – for example, by contrasting 'old' European countries with newly associated transition states from Central and Eastern Europe.

In this latter regard, SHARE (wave 2) has already included some Central and Eastern European countries (Czechia and Poland). Moreover, newly established ageing studies in rapidly developing countries (such as South Korea and China) will allow further comparisons (Meijer et al., 2008). In addition, comparable datasets are available from England (English Longitudinal Study of Ageing [ELSA]; Marmot et al., 2002) and the United States (Health and Retirement Study [HRS]; Juster and Suzman, 1995).

Along these lines, we started preliminary analyses in which we categorised the countries of the first wave of the SHARE study together with England (based on respective data from ELSA; Bank et al., 2006) into three groups: social-democratic (Denmark, Sweden), liberal (England, Switzerland) and conservative (the remaining countries of SHARE). Multivariate logistic regressions were run for each of the three categories where odds ratios of depressive symptoms were estimated according to quality of work. As shown in Figure 5.1, low quality of work was defined in terms of scoring high on the summary measure of imbalance between effort and reward and in terms of scoring high on the measure of low control at work (upper tertiles).

Figure 5.3 indicates that in each type of welfare system, low quality of work is associated with a higher relative risk of experiencing depressive symptoms. However, odds ratios in both regards (effort-reward imbalance and low control) were lowest in the social-democratic countries, intermediate in conservative countries and highest in countries with a liberal welfare state regime. The odds ratios in Figure 5.3 are adjusted for socioeconomic status, age and gender.

Figure 5.3: **Associations between low quality of work (yes= highest tertile effort-reward ratio or low control; no= lower tertiles) and depressive symptoms: Odds ratios adjusted for socioeconomic position (education, income), age and gender**

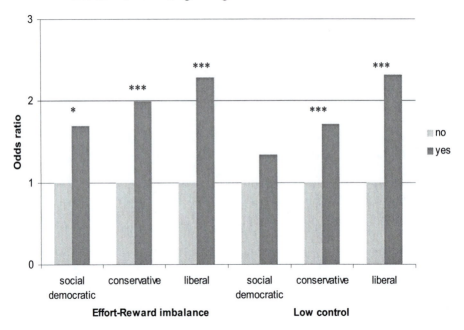

*p < 0,05; ** p< 0,01; *** p < 0,001

As stated, these are preliminary results that need to be elaborated more systematically in future research in order to elucidate the impact of larger socio-political contexts, such as welfare regimes, on the well-being of socially productive older people across Europe. It may be more difficult to study these associations in case of informal work than in case of paid work and employment. The main reason for this difficulty has to do with the fact that the welfare typology is more explicitly based on labour-market-related transfer characteristics than on characteristics of welfare services, such as social services and healthcare services provided to the elderly. For instance, a recent study demonstrated that the welfare regimes vary negligibly with regard to healthcare expenditures, and that the correlation between the two components (transfers and welfare services) is generally weak (Jensen, 2008; Dahl et al., 2006). Thus, it seems unlikely that variations will be found in the strength of association between quality of informal work (volunteering, for example) and well-being according to type of welfare regime, similar to those reported above (Figure 5.3).

A further step for future research on this topic concerns the extension of information on occupational histories of older people – in particular, duration and intensity of exposure to a health-adverse work environment. A first step in this direction has been initiated in the ELSA and SHARE studies, where a questionnaire module assesses in more detail various occupational trajectories over the life course. Similarly, trajectories of informal socially productive activities need to be monitored in population groups outside the labour market. It will be important to study effects of transitions (retirement, for example) and disruptive life events on associations between activities and well-being. The final section of this chapter discusses the policy implications of current evidence, and points to unresolved questions whose answers will rely, in part, on future research evidence along the lines mentioned above.

4   Policy implications of current evidence and open questions

This chapter set out to analyse socioeconomic and psychosocial determinants of well-being in early old age in European countries. The aim was to emphasise the importance of the social opportunity structure in supporting and strengthening the health and well-being of older people through their participation in rewarding and control-enhancing productive activities. Two such opportunity structures were considered, the formal labour market (offering varying degrees of quality of work and employment) and the informal market of unpaid work (volunteering, informal help, caring). In either case, the theoretical notions of autonomy and rewards resulting from reciprocal exchange were applied to explore variations in well-being.

The results indicate that belonging to a higher socioeconomic status group and being socially productive in a rewarding and control-enhancing formal (paid work) or informal (volunteer) social role is associated with greater well-being.

This holds true for different indicators of well-being (depressive symptoms, self-rated health, quality of life), and associations are consistent in cross-sectional and longitudinal analyses. An additional finding points to relevant variations according to type of country under study (for instance, a North-South European gradient) or according to type of welfare-state regime (social-democratic, conservative, liberal).

Despite the fact that some of these results (mainly based on data from the first two waves of the SHARE study) deserve further analysis, several policy implications become obvious at this stage of available evidence. These implications differ between formal and informal social roles, and are more explicit in the former case. In the introductory remarks, three related policy questions were raised: What needs to be done to retain as many older people as possible in work – either employed or self-employed? What might be done to reduce the proportion of employees with early exit from the labour market? What measures might be taken to better protect and improve the health and well-being of middle-aged to early old-aged working men and women?

Answers to these questions are obviously complex and involve different levels of activity of shareholders, organisations and policy-making bodies. Yet, a common feature of such activities concerns measures that aim at improving the quality of work. In the context of our research, the following recommendations are examples of theory- and evidence-based policy implications:

- Improve monitoring activities of health-adverse working conditions (including traditional physical and chemical hazards and, more recently, psychosocial hazards) at national and regional levels, as well as at the level of branches, occupations and firms or organisations;
- Implement more comprehensive occupational health and safety measures that include the promotion of healthy work;
- Monitor occupational high-risk groups, particularly those employed in precarious work, temporary and irregular work, and those working in risky jobs (transport workers, construction workers, employees in emergency services, for example);
- Increase the flexibility of work-time arrangements, including broader opportunities for part-time work and continued training, as well as 'flexicurity' models of occupational careers;
- Secure fair pension- and retirement arrangements in relation to both lifetime contributions to the labour market and major shocks (long-term unemployment, forced early retirement, disability pension);
- Design and implement measures of organisational and personnel development that are instrumental in increasing control at work and in providing

fair rewards in return for effort expended; these measures concern separate organisations as well as larger bodies of branches, stakeholder associations, trade unions or even national and transnational legislation.

This latter recommendation could be further specified by pointing to models of good practice that are already available in the context of European-wide initiatives (Breucker, 2000). Measures include the reorganisation of division of work, with the aim of developing more complete job task profiles (job enlargement, job enrichment, for example) and more adequate promotion prospects (including job security, more flexible forms of remuneration and non-monetary gratifications, enhanced leadership training and the development of a culture of trust, fairness, and transparency at organisational level). These measures could be tailored further towards specific age groups – particularly middle-aged and early old-aged groups, in order to maintain their health and well-being, and to enhance their motivation to stay at work.

Less-elaborate policy measures concern the realm of informal work. The following policy questions were addressed in the introductory section: What could be done to enable retired people to continue or to initiate socially productive activities (such as volunteering, being engaged in informal help or caring for a sick or disabled person)? How might the proportion of socially productive early old-aged people be augmented? What are the costs and benefits of extending respective opportunities and incentives?

Two broad answers are obvious. First, the opportunity structure of informal work for third-age population groups must be developed in a pro-active manner. This includes the creation of new social roles in the context of an emerging civil society, the liberalisation of legal restrictions including tax allowance, and a change of societal attitudes and views of the role of ageing in the life course (Erlinghagen and Hank, 2008). In all of these instances, lower socioeconomic status groups are a primary target group, as they are largely excluded from opportunities to take part in socially productive activities in early old age. The second answer points to the need of developing quality measures of informal work – particularly, of providing informal work roles with options of reward that entail the experience of recognition and esteem, and with options of control that entail the experience of autonomy and self-efficacy. Our results have demonstrated that the beneficial effects of being socially productive in an informal role to a large extent are contingent on the presence of these characteristics.

Whereas these policy measures are supported by scientific results derived from currently available datasets, additional policy challenges will require new data and further scientific work. Open questions to be addressed concern the role of both formal and informal work in ageing societies. For instance, it is important to know more exactly how national social policies (welfare typologies) and socioeconomic conditions (macro-indicators) affect the well-being of different groups of older people. So far, our focus has mainly been on regularly employed segments of the population, and on those who experience opportunities to engage in informal work. Little is known about homemakers, older people exposed to precarious work, migrant workers or the long-term unemployed. And even within the majority of employed people, the impact of significant changes of work and employment due to technological and economic globalisation has not been well explored. With recent progress in information technology and automatisation, the traditional separation of the spheres of work and home is vanishing. Working at home, telework, participation in virtual networks, and an unprecedented degree of flexibility in local and temporal work arrangements contribute to this process. Moreover, due to economic globalisation and a related segmentation of the labour market, income inequalities are increasing – with far-reaching consequences for economic standing and quality of life in old age.

A further open question concerns the extent to which health and well-being among older persons are determined by opportunities and achievements in midlife – both within and outside work. Longitudinal data mirroring trajectories of formal and informal productive activities are needed to explain the development of large social inequalities in health and well-being that persist into old age. Based on these explanations, targeted policy measures can be developed that aim at reducing the social gradient of health.

In conclusion, the socioeconomic and socio-emotional consequences of participating in rewarding and control-enhancing formal and informal productive activities for healthy ageing are far-reaching. Strengthening these conditions through targeted policy measures seems to be a promising approach to tackling the economic-, social- and health-related challenges of ageing societies.

References

Antoniou, A.S. and C.L. Cooper (eds) (2005) *Research Companion to Organizational Health Psychology* (Cheltenham: Edward Elgar Publishers).

Banks J., E. Breeze, C. Lessof and J. Nazroo (eds) (2006) *Retirement, Health and Relationships of the Older Population in England: The 2004 English Longitudinal Study of Ageing* (London: Institute for Fiscal Studies).

Blane D., G. Netuveli and M. Bartley (2007) 'Does Quality of Life at Older Age Vary with Socioeconomic Position?' *Sociology* 41, 717-726.

Blekesaune M. and P.E. Solem (2005) 'Working Conditions and Early Retirement. A Prospective Study of Retirement Behaviour', *Research on Ageing* 24: 3-30.

Börsch-Supan A., A. Brugiavini, H. Jürges, J. Mackenbach, J. Siegrist and G. Weber (eds) (2005) *Health, Ageing and Retirement in Europe* (Mannheim: Mannheim Research Institute for the Economics of Ageing).

Börsch-Supan A., A. Brugiavini, H. Jürges, A. Kapteyn, J. Mackenbach, J. Siegrist and G. Weber (eds) (2008) *Health, Ageing and Retirement in Europe – Starting the Longitudinal Dimension.* (Mannheim: Mannheim Research Institute for the Economics of Ageing).

Breeze E., A.E. Fletcher and D.A. Leon M.G. Marmot, R.J. Clarke and M.J. Shipley (2001) 'Do Socioeconomic Disadvantages Persist into Old Age? Self-reported Morbidity in a 29-year Follow-up of the Whitehall Study', *American Journal of Public Health* 91, 1126-29.

Breucker G. (ed.) (2000) *Towards Better Health at Work. Successful European Strategies* (Bremerhaven: Wirtschaftsverlag NW).

Brugiavini A., E. Croda and F. Mariuzzo (2006) 'Labour Force Participation of the Elderly: Unused Capacity?' in A. Börsch-Supan, A. Brugiavini, H. Jürges, J. Mackenbach, J. Siegrist and G. Weber (eds) *Health, Ageing and Retirement in Europe* (Mannheim: Mannheim Research Institute for the Economics of Aging), pp. 236-240.

Cartwright S. and C.L. Cooper (eds) (2008) *The Oxford Handbook of Organizational Wellbeing* (Oxford: Oxford University Press)

Dahl E., J. Fritzell, E. Lahelma, P. Martikainen, A. Kunst and J. Mackenbach (2006) 'Welfare State Regimes and Health Inequalities' in J. Siegrist and M. Marmot (eds) *Social Inequalities in Health. New Evidence and Policy Implications* (Oxford: Oxford University Press), pp. 193-222.

Dewey, M.E. and M.J. Prince (2005) 'Mental Health' in A. Börsch-Supan A. Brugiavini, H. Jürges, J. Mackenbach, J. Siegrist and G. Weber (eds.) *Health, Ageing and Retirement in Europe* (Mannheim: Mannheim Research Institute for the Economics of Ageing), pp.108-117.

Dragano, N. (2007) *Arbeit, Stress und Krankheitsbedingte Frührente* (Work, Stress and Disability Pension) (Wiesbaden: VS Verlag für Sozialwissenschaften).

Dubos, R. (1969) *Man, Medicine, and Environment* (New York: Mentor).

Emmerson, C. and G. Tetlow (2006) 'Labour Market Transitions' in J. Banks, E. Breeze, C. Lessof and J. Nazroo (eds) *Retirement, Health and Relationships of the Older Population in England* (London: Institute for Fiscal Studies), pp. 41-82.

Erlinghagen M. and K. Hank (2006) 'The Participation of Older Europeans in Volunteer Work', *Ageing and Society* 26, 567-84.

Erlinghagen M. and K. Hank (eds) (2008) *Produktives Altern und Informelle Arbeit in Modernen Gesellschaften.* (Productive Ageing and Informal Work in Modern Societies) (Wiesbaden: VS Verlag für Sozialwissenschaften).

Esping-Andersen, G. (1990) *The Three Worlds of Welfare State Capitalism* (Cambridge: Cambridge University Press).

Gouldner, A.W. (1960) 'The Norm of Reciprocity', *American Sociological Review* 25, 161-178.

Gruber, J. and D. Wise (1999) *Social Security and Retirement around the World* (Chicago: University of Chicago Press).

Hagen, K.B., H.H. Holte, C. Tambs and T. Bjerkedal (2000) 'Socioeconomic Factors and Disability Retirement from Back Pain', *Spine* 25, 2480-87.

Haidt, J. and J. Rodin (1999) 'Control and Efficacy as Interdisciplinary Bridges', *Review of General Psychology* 3, 317-37.

Huisman, H., A.E. Kunst, and O. Andersen, M. Bopp, J.-K. Borgan, C. Borrell, G. Costa, P. Deboosere, G. Desplanques, A. Donkin, S. Gadeyne, C. Minder, E. Regidor, T. Spadea, T. Valkonen and J. P. Mackenbach (2004) 'Socioeconomic Inequalities in Mortality among Elderly People in 11 European Populations', *Journal of Epidemiology and Community Health* 58, 468-75.

Hyde, M.R., R. Wiggins, P. Higgs and D. Blane (2003) 'A Measure of Quality of Life in Early Old Age: The Theory, Development and Properties of a Needs Satisfaction Model (CASP-19)', *Aging and Mental Health* 7, 186-94.

Jensen, C. (2008) 'Worlds of Welfare Services and Transfers', *Journal of European Social Policy* 18, 151-62.

Juster, F.T. and R. Suzman (1995) 'An Overview of the Health and Retirement Study', *Journal of Human Resources* 30, 7-56

Karasek, R. and T. Theorell (1990) *Healthy Work* (New York: Basic Books).

Karasek, R., C. Brisson, N. Kawakami, I. Houtman, C.P. Bongers and C. Amick (1998) 'The Job Content Questionnaire (JCQ): An Instrument for Internationally Comparative Assessments of Psychosocial Job Characteristics', *Journal of Occupational Health Psychology* 3, 322-355.

Kivimäki, M., M. Virtanen, A. Elovainio, A. Kouvonen and J. Vahtera (2006) 'Work Stress in the Etiology of Coronary Heart Disease – A Meta-analysis', *Scandinavian Journal of Work, Environment and Health* 32, 431-442.

Krokstad, S., R. Johnsen and S. Westin (2002) 'Social Determinants of Disability Pension: A 10-year Follow-up of 62.000 People in a Norwegian Country Population', *International Journal of Epidemiology* 31, 1183-91.

Laslett, P. (1989) *A Fresh Map of Life: The Emergence of the Third Age* (London: Weidenfeld and Nicolson).

Mackenbach, J.P. and M. Bakker (eds) (2002) *Reducing Inequalities in Health. A European Perspective* (London: Routledge).

Marmot, M., J. Banks, R. Blundell, C. Lessof and J. Nazroo (2002) *Health, Wealth and Lifestyles of the Older Population in England: The 2002 English Longitudinal Study of Ageing* (London: Institute for Fiscal Studies).

Marmot M. and R. Wilkinson (eds) (2006) *Social Determinants of Health* (Oxford: Oxford University Press).

Marmot, M., J. Siegrist and T. Theorell (2006) 'Health and the Psychosocial Environment at Work' in M. Marmot and R. Wilkinson (eds) *Social Determinants of Health* (Oxford: Oxford University Press), pp. 97-130.

McMunn, A., E. Breeze, A. Goodman, J. Nazroo and Z. Oldfield (2006) 'Social Determinants of Health in Older Age' in M. Marmot and R. Wilkinson (eds) *Social Determinants of Health* (Oxford: Oxford University Press), pp. 267-296.

Meijer, E., G. Zamarro and M. Fernandes (2008) 'Overview of Available Aging Data Sets' in A. Börsch-Supan, A. Brugiavini, H. Jürges, J. Mackenbach, J. Siegrist and G. Weber (eds) *Health, Ageing and Retirement in Europe-Starting the Longitudinal Dimension* (Mannheim: Mannheim Research Institute for the Economics of Ageing).

Parent-Thirion, A., E.F. Macias, J. Hurley and G. Vermeylen (2007) *Fourth European Working Conditions Survey* (Luxemburg: Office for Official Publications of the European Communities).

Rowe J.W. and R.L. Kahn (1997) 'Successful Aging', *The Gerontologist* 37, 433-440.

Salomon, L.M. and S.W. Sokolowski (2003) 'Institutional Roots of Volunteering. Towards a Macro-structural Theory of Individual Voluntary Action' in P. Dekker and L. Halman (eds) *The Values of Volunteering* (New York: Kluwer Academics), pp. 71-90.

Schultz, W., P. Dayan and P.R. Montague (1997) 'A Neural Substrate of Prediction and Reward', *Science* 275, 1593-99.

Siegrist, J. (2005) 'Social Reciprocity and Health: New Scientific Evidence and Policy Implications', *Psychoneuroendocrinology* 30, 1033-38.

Siegrist, J. (2008) 'Job Control and Reward: Effects on Well-being' in S. Cartwright and C.L. Cooper (eds) *The Oxford Handbook of Organizational Wellbeing* (Oxford: Oxford University Press), pp 109-132.

Siegrist, J., O. vd Knesebeck and C.E. Pollack (2004) 'Social Productivity and Well-being of Older People: A Sociological Exploration', *Social Theory and Health* 2, 1-17.

Siegrist, J., D. Starke, T. Chandola, I. Godin, M. Marmot, I. Niedhammer and R. Peter (2004) 'The Measurement of Effort-reward Imbalance at Work: European Comparisons', *Social Science & Medicine* 58, 1483-1499.

Siegrist, J. and M. Marmot (eds) (2006) *Social Inequalities in Health: New Evidence and Policy Implications* (Oxford: Oxford University Press).

Siegrist, J., M. Wahrendorf, O. vd Knesebeck, H. Jürges and A. Börsch-Supan (2007) 'Quality of Work, Well-being, and Intended Early Retirement of Older Employees – Baseline Results from the SHARE Study', *European Journal of Public Health* 17, 62-68.

Siegrist, J. and M. Wahrendorf (2008) 'Participation in socially productive activities and quality of life in early old age: Findings from SHARE', *Journal of European Social Policy* 19, 317-326..

Skinner, E.A. (1996) 'A Guide to Constructs of Control', *Journal of Personality and Social Psychology* 63, 407-14.

Smith, J.P. (1999) 'Healthy Bodies and Thick Wallets: The Dual Relation between Health and Economic Status', *Journal of Economic Perspectives* 13, 145-66.

Stansfeld, S. and B. Candy (2006) 'Psychosocial Work Environment and Mental Health – A Meta-analytic Review', *Scandinavian Journal of Work, Environment and Health* 32, 443-462.

Stattin, M. and B. Järvholm (2005) 'Occupational Work Environment and Disability Pension: A Prospective Study of Construction Workers', *Scandinavian Journal of Public Health* 33, 84-90.

Theorell, T. (2008) 'After 30 Years with the Demand-control-support Model – How is It Used today?' *Scandinavian Journal of Work, Environment and Health* Special Issue 6, 1-184.

Van Rossum, C.T.M., M.J. Shipley, H. van de Mheen, D.E. Grobbee and M.G. Marmot (2000) 'Employment Grade Differences in Cause-specific Mortality. A 25-year Follow-up of Civil Servants from the First Whitehall Study', *Journal of Epidemiology and Community Health* 54, 178-84.

Von dem Knesebeck, O., M. Wahrendorf, M. Hyde and J. Siegrist (2007) 'Socio-economic Position and Quality of Life among Older People in 10 European Countries: Results of the SHARE Study', *Ageing & Society* 27, 269-84.

Wahrendorf, M., O. vd Knesebeck and J. Siegrist (2008a) 'Social Productivity and Quality of Life-first Prospective Results' in A. Börsch-Supan, A. Brugiavini, H. Jürges, J. Mackenbach, J. Siegrist and G. Weber (eds) *Health, Ageing and Retirement in Europe-Starting the Longitudinal Dimension* (Mannheim: Mannheim Research Institute for the Economics of Ageing).

Wahrendorf, M., C. Ribet, M. Zins and J. Siegrist (2008b) 'Social Productivity and Depressive Symptoms in Early Old-Age Results from the GAZEL Study', *Aging & Mental Health* 12, 1-7.

Wahrendorf, M. (2009) *Soziale Produktivität und Gesundheit im Höheren Lebensalter-Vergleichende Untersuchungen (Social Productivity and Health in Early Old Age-Comparative Investigations)* (Berlin: Lit Verlag)

World Health Organization (2001) *International Classification of Functioning* (Geneva: WHO).

Socioeconomic and Psychosocial Determinants of Well-being in Early Old Age

Comments by Jim Ogg

The chapter by Siegrist and Wahrendorf makes an important contribution to the growing body of evidence-based research supporting social policies that promote well-being. The focus is on the interaction between goal-orientated behaviour, socially productive activities and positive well-being. Simply stated, having goals and meaningful objectives in life produces positive health outcomes. By creating the opportunity structures within which individuals can maximise their potential, social policies can have a positive impact on promoting well-being. There is a further context to the chapter – namely, that goal-oriented behaviours should be socially productive in order to have a maximum effect on outcomes of positive well-being. Efforts that are rewarded, in a socially productive context, thus promote positive health outcomes.

There are clear policy implications arising from the chapter's findings: improving the quality of the work environment, for example, is crucial to promoting well-being. The monitoring of high-risk groups is proposed. The introduction of flexible working arrangements is another possible measure. Also important is that workers receive their due reward – not simply in terms of adequate salaries and pensions, but also through a series of non-monetary measures including the development of cultures of trust, fairness and transparency within the workplace. As far as the voluntary sector is concerned, more work needs to be done to open the many pathways of informal helping and to harness these socially productive activities to structures that reward efforts and enhance self-esteem.

This said, we should be aware that the concept of well-being has an 'amorphous, multidimensional and complex nature' (Walker, 2005: 2), and many people with similar but different ideas cluster under the same banner.

It is sometimes difficult to untangle the various ideas and approaches, and determine their relevance for public policy. Perhaps more importantly, there remains a precarious relationship between the scientific exploration of quality of life and what individuals actually experience. This means that, despite the usefulness of the concept of quality of life as a central tenant of social policy, it may have its limitations.

Adding to the complexity of the concept is the cross-national context of the empirical data used to research quality of life in early old age. We need to be careful about how groups of older people within different European settings interpret and respond to self-assessed subjective areas of their lives. Evidence from the World Values Survey suggests that terms such as 'happiness' and 'satisfaction with life' generally mean the same thing in different languages, and can therefore be compared (Layard, 2005). So we can be reasonably assured that the same applies to the measures of quality of life contained within the SHARE survey. However, there is a lot more noise around concepts such as informal work, volunteering, caring and helping others. These concepts seem to mean different things, depending on social class and national setting. Take, for example, the finding that rates of volunteering are highest among Northern Europeans who themselves are from higher socioeconomic status groups. This may say more about how informal help is structured than reflect any real differences between Northern and Southern Europe. Many people who are routinely involved in everyday helping activities do not see these activities as helping per se – and thus they can be lost in large surveys.

These conceptual difficulties, together with the broad scope of the chapter, inevitably lead to some caution being applied to the findings. The need for caution becomes amplified, however, when interpreting the policy implications that arise from the hypotheses and findings of the research. The findings and policy implications outlined in the chapter that are based on the relationship between high-quality work environments for people in the labour market and positive outcomes for quality of life make good sense. So, too, does the general theme of the positive quality-of-life outcomes to be gained by people remaining active in later life. But perhaps less convincing is the idea that runs through the chapter linking socially productive activities and well-being.

According to the authors, key socially productive activities are paid work, volunteering, informal help and caring. Being involved in these activities, under the right circumstances, enhances well-being. Individuals in later life should therefore be encouraged to do them. There is good evidence to suggest a positive association of well-being with these activities (with, perhaps, the exception of caring – for which there are strong indications of the adverse effects on the psychological health of the carers). But I am slightly uneasy with the prescriptive undertones that are inevitably present in policy measures that promote socially productive activities in later life. Doing good deeds for the benefit of society, as well as oneself, has of course been the subject matter of many philosophers, not to mention theologians. The maxim of the seventeenth century utilitarianist Jeremy Bentham was, "I ought do that act which will bring about the greatest happiness for the greatest number of persons". But happiness, pleasure, positive well-being, or whatever we call it, can also be gained through activities that are not necessarily socially productive, particularly in later life. Continuing education (or 'learning for learning's sake') is not an activity that one associates directly with being socially productive, yet there is good evidence that it enhances well-being. As Bertrand Russell said, 'there is much pleasure to be gained from useless knowledge'. The same could be said about leisure pursuits that are often associated with the end of paid work in later life – perhaps even more so with the current generation of mobile baby-boomers who are in their sixties. So it is questionable that an activity should be socially productive in order to maximise well-being.

These observations, however, should not diminish the important message contained within this chapter: namely, that the links between activities and well-being are well-established, that the context within which activities are undertaken is all-important, and that the changing landscape of ageing populations presents major challenges for the well-being of citizens. To quote the authors, 'the socioemotional consequences of participating in rewarding and control-enhancing formal and informal productive activities for healthy ageing are far-reaching'.

References

Layard, R. (2005) *Happiness: Lessons from the New Science* (New York: The Penguin Press).
Walker, A. (2005) 'A European Perspective on Quality of Life in Old Age', *European Journal of Ageing*, 2: 2-12.

Socioeconomic and Psychosocial Determinants of Well-being in Early Old Age

Comments by Janneke Plantenga

In their contribution on sociological and psychosocial determinants of well-being in early old age, Siegrist and Wahrendorf emphasise the importance of goal-oriented agency (particularly, the importance of paid work and voluntary, or informal, work) on health and well-being. It appears that people who are engaged in socially productive activities experience better health and well-being compared to those who are not so engaged – especially if these productive activities are compensated by appropriate rewards, and if these activities provide a measure of control and autonomy. Indeed, the authors provide extensive evidence that low quality of work (in terms of an imbalance between effort spent and rewards received, and in terms of low control) is associated with reduced health and well-being.

Their basic claim, that social productivity in high-quality work (with 'quality' defined in terms of rewards and control) is associated with better well-being, seems to be rather generally applicable, but Siegrist and Wahrendorf claim that issues of reward and control are especially relevant in early old age, as other options of agency, control and reward become, at that point, less frequent and less pronounced (with reference to work roles, family roles and civic roles). As boosting the participation rate of older people is becoming increasingly important against the background of Europe's rapidly ageing population, the authors' suggestion to focus on high-quality work seems extremely relevant.

Whereas Siegrist and Wahrendorf focus on reward and control in rather general terms, it seems important to take the analysis a step further, and to differentiate between intrinsic and extrinsic rewards, thereby linking the reward structure to employee motivation and employee governance. People do not work just for the money; they are also motivated by the work itself (Deci,

1975; Frey, 1997). In most jobs, both types of reward form an important source of motivation. Extrinsic rewards (such as financial rewards and status) trigger the extrinsic motivation of the employee, while intrinsic rewards (such as having an enjoyable job and opportunities for personal development) trigger intrinsic motivation. Economists tend to concentrate on extrinsic rewards and motivation, but there is evidence that work governed by intrinsic motivation is associated with better mental and physical health than work governed by extrinsic motivation. Employees who are characterised by a high level of intrinsic motivation indicate higher work- and life satisfaction than employees who are characterised by a high level of extrinsic motivation (Frey, 1997). Employers will also do well to take intrinsic motivation into account, as intrinsically motivated employees have a higher learning capacity than extrinsically motivated employees. Furthermore, laboratory experiments suggest that cognitively difficult tasks are better solved by intrinsically motivated employees. Finally, intrinsically motivated employees are less costly when it comes to maintaining discipline. They work well, regardless of whether the employer is monitoring their performance, because the work as such is fulfilling (Gallie, 2007). The conclusion therefore seems to be that it is important to look at both intrinsic and extrinsic rewards and motivation.

With regard to older employees, we may presume that especially those with low-paid jobs run the risk of receiving few rewards – intrinsic or extrinsic – from their job. Wages are low, status is low, the job is not enjoyable and opportunities for personal development are limited. In such cases – when jobs are no longer rewarding – early retirement seems the optimal option. A more interesting hypothesis might involve our focussing not only on the level of intrinsic and extrinsic motivation as such, but rather at the changing weight. Older people might not be that easily motivated purely in terms of extrinsic motivation. Whereas young people still strive for higher financial rewards, older people have come to terms with a certain level of income; they have raised their children, bought a house, own a car, and so on. In short, the increase in marginal utility of yet another euro is rather limited for older people. If this is indeed true, especially in the case of older employees, then increasing their rewards by increasing their intrinsic motivation is extremely important. This has special relevance for the organisation of work. The literature on task characteristics claims that intrinsic motivation at work 'may actually have more to do with how tasks are designed and managed than with the personal dispositions of the people' (Hackman and Oldham, 1980: 76-77). Over the years, much has been learned about how work can be organised and how employees can be governed in order to create high levels of intrinsic rewards – and thus high levels of intrinsic motivation. So, depending on the type of employee governance, various packages of extrinsic and intrinsic rewards might emerge, each having a different impact on intrinsic and extrinsic motivation, well-being and health.

This brings me to issues of further research. Siegrist and Wahrendorf refer to the role of national welfare state regimes as a promising new pathway. Social-democratic welfare states may indeed score lower than liberal welfare state regimes on the relation between low quality of work and depressive symptoms. Yet, as also suggested by the authors, it is a long way (both methodologically, conceptually and empirically) to translate different sociopolitical contexts into an independent variable that is somehow connected to outcomes in terms of health and well-being. Instead of focusing on macro concepts, my preference is for continuing research at the meso- and micro levels, focussing on the type of employee governance – for example, in terms of working time regime, internal and external flexibility, the extent of teamwork, and the reward- and career system. This would also engage the employer – an angle that is largely missing in the analysis of Siegrist and Wahrendorf, given their focus on 'societal opportunity structures' and 'welfare institutions'. It seems likely, however, that it is chiefly the organisation of work, and the type of employee governance by the employer that makes a difference in terms of the quality of work – and thus in terms of employee well-being.

References

Deci, E.L. (1975) *Intrinsic Motivation* (New York: Plenum Press).
Frey, B.S. (1997) *Not Just for the Money. An Economic Theory of Personal Motivation* (Cheltenham: Edgar Elgar).
Gallie, D. (2007) *Employment Regime and the Quality of Work* (Oxford: Oxford University Press).
Hackman, J.R. and G.R. Oldham (1980) *Work Redesign* (Reading, MA: Addison Wesley).

6

Social Networks

Martin Kohli / Harald Künemund

Introduction

Social networks are constituted by socially interacting units or actors at different levels of aggregation (individuals, organisations or societies). Such networks may be defined and qualified by their density and cohesion, the diffusion of information within them, the multiplexity of relationships, or the amount and substantive content of exchanges. This chapter focuses on the social networks – the social connectedness – of elderly individuals: on their relationships with others (their family members, friends, neighbours, colleagues or service providers), and on the content of their exchanges with these others.[1]

Maintaining social connectedness through the transitions of later life (fraught with the potential for isolation) is an important prerequisite for 'successful ageing' (see Rowe and Kahn, 1998; Kohli et al., 2009; Sirven and Debrand, 2008). Despite frequently raised concerns about a lack of social integration of the elderly in modern societies (see Cornwell et al., 2008 and de Jong Gierveld and Havens, 2004 for overviews of the discussion), there is now a broad body of literature that emphasises the ongoing integration of the elderly into family networks (Attias-Donfut, 1995) and into networks of more general social participation (Kohli and Künemund, 1996).

Social networks are thus crucial for the well-being of elderly individuals (Litwin, 1996). Such networks provide moreover a range of benefits for age-

---

1 This is not the place for a general formal analysis of social networks, a burgeoning field that has spawned its own theoretical and methodological approaches as well as a broad research literature. In fact, the lack of articulation between general network analysis and the analysis of social networks of the elderly is an obvious knowledge gap (see section 3).

ing societies: they are a source of support for persons in need, they are a site of productive acitivities of the elderly, and they organize social participation in community affairs. By this, they also contribute to a reduction in public expenditure (European Commission, 2006; Johnson et al., 2007). Social networks are thus a key issue for policies that address ageing at both the individual and the societal level.

The conceptualisation of social connectedness at older ages has changed over time. Originally, the emphasis was on activities (as such) – on the roles still available to the elderly (Rosow, 1974), and on the properties of the social fields in which they participate (Kohli et al., 1993; Kohli and Künemund, 1996). Recent years have seen a shift away from such conceptualisations that focus on roles and activities toward more network-oriented constructs (Cornwell et al., 2008, 2009; Kohli et al., 2009). The network approach positions itself against both the Parsonian emphasis on the normative integration of society and the atomising emphasis of much of survey research. The more recent survey work on intergenerational relationships and exchanges also goes beyond treating the respondents as atomised individuals, and thus incorporates (some of) the network dimensions (Kohli, 1999; Künemund and Hollstein, 2000). A third approach that has recently gained currency is that of social capital, a term that is used to refer to both activities and networks: to *'the way in which people participate in their society and the forms of social bonding that take place'* (Pichler and Wallace, 2007: 423). The approach has raised a great deal of interest in sociology and political science (Putnam, 2000) as well as in economics (Glaeser et al., 2002).

The three approaches (*activity*, *network* and *social capital*) are often treated as mutually exclusive alternatives. In our view, this is not appropriate. The concept of 'activity' has to do with the opportunities and demands for individuals to be socially productive, while the concept of 'network' focuses on the social relationships that accompany such opportunities and demands. The concept of 'social capital' implies that activities and networks yield certain profits or dividends from which individuals can draw, such as contacts and information that facilitate access to a better position in the labour market or help in situations of need. Our examination of the modes individuals employ to keep socially connected through the later stages of the life course uses the three approaches in articulation with each other.

1 Policy questions

Social networks have been shown to buffer the effects of negative life events such as spousal bereavement (Ferraro, 1984; Li, 2007), to reduce mortality (Moen et al., 1989; Musick et al., 2004; Litwin and Shiovitz-Ezra, 2006) or to serve as a social protection mechanism (Lyberaki and Tinios, 2005; Wall et al., 2001). These networks are thus crucial for well-being in old age. They mediate a range of activities and support, such as participation in volunteering and societal affairs, emotional support, practical help with everyday activities, and personal care in situations of dependency. Beyond personal well-being, such support is also important in economic terms (for the prevention or delay of costly institutional placement, for example), as well as in sociological terms (for the social inclusiveness and moral cohesion of societies).

The discussion of social support in old age has long been dominated by a controversy between two models, that of hierarchical compensation and that of task specificity. The *hierarchical compensation model* (Cantor, 1979) states that – regardless of the nature of the task – elderly people in need of help will turn to the next available kin. Partners are the first to be responsible for help, followed by daughters, daughters-in-law, sons, sons-in-law, other relatives, and (only at the end of this hierarchy) friends, neighbours and other non-relatives. If a specific relation is missing, the next person in the hierarchy is regarded as being responsible. The *task-specificity model* (Litwak, 1985), alternatively, states that persons with different relations to the elderly individual are considered appropriate for specific tasks, depending on their capabilities and competencies. Consequently, non-kin persons such as neighbours and friends can be more important, depending on the nature of the task needed – although the model would not challenge the claim that the more-demanding types of help, such as personal care or heavy monetary support, are usually attended to by family.[2]

The same point is made by the (much smaller) literature on non-kin relationships in old age (de Jong Gierveld and Perlman, 2006). It has repeatedly been shown that networks become smaller and more family-centred in older age groups – a phenomenon that seems indeed to be an age- rather than a

---

2   Although these models were usually discussed as if they were contradictory (Crohan and Antonucci, 1989: 137f.; Messeri et al., 1993; Lang, 1994: 61; Kinney, 1996: 671; Broese van Grenou and van Tilburg, 1997: 24f.), it may be argued that they are not. While the *hierarchical compensation model* asks to whom one would turn in case of need, the *task-specific model* asks who is in fact helping the elderly. In that view, it is plausible that elderly people prefer to ask their next of kin in the first place, but receive help from other sources as well (Künemund and Hollstein, 2000).

cohort effect (Wagner and Wolf, 2001). It may be explained by shrinking opportunity structures with age, while norms of parental and filial solidarity remain in place. There are also evolutionary explanations for this solidarity, arguing that kinship orientation, especially from parents towards offspring, is genetically determined as a result of evolved psychological adaptation (Neyer and Lang, 2003). Other explanations for the trajectory of social support refer to attachment, emotional closeness, or the 'convoy' of supportive relationships in the life course (Kahn and Antonucci, 1980). Regardless of which explanation is correct, the descriptive evidence is clear: When the elderly give or receive support of any kind, this occurs most often in the couple or in the generational lineage. Besides spouses, the children and children-in-law of the elderly are the foremost addressees of help – given and received. For these reasons, the following sections give particular weight to kin networks.

There are three current policy issues where intergenerational family linkages play a crucial role (see Kohli et al., 2010). The first has to do with life-course risks such as unemployment or divorce. These risks are increasing, while at the same time the coverage provided by public transfers is decreasing, due to current welfare state retrenchments. Financial transfers and social support by parents can help adult children cope with the income loss and turbulence that typically attend such events. The second policy issue has to do with ensuring demographic reproduction while keeping up female labour-force participation, or (as it is usually stated at the individual level) reconciling parenthood and employment (which increasingly becomes a precondition for women to be willing to engage in parenthood). The time given by the older generation in terms of grandparenting activities, particularly in countries with a weak provision of public childcare services, is possibly the most important reconciliation measure available to many young dual-earner families with pre-school-age children (Attias-Donfut and Segalen, 1998). The third issue is providing care for the dependent elderly. The family has traditionally been the key provider of such care (more so in the European South than in the North)[3], and the impending 'care crisis' (Anttonen et al., 2003) engendered by rising demand and costs

---

3   Surveys of the living situation of older persons and time-use studies consistently show that the majority of care is provided informally (above all within families), usually in a range of 80% plus of hours of care provided (see Sundström et al., 2002; OECD Health Project, 2005). Most of this time is spent on lower-level care, such as help with instrumental activities of daily living. But family carers also extend care to many older persons with the highest care needs (such as dementia patients), for whom informal care is often the most important source of support.

threatens to overburden the public (state and market) system of care service provision and institutional care.

Intergenerational family networks thus function as insurance for children's life-course risks, as support for children's parenting, and as care for the dependent elderly. At the same time, these networks are an important source of generational integration. Contemporary societies are highly age-graded and age-segregated, and thus present a risk of intergenerational conflict and warfare. Intergenerational family networks create emotional and material linkages and help to equalise the disparities between generations (Kohli, 1999).

But will families still be able to perform? The threats for network effectiveness come first of all from the current demographic shifts. Increasing generational co-longevity and decreasing numbers of siblings and children combine to create 'beanpole families'. Increasing proportions of singles, both among the elderly and among their children, reduce the supply of carers. Increasing rates of divorce and remarriage produce 'blended families'. As a consequence, we may predict a higher potential for parental support and transfers to each adult child, but smaller and less reliable support networks for the elderly.

But other societal changes have also impacted network effectiveness. The historical shift of responsibility from the family to public social security – with respect, for example, to income (from children and savings to pension systems) or care (from the family to the state or community) – may have resulted in a general decline of private intergenerational solidarity ('crowding out'; Künemund and Rein, 1999). Cultural individualisation results in a diminished sense of obligation towards other family generations and more legitimacy of personal choice. The increasing labour-force participation of women and their higher geographical mobility make them less available for family services. The reduction in welfare state spending for the elderly – in the shape of lower and later pensions, in particular – diminishes their ability to give money and time to their offspring. In addition to the smaller support networks, there may thus also be less willingness and ability to provide help.

Public policy may thus have to cope with both rising demand for and shrinking supply of kin network support. It will be critical to support the networks themselves so that they remain viable and productive in spite of the foreseeable risks for their functioning (see below).

2   Major progress in understanding

2.1   *Theoretical traditions*

The traditional view of population ageing promotes the idea of the elderly as (only or mainly) a burden on society (in terms of both income and care needs). Much of the discourse on the new challenges posed by ageing populations to contemporary welfare systems is based on the assumption that greater longevity implies a higher financial demand for pensions and a higher care demand from families and public services – at a time when the proportion of 'producers' (those in the labour force and those able to give care) is shrinking. The extent to which these, often catastrophic, predictions will become reality depends on several factors, among them (i) how people will be ageing: in other words, the extent to which the increasing life expectancy will be accompanied by an improvement in the health of the elderly population (compression of morbidity); (ii) when people will move into retirement, and thus no longer have earnings that cover their financial needs, (iii) what social networks will be available for supporting the elderly, and (iv) the extent to which elderly people will themselves remain productive in their social networks.

This chapter focuses on the third and fourth of these factors. The third factor represents the traditional view of the elderly as receivers of support from the younger generations, and thus *as a social problem to be solved*. The fourth factor addresses the opposite view of the elderly as supporters of others, and thus *as a social resource* – a view that has been gaining ground since the late 1980s (Herzog et al., 1989; Coleman, 1995; Künemund, 1999; Erlinghagen and Hank, 2008). We should of course not fall into the trap of an exaggerated gerontological optimism by claiming that old age is only about productivity and not also about dependency and need for care. But until recently, the latter aspect has been unduly exaggerated at the expense of the former.

That the elderly are at risk of social isolation has been one of the founding ideas of modern ageing research. The explanations given for this risk and its evaluation in terms of positive or negative outcomes have changed over time. The early activity theories proposed that the act of taking over new roles and relations is important for successfully adapting to retirement (Cavan et al., 1949). Social barriers were seen as a major reason for fewer activities and shrinking social networks in old age. As an alternative perspective, Cumming and Henry (1961) proposed their disengagement theory of ageing. They shared the view that as people approach old age, they gradually withdraw from social roles and relations; this 'disengagement', however, was something they held to be a functionally necessary process of the individual retreating from society and the society releasing its hold on the individual – necessary in order to minimise the disruption caused by the individual's eventual physical decline and death. In a less optimistic vein, Burgess (1960) spoke of ageing as a 'roleless role', and thus a life phase whereby individuals were at risk of losing their social connectedness. Many authors asserted this to be a consequence of societal modernisation, with its devaluation of the productive capacities of the elderly and its dissolution of family solidarity beyond the nuclear household (Cowgill and Holmes, 1972).

More recent approaches have focused on preferences and routines that develop earlier in life, so that a lifespan perspective seems more appropriate. This applies also to the establishment of social networks (Kahn and Antonucci, 1980). Here, the fact that network size decreases with age is partly explained by the increasing probability of a death of parents, partners, siblings and friends. The theory of socioemotional selectivity (Carstensen, 1991) explains these changes by changes in individual time perspective: People carefully choose with whom to spend their remaining lifetime, and drop less important relations. This explanation has recently been used to explain the reduction of volunteer activity in old age as well (Hendricks and Cutler, 2004).

While most studies concur that network size decreases with age, this does not mean social isolation. With regard to family integration, the literature today concludes that intergenerational family bonds usually remain strong throughout adulthood and old age – and may even become more important for well-being over the life course than nuclear family ties (Bengtson, 2001). This has been shown for several Western countries such as France (Attias-Donfut, 1995), Germany (Kohli, 1999) and the US (Bengtson, 2001). These bonds range across different dimensions that have been conceptualised as dimensions of intergenerational 'solidarity': affectual, associational, consensual, functional, normative and structural (Szydlik, 2000; Bengtson, 2001). It is especially functional solidarity – the giving and receiving of support – that has drawn the most attention in recent years. The stylised results may be summarised in the following points (Kohli et al., 2010):

- Adult children and their elderly parents live close to each other (although seldom in the same household), feel close to each other emotionally, have frequent contact with each other, and mutually support each other with several types of help.

- Financial transfers and social support are (still) frequent and substantial, they occur mostly in the generational lineage, and their net flow is mostly downward, from parents to children.[4]
- Financial transfers *inter vivos* are complemented by bequests. *Inter vivos* transfers often go to children in need ('altruism'), while bequests are distributed equally among all children.

The protracted neglect of the socially productive activities of the elderly has now been redressed as well (Erlinghagen and Hank, 2008; Künemund, 2001). Such activities range from the continuation of gainful work to voluntary work of various sorts, and may even include 'hobby work' (Kohli et al., 1993; see also the chapter by Siegrist and Wahrendorf in this volume). For most of the elderly, labour-market issues – which have been at the origin of social network research (Granovetter, 1973, 1974) – are no longer relevant. But the importance of social connectedness through participation in activities and inclusion in networks has come into sharper focus for other issues as well.

2.2 Methodological issues

As mentioned previously, there is mounting evidence for the beneficial effects of social networks in old age on a variety of individual life outcomes. Claiming a positive impact of social networks on health and survival requires a study design that is able to differentiate between the impact of being part of a network and the selection into network membership on the basis of personal characteristics (including genetic ones) associated with positive health and survival outcomes. To explore genetic characteristics, twin designs are especially appropriate. A twin study in Denmark showed that health and mortality effects could partly be attributed to genetic factors, but that in addition to these, being in a network of family and friends also significantly contributed to survival (Rasulo et al., 2005). Also needed are longitudinal studies, such as the one by Lum and Lightfoot (2005) based on the 1993 and 2000 panels of the US *Asset and Health Dynamics Among the Oldest Old (AHEAD)* study. Their findings provide empirical support to earlier claims that doing volunteer work slows the decline in self-reported health- and functioning levels, slows the increase in depression levels, and decreases mortality rates. Some of these benefits – especially those linking activity with a reduction in mortality – seem almost too good to be true. They obviously hinge on the range and validity of the control factors that have been incorporated into the study design. The theoretical plausibility

---

4   See the calculation of net balance of monetary and time support by Litwin et al. (2008).

of the findings, in turn, depends on the availability of an appropriate causal mechanism. As an example, the mechanism by which productive activity and other forms of social participation affect mortality may be through enhancing the positive self-perceptions of ageing, which have been shown to increase longevity (Levy et al., 2002).

Since personal networks – in other words, the social capital of individuals – can hardly be observed directly, data are usually gathered by means of personal interviews or self-administered questionnaires. Numerous different techniques can be used to identify the members of a personal network. From a broader perspective, two types may be differentiated (Antonucci et al., 1990): role-relation methods and exchange methods. While role-relation methods use the type of network partners (for example, parents, children, neighbours or friends) as a starting point to collect information on these network members and the relationships with them, exchange methods start with the – factual or potential – substantive content of relationships (for example, exchanging goods, taking meals, or sharing close emotions), and then ask for information on the people with whom these exchanges and activities typically take place.[5] Some newer studies (such as the *German Aging Survey* or *SHARE*) use both strategies simultaneously by backing up the exchange network approach with other kinds of pertinent information (for example, on all existing children – whether or not they appear in the exchange network). However, since most of the support exchange takes place within the family, both approaches often produce similar results (see Kogovšek and Hlebec, 2008). This is especially the case in older age groups, as close relationships become more important with increasing age (Carstensen et al., 1999) and as the number of members of one's 'convoy' across the life course decreases (Kahn and Antonucci, 1980).

2.3   Social changes and network quality

The crucial question is whether social networks are still able to perform – and will be so in the future. There is, for instance, a potential dilemma between care and paid employment both at the individual and at the societal level. Johnson and Lo Sasso (2000) examine whether the rising labour-force participation rates of married women interfere with care-giving for frail elderly parents. Their results for the US (based on the Health and Retirement Study, the HRS) indicate that time spent helping parents substantially reduces labour supply for both women and men. This is especially acute for those in the position of

---

5   In psychology, the latter is sometimes conceptualised as a distinct third strategy (referred to as the *affective network approach*).

the 'sandwich generation' (Künemund, 2006) – those with a double obligation of care for dependent parents and children. A 'hard' sandwich position – having to care for both dependent generations while simultaneously being in the labour force – is rare. In Germany in 2002, only 3% of the women aged 40 to 44, 8% of those aged 45 to 49, and 3% of those aged 50 to 54 cared for an elderly family member while having children at home or caring for grandchildren and simultaneously participating in the labour force at least one hour per month (Künemund, 2006). But these low numbers may already be partly the result of a withdrawal from the labour force due to family care obligations.[6] With the rising labour-force participation of women, and the extension of working life through a later retirement age, the potential time crunch is likely to become harsher. The mid-lifers and young elderly will be faced with the choice between foregoing care and reducing or abandoning employment – even at the expense of the heavy penalties for early exit from the labour force in terms of pension levels that are now being instituted in many European countries. At the societal level, the dilemma is between increasing the labour-force participation of those beyond 55 (as required by the Lisbon Agenda of the EU) and increasing the demand for public care services and institutions. The dilemma also applies to grandparenting. While grandparents may be willing to give substantial amounts of time to the task of caring for their grandchildren – thus allowing their daughters or daughters-in-law to combine parenthood and engagement in the labour force – this may interfere with their own employment.

This creates a need for new arrangements between employment and care – not only for young parents, but for the younger old as well (for example, through the availability of part-time work and leaves or sabbaticals). Family care work also needs to be supported by public policy, both in terms of financial subsidies (which will remain much less costly than publicly funded institutional care) and in terms of services to help the helpers. Another principle for public

---

6   Note that the numbers reported in the literature diverge hugely. Between one and 80% of the population have been identified as sandwiched adults. The divergence is a consequence of different concepts, operationalizations and samples. In Germany, more than 80% of the men and women aged 40 to 44 have at least one relative of both younger and older generations, but most of the parents of these sandwiched adults are not in actual need of personal care, nor are most of their children. Furthermore, many studies do not find a negative relationship to well-being (Spitze et al., 1994; Penning, 1998; Ward and Spitze, 1998; AARP, 2001). There is already a tradition of calling the metaphor of the sandwich generation a modern myth (Loomis and Booth, 1995; Rosenthal et al., 1996; Hörl and Kytir, 1998; Höpflinger and Baumgärtner, 1999; Putney and Bengtson, 2001). There can be no doubt, however, that the competing demands from work and family – especially with regard to caring for disabled family members – take a heavy toll on some women (Brody, 1990; Nichols and Junk, 1997).

policy is that it needs to be conceptualised as generational policy, being aware that provisions (or their withdrawal) for one generation affect the welfare of all other generations.

Taking into account the risks for network effectiveness discussed above (which stem from demographic shifts as well as from societal change), we may conclude that social networks of the future elderly will be very different from those of elderly people today – and this is one of the main research topics that should receive more scientific attention.

Network effectiveness is also a function of the quality of ties. How the close network ties should be evaluated is controversial. Intergenerational relationships in the family are inherently ambivalent (Pillemer and Lüscher, 2004); the balance between solidarity and autonomy has to be continually negotiated among the participants, and under contemporary conditions of individualisation this is no easy matter. Close relationships need not be an unmitigated blessing; they can also be a source of conflict, anxiety and frustration. The family itself may generate conflicts (inheritances, for example). Network obligations may become burdensome, as in the case of sustained family care for dependent elderly parents, which may restrict caretakers' own life plans. However, open intergenerational conflicts occur in only a small minority of families (Szydlik, 2002)[7] and are less often present in older age groups than in younger ones (Akiyama et al., 2003). The individual and social determinants of relationship quality in old age need to be examined in greater detail.

Another controversial point is partnership disruption. While many studies confirm the view that current family changes pose a risk for the support potential available to future cohorts of elderly people (Pezzin et al., 2008), some evidence indicates that the increase in divorce has not resulted in less support in old age (Glaser et al., 2008). A key for understanding these partly counter-intuitive results may be the interplay between family changes and increasing resources of individuals (education, health, wealth, social networks) and society (welfare state spending, availability of community- and market-based services). These resources help to improve intergenerational relationships and to make family support less burdensome (Künemund, 2008). Again, better data – especially longitudinal data – is needed to address these questions (van der Pas et al., 2007).

A crucial question for policy is whether the available family potential can be activated in times of need. A number of factors may weaken the reliability

---

7   Among the respondents in the German Aging Survey (1996) – the German population aged 40-85 – 11% said they were in conflict with a close person in an intergenerational family relation (Szydlik, 2002).

of these family ties, among them (a) geographical distance, which is increasing due to increasing mobility; and (b) a shrinking willingness to help, in the sense of fading norms of solidarity and responsibility. As an alternative, we need to assess (c) the availability of non-kin social networks that may fill in the gap.

(a) Co-residence among adult family generations has decreased massively in all Western societies. Today, among the Europeans above the age of 80 who have at least one living child, only 17% live together with a child in the same household (Kohli et al., 2008; Hank, 2007).[8] However, by extending the boundaries of 'togetherness' the situation turns out to be very different. If one includes parents and children living not only in the same household but also in the same house – a pattern that is very common in Germany and Austria – the proportion rises from 17 to 32%. By including the neighbourhood (less than 1 km away), the proportion rises to 53%. 84% have a child living not farther away than 25 km. The preference today seems to be for 'intimacy at a (small) distance' – small enough so that relations of exchange and support may function easily across the boundaries of the separate households.

The increase in the proportion of elderly people living alone does therefore not imply that they have been left alone by their children. Unlike their own parents or grandparents, they are not forced in any way by economic necessity to co-reside, and this may have resulted in even better generational relationships (Kohli, 1999; Künemund and Rein, 1999). Findings on the frequency of contact, emotional closeness and the exchange of support confirm this view (Kohli et al., 2005). Adult generations in families – even in countries with comparatively weaker family traditions and larger geographical distances – remain closely linked. Contact with the most contacted child is daily for 42 and 45% in Denmark and Sweden, respectively, and for between 47 and 55% in the countries of continental Western Europe; the Mediterranean countries stand out, with between 84 and 86%. In all countries, 70% or more have contact at least several times a week; in the Mediterranean countries, this amounts to 95% or more. There are those who have no contact at all with their living child or children, but in no country do they comprise more than 1%. In the older age groups contact is less frequent – but even among those over 80 years of age between three-fifths (in Switzerland) and more than nine-tenths (in Mediterranean countries) are in contact with a child daily or several times a week. A

---

8   These and the subsequent findings are based on the second wave of the *Survey of Health, Ageing and Retirement in Europe (SHARE)*, with data collected in 2006 in 14 countries (Sweden, Denmark, Germany, the Netherlands, Belgium, France, Switzerland, Austria, Poland, Czech Republic, Spain, Italy, Greece and Israel).

recent study of longitudinal cohort data in the Netherlands between 1992 and 2002 confirms the view that support exchange and contact have held their own: '*Macrostructural changes have had less destructive influence on parent-child relationships than we initially thought*' (van der Pas, 2007: 271).

It is obvious that geographic mobility increases in line with labour-market flexibility. But to date, it is less well known whether at least one of the children (most likely the youngest; see Konrad et al., 2002) stays within close distance to their parents, or whether 'trailing' parents relocate to move closer to one of their children (Attias-Donfut and Renaut, 1994). Such 'trailing' has its price. Current findings suggest that elderly persons in need of help will try to keep living as independently as possible in order not to 'burden' their children (Lewinter, 2003), but will turn to their next of kin when this is no longer possible. Providing the elderly with sufficient resources to live on their own so that they can age in place is therefore one of the main policy goals here (OECD Health Project, 2005).

(b) Of special interest is the articulation between private and public provision of services and resources. In the conventional story of modernisation, the emergence of the nuclear family and that of the public old-age security system were seen as parallel and mutually reinforcing processes. The basic assumption was that the development of the welfare state would crowd out the private support within families. Recent evidence, however, points to the opposite conclusion: Welfare state provisions, far from crowding out family support, enable the family, in turn, to provide new intergenerational support and transfers (Kohli, 1999; Künemund and Rein, 1999; Lowenstein and Ogg, 2003). For example, when the elderly have sufficient financial resources from state pensions, there is less need for their kin to provide monetary support. Thus, monetary transfers from children to their elderly parents may indeed be crowded out (Reil-Held, 2006). But the level of services provided by the children may be expected to increase: parents' increasing resources allow them to be not only better able to buy missing services on the market, but also in a better position to initiate an exchange with their children; they can accept help offered to them more easily, as they have something to reciprocate; the quality of the parent-child relationship is improved when there is no need to financially support the aged parents and when public services support the relationship; or some children offer help in expectation of later transfers (Künemund and Rein, 1999). When the level of concrete public services increases, then the family may be expected to provide more of other services, such as emotional support. The overall outcome of this process of substitution is probably more, not less, family solidarity. The available empirical findings do not support the crowding-out hypothesis, and partly support the reverse: namely, that the relationship between the state and the family is a process of 'crowding in': Generous welfare systems give resources to the family that help to increase rather than undermine its own solidarity among its members.

The above discussion allows us to conclude that existing public retirement pension arrangements currently support not only the elderly themselves, but through them, also their descendants, which improves the quality of intergenerational relations within the family – with the likely result of even stronger family solidarity. Reductions in welfare state spending could therefore result in less family solidarity. To date, several studies have confirmed this view (Attias-Donfut, 2000; Lund, 2002; Knijn, 2004; Motel-Klingebiel et al., 2005; Armi et al., 2008; Kalmijn and Saraceno, 2008). Nevertheless, these findings are often overlooked in the debate on generational equity (see Kohli, 2006); the idea of a general crowding out of the family is still prominent among politicians and the broader public.

(c) The patterns of non-kin networks are much less often subject to scientific research. Of importance here are friendships and privately organised groups as well as the weak (and sometimes strong) ties that are connected to social activities in associations, ranging from active involvement and volunteering to long-term membership.

2.4   Social participation

Withdrawal from the labour force may be expected to result in diminished social networks, as far as interactions with work colleagues are concerned. But the transition to retirement is not paralleled by withdrawal from social participation in general. The term 'productive ageing' is sometimes used to refer to continued labour-force participation, but this is overly restrictive; it should rather be used to point out that activity and social engagement of any kind play a major role in old age as well – a fact that is often neglected in the discourse of intergenerational equity. Several attempts have also been made to estimate the value of productive activities in economic terms (Herzog et al., 1989; Herzog and Morgan, 1992; Coleman, 1995; Künemund, 1999). The results obviously differ according to how productive activity is defined.

Economic perspectives generally focus on remunerated activities or those producing valued goods and services that would otherwise have to be purchased in the market – for the most part, paid work and some dimensions of household production. Sociological perspectives add voluntary activities, child-care, personal care, informal help and other activities (to the extent that they are of value to others). A broader definition may also take into account certain activities that help to maintain the ability of an individual to be productive (further education, for example). Finally, psychological perspectives include activities that help to maintain the emotional and motivational preconditions for productivity, for example, the successful adaptation to age-related losses (Staudinger, 1996; O'Reilly and Caro, 1994, provide an excellent overview of definitions of productivity). In this perspective, individual well-being also becomes a productive resource. According to the definition chosen for analysis, either only a minority of older people or nearly all of them are productive (and those who are not can hardly be accused of their 'unproductivity' – in most cases we would expect that it is not a voluntary decision to be unproductive in psychological terms).

Using a sociological definition, we find that older people are productive to a remarkable degree. In Germany, for example, older people are seldom active on the labour market (3% of the 70-85 year-olds), but 16% have been active in childcare, 8% in personal care, 7% in voluntary activities, and 18% in informal help within the last three months. Taken together, 39% of this age group are 'productive' in a way that produces valued goods and services (Künemund, 1999). Furthermore, 39% support their children and grandchildren by substantial intergenerational material transfers (Kohli et al., 2000). The total number of hours of productive activity solely in volunteerism, childcare and personal care in the population aged 60 to 85 is nearly 3.5 billion hours of (usually unpaid) production of valued goods and services a year. The monetary value of these activities – calculated by the mean net hourly wage rate of workers in the non-profit sector (welfare associations or political parties, for example) – has been estimated to be equivalent to about 21% of the public pensions in Germany (Künemund, 1999).

Studies in other countries, such as the United States, come up with roughly similar numbers. More than one-third of Americans aged 65 or over report voluntary activities, more than 50% informal help (Herzog et al., 1996: 327). Coleman (1995) estimated the value of productive activities – care and volunteerism – to be almost 2% of US GDP in 1990.

It is evident that demand structures play a major role with respect to age-group differences in these activities. Childcare becomes less relevant in old age because the grandchildren have by then reached an age at which they no longer need care. Many aggregate group differences in levels of childcare – such as the difference between East and West Germany – mainly reflect the difference in the existence of children. Personal care, on the other hand, declines with age mainly because fewer parents are left. For voluntary activities, demographic opportunity structures are less important. The decrease of volunteerism with advancing age – to the extent that it really represents an age rather than a cohort process – may result from self-initiated disengagement on the part of the elderly; our findings suggest, however, that the organisations of this sector have also implemented some informal age limits (Kohli and Künemund, 1996).

International comparisons show that even the definitions of 'voluntary activity' and 'informal help' differ among countries – and so do the opportunities, necessities and constraints for such activities. The numbers are therefore not strictly comparable. They do show, however, that older people are not merely passive recipients of pensions and welfare. In addition to the economic value of their activities, the elderly also contribute productively to society in many other ways – not least in terms of social integration and cohesion. The psychological aspects include benefits in terms of self-efficacy and life satisfaction, which in turn are related to health and longevity.

The positive impact of networks of productive activities would show up even more strongly if reliable and comparable data were available on new forms of engagement and volunteering beyond formal organisations (such as social movements, self-help groups or self-organised learning networks).

By any definition, however, such productive networks comprise, for most people, only a small part of everyday life in retirement – and a large share of older people are not involved at all. Despite the substantial productivity outlined so far, much of life in retirement seems to be oriented towards leisure and consumption. But even here, social networks may play an important and hitherto neglected role. While many of these consumptive activities may be solitary, others are embedded in regular and stable network structures, such as informal groups playing cards or bowling together, or just meeting for purposes of sociability (Putnam, 2000; Brauer, 2005). To date, reliable knowledge about these types of social integration into private (mostly non-kin) networks is extremely rare. For Germany, we have shown that about 40% of the 40- to 85 year-olds are regularly involved in at least one such setting (Künemund, 2001). Further research is needed here to identify, and possibly strengthen the potentials of these informal social networks. The latter may, in turn, become productive in a wider sense – for instance, in rural communities where formal community services and supports are increasingly lacking.

3 Remaining gaps in knowledge: main challenges

Research on social networks is increasingly important in ageing societies. As the preceding sections have shown, the basic patterns of family networks in old age and of the exchanges within them have by now been well established. The same is true for some of the networks of productive activities outside the family. In order for policy to be adequately targeted, however, more specific research will need to address a number of substantive topics that have been neglected so far:

- The role of *wider kinship networks* (for example, the demographic potential and relevance in case of need of siblings or in-laws).
- The needs of *special groups* (the childless or divorced, for example) and the availability of compensatory networks.
- The extent and effect of *gender differences* in networks and exchange activities (the incidence of caring men, for example, or the effects of caring on the couple's division of formal and informal work).
- The incidence, strength and reliability of *informal sociability and solidarity* in situations of need.
- The extent and consequences of *regional disparities* (urban-rural differences in ageing and service provision, for example).
- The impact of *ethnic or religious disparities* (with respect to family networks, for example).[9]
- The way in which *socio-economic disparities* affect both formal and informal networks.[10]
- The functioning of *kin and non-kin networks of elderly migrants* (both for those having migrated earlier and those migrating after retirement; 'amenity migration').

The networks of elderly migrants merit special attention. Immigrants in most European countries are younger than the native populations, and this may explain the fact that research has been late to address elderly migrants. But immigrant populations now age rapidly and present new challenges for policies of ageing. There is some evidence that the patterns of exchange in family networks of immigrants differ systematically from those of natives (Baykara-Krumme, 2008). Networks of immigrants often remain transnational and may function as channels of remittances from host to home societies. Another impor-

---

9   See Litwin's typological approach to comparison of ethnic groups in Israel (Litwin, 2004).
10   Findings so far tend to show that upper classes have higher levels of formal social capital, especially through associational networks, while informal contacts are not so clearly stratified by class (Pichler and Wallace, 2009).

tant phenomenon that gives rise to special transnational network types is the migration of retirees from Northern and Western Europe to the Mediterranean countries in search of more amenable lifestyles (King et al., 2000; Braun and Arsene, 2009).

There is, moreover, a range of more systematic problems to be addressed:

- *The comparative dimension:* Europe can be considered a 'laboratory' for assessing the impact of structures, cultures and institutions (Kohli, 2004). There is a well-developed research literature on welfare states and family regimes, but few studies have focused on social networks. Comparisons are needed among European societies, but also with the other advanced societies in North America and East Asia that face similar challenges of population ageing.

- *The life-course dimension:* A key question here is how networks are differentially important and provide different benefits in different periods of the life course. On the one hand, networks may be less important for the elderly because there is no longer a need for them in processes of job search or partner search (which has been the focus of much of early social network analysis). On the other hand, they may become more critical because the elderly are no longer integrated into the formal organisations of the workplace, and have higher needs of security and support. There is some research on the life-course changes of network size and composition (Wagner and Wolf, 2001, for example). But more systematic analyses of network patterns and network salience over the life course are required to overcome the narrow focus on specific periods such as older age (or adolescence or early adulthood, for that matter).

- *The longitudinal dimension:* While cross-sectional associations between the dimensions of network structure and participation and their antecedents and consequences can yield more or less plausible interpretations, reliable knowledge about trends (age-period-cohort matrix) and about causal processes presupposes longitudinal data. A number of single-country longitudinal surveys (especially the large household panels) are available for this purpose, but comparative panel databases – such as *SHARE* – need to be expanded. Changes over time in network composition and in the support and exchange relations of elderly people need to be described and analysed in greater depth (Peek et al., 1997).

- *Integration of substantive research on social networks of the elderly with the formal approaches of general network analysis:* Network analysis shifts the attention from dyadic relationships to indirect ties, third-party effects, and other more complex patterns of ties between multiple actors that have profound consequences for their behaviour. Furthermore, these ties between multiple actors lead to feedback patterns and phenomena that can be studied only longitudinally.

- *Trends in technology:* The chapter has not gone into how new information technologies facilitate social interaction, or into social networks in terms of 'web 2.0' internet communities. The current elderly are not yet much affected by networking developments such as computer-mediated friendships, chatting or discussion groups; the 'digital divide' is not least a cohort divide. But it is obvious that these new technologies will in many ways alter our social relations, including those of the coming cohorts of elderly people. These trends are important not only from an individual point of view, but also with respect to interfaces between formal and informal support (healthcare systems, service providers etc.). Serious research in this new field is still rare and should be expanded.

4 Research infrastructures and networks

National research infrastructures for the study of social networks among the elderly population are fairly well developed in many European countries. Sizable research institutes are active in this field, often clustered around large surveys such as the *English Longitudinal Study of Ageing (ELSA)*, the French *Three-Generation Study* or the *German Aging Survey*. Family networks are also addressed by the large demography institutes, whether directly state-organised (such as the French *INED* or the Dutch *NIDI*) or as part of the academic research environment (such as the German *Max Planck Institute for Demographic Research* or the *Vienna Institute of Demography*).

The large household panel studies that have partly been operative for more than two decades – such as the *German Socio-Economic Panel* (since 1984), the *British Household Panel* or the *Swiss Household Panel* – provide another important infrastructure for studying social networks. At the European level, an attempt was made to create a *European Community Household Panel*. This was discontinued and partly replaced by the *European Union Statistics on Income and Living Conditions (EU-SILC)*, which is, unfortunately, much more limited in its network information.

An early cross-national survey with some network information is the *International Social Survey Program (ISSP)*. It has been especially useful through some of its special topical modules that change yearly. A more recent instrument that has become the main rallying point for researchers in this field is the *Survey of Health, Ageing and Retirement in Europe (SHARE)*, a longitudinal survey started in 2004, and now in its fourth wave of data collection. It has become part of the *European Strategy Forum on Research Infrastructures (ESFRI)*, a major multidisciplinary EU research platform beyond the Framework Programmes. Another large social science infrastructure project now part of *ESFRI* is the *European Social Survey (ESS)*, with a focus on political behaviour and attitudes but also some information in its yearly changing modules that may be relevant for network research.

5   Required research infrastructures, methodological innovations and data

In order to address the gaps in knowledge and the problems outlined above, the existing survey data are often not sufficient. While *SHARE* now allows for describing and analysing social networks of the elderly in at least some detail, many dimensions of social networks – such as changes over time or differences among cohorts – have only been studied, if at all, in national contexts and with cross-sectional data. The main task for survey data collection is to integrate different types of transfers and support for various groups with multiple relationships in a biographical approach to network structure and exchange. This requires a longitudinal dataset (which obviously takes time to mature and bear fruit). Retrospective designs are problematic in this field.[11]

Time-use studies represent a well-tested way to describe and analyse the patterns of activities in which individuals engage. While they usually do not contain enough information on role relations and exchange processes, they can be extended in these dimensions, which allows them to serve as a basis for network analyses as well.

Also needed are more detailed comparative qualitative studies. Particularly promising are comparative qualitative *panel* studies, although their methodology still needs to be developed (as it is much less standardised and depends more on specific national research traditions). A better understanding of normative obligations and their change, for example, is a precondition for not only the construction of reliable survey research instruments but also for the

---

11   Standardised retrospective questions on the change of network size and composition are subject to serious bias, so that this way of collecting data on network change over the life course is limited. Qualitative procedures such as detailed biographical narrations may be better suited to reveal the structure of social relations earlier in life.

development of better theories. While many basic facts are well-known – the shrinking network size and increasing dominance of the family with advancing age, for example – their explanation is still not satisfactory. Explanation requires interdisciplinary accounts that combine, for example, sociological and psychological theories; these accounts may also find their starting point in detailed qualitative biographical data.

For some of the topics identified above as needing further research, there exist special comparative surveys. Examples are political participation (surveyed by the *European Social Survey*) or changing family forms (surveyed by the *Gender and Generations Survey*). In other fields, such as volunteering, national surveys are currently being extended to include most or all other European countries. Care should be taken to standardise the core instruments across surveys, so that the results of the respective studies can be compared and linked. The variables for political orientation and participation in a survey oriented towards other topical fields (*SHARE*, for example) should be closely modelled after those in a specialised survey (such as the *ESS*). However, allowing for full micro-level analyses requires incorporation of as many variables as possible in the same survey instrument. As an example, *SHARE* should incorporate more key variables of participation in political networks and in those of voluntary work. One possibility for doing this is to have a smaller share of core questions repeated annually and a larger space allowed for changing modules, as successfully practiced by both the *ISSP* and the *ESS*. Changing modules may also be used for surveying special situations such as men as caregivers or exchange relations with siblings and other kin.

Similarly, special groups such as migrants – too small to be sufficiently picked up by most general population surveys – need specific comparative surveys that have appropriate sampling frames and also instruments tailored to the particular living situations and contexts of these groups. Here again, however, making full use of the potential of micro-level analysis requires incorporation of these groups in one and the same survey together with the majority groups (for example, by over-sampling them – if not on a regular basis, then at least in some years).

References

Akiyama, H., T.C. Antonucci, K. Takahashi and E.S. Langfahl (2003) 'Negative Interactions in Close Relationships across the Life Span', *Journal of Gerontology* 58B, P70-P79.

Albertini, M., M. Kohli and C. Vogel (2007) 'Intergenerational Transfers of Time and Money in European Families: Common Patterns - Different Regimes?' *Journal of European Social Policy* 17, 319-334.

American Association of Retired Persons (AARP) (2001) *In the Middle: A Report on Multicultural Boomers Coping with Family and Aging Issues* (Washington: AARP).

Antonucci, T.C. and K.C.P.M. Knipscheer (1990) 'Social Network Research: Review and Perspectives' in K.C.P.M. Knipscheer and T.C. Antonucci (eds) *Social Network Research: Substantive Issues and Methodological Questions* (Amsterdam: Swets and Zeitlinger), 161-175.

Anttonen, A., J. Baldock and J. Sipilä (eds) (2003) *The Young, the Old and the State: Social Care Systems in Five Industrial Nations* (Cheltenham: Elgar).

Armi, F., E. Guilley and C.J. Lalive d'Epinay (2008) 'The Interface between Formal and Informal Support in Advanced Old Age: A Ten-year Study', *International Journal of Ageing and Later Life* 3(1), 5-19.

Attias-Donfut, C. (1995) 'Le double circuit des transmissions' in C. Attias-Donfut (ed.) *Les solidarités entre générations. Vieillesse, familles, État.* (Paris: Nathan), 41-81.

Attias-Donfut, C. (2000) 'Familialer Austausch und soziale Sicherung' in M. Kohli and M. Szydlik (eds) *Generationen in Familie und Gesellschaft.* (Opladen: Leske + Budrich), 222-237.

Attias-Donfut, C. and S. Renaut (1994) 'Vieillir avec ses enfants - Corésidence de toujours et reco-habitation., *Communications* 59, 29-53.

Attias-Donfut, C. and M. Segalen (1998) *Grands-parents. La famille à travers les générations.* (Paris: Éditions Odile Jacob).

Baykara-Krumme, H. 2008. *Immigrant Families in Germany: Intergenerational Solidarity in Later Life* (Berlin: Weißensee).

Bengtson, V.L. (2001) 'Beyond the Nuclear Family: The Increasing Importance of Multigenerational Bonds', *Journal of Marriage and Family* 63, 1-16.

Brauer, K. (2005) *Bowling Together: Clan, Clique, Community und die Strukturprinzipien des Sozialkapitals* (Wiesbaden: VS Verlag für Sozialwissenschaften).

Braun, M. and C. Arsene (2009) 'The Demographics of Movers and Stayers in the European Union' in E. Recchi and A. Favell (eds) *Pioneers of European Integration: Citizenship and Mobility in the EU* (Cheltenham: Edward Elgar), 26-51.

Brody, E.M. (1990) *Women in the Middle: Their Parent-care Years* (New York: Springer).

Broese van Grenou, M. and T. van Tilburg (1997) 'Changes in the Support Networks of Older Adults in the Netherlands', *Journal of Cross-Cultural Gerontology* 12, 23-44.

Burgess, E.W. (1960) 'Aging in Western Culture' in E.W. Burgess (ed.) *Aging in Western Societies* (Chicago: University of Chicago Press), 3-28.

Cantor, M.H. (1979) 'Neighbors and Friends: An Overlooked Resource in the Informal Support System', *Research on Aging* 1, 434-463.

Carstensen, L.L. (1991) 'Socioemotional Selectivity Theory: Social Activity in Life-span Context', *Annual Review of Gerontology and Geriatrics* 11, 195-217.

Carstensen, L.L., D.M. Isaacowitz and S. Turk-Charles (1999) 'Taking Time Seriously: A Theory of Socioemotional Selectivity', *American Psychologist* 54, 165-181.

Cavan, R.S., E.W. Burgess, R.J. Havighurst and H. Goldhamer (1949) *Personal Adjustment in Old Age* (Chicago: Social Science Research Associates).

Coleman, K.A. (1995) 'The Value of Productive Activities of Older Americans' in S.A. Bass (ed.) *Older and Active: How Americans over 55 are Contributing to Society* (New Haven: Yale University Press), 169-203.

Cornwell, B., E.O. Laumann and P.L. Schumm (2008) 'The Social Connectedness of Older Adults: A National Profile', *American Sociological Review* 73, 185-203.

Cornwell, B., L. Philipp Schumm, E.O. Laumann and J. Graber (2009) 'Social Networks in the NSHAP Study: Rationale, Measurement, and Preliminary Findings', *Journal of Gerontology: Social Sciences*, 64B (Supplement 1), i47-i55.

Cowgill, D.O. and L.D. Holmes (eds) *Aging and Modernization* (New York: Appleton-Century-Crofts).

Crohan, S.E. and T.C. Antonucci (1989) 'Friends as a Source of Social Support in Old Age' in R.G.Adams and R. Blieszner (eds) *Older Adult Friendship: Structure and Process* (Newbury Park: Sage), 129-147.

Cumming, E. and W.E. Henry (1961) *Growing Old: The Process of Disengagement* (New York: Basic Books).

de Jong Gierveld, J. and B. Havens (2004) 'Cross-national Comparisons of Social Isolation and Loneliness: Introduction and Overview', *Canadian Journal on Aging* 23, 109-113.

de Jong Gierveld, J. and D. Perlman (2006) 'Long-standing Non-kin Relationships of Older Adults in the Netherlands and the United States', *Research on Aging* 28, 730-748.

Erlinghagen, M. and K. Hank (eds) (2008) *Produktives Altern und informelle Arbeit in modernen Gesellschaften. Theoretische Perspektiven und empirische Befunde* (Wiesbaden: VS Verlag für Sozialwissenschaft).

European Commission (2006) *The Impact of Ageing on Public Expenditure: Projections for the EU-25 Member States on Pensions, Healthcare, Long-term Care, Education and Unemployment Transfers (2004–50). Report Prepared by the Economic Policy Committee and the European Commission (DG ECFIN).* (Brussels: European Communities).

Ferraro, K.F. (1984) 'Widowhood and Social Participation in Later Life: Isolation or Compensation?' *Research on Aging* 6, 451-568.

Fokkema, T., S. ter Bekke and P.A. Dykstra (2008) *Solidarity between Parents and their Adult Children in Europe* (Amsterdam: KNAW Press).

Glaeser, E.L., D. Laibson and B. Sacerdote (2002) 'An Economic Approach to Social Capital', *Economic Journal* 112, 437-458.

Glaser, K., C. Tomassini and R. Stuchbury (2008) 'Differences over Time in the Relationship between Partnership Disruptions and Support in Early Old Age in Britain', *Journal of Gerontology* 63B, S359-S368.

Granovetter, M.S. (1973) 'The Strength of Weak Ties', *American Journal of Sociology* 78, 1360-1380.

Granovetter, M.S. (1974) *Getting a Job: A Study of Contacts and Careers* (Cambridge: Harvard University Press).

Hank, K. (2007) 'Proximity and Contacts between Older Parents and their Children: A European Comparison', *Journal of Marriage and Family* 69, 157-173.

Hendricks, J. and S.J. Cutler (2004) 'Volunteerism and Socioemotional Selectivity in Later Life', *Journal of Gerontology* 59B, S251-S257.

Herzog, R.A. and J.N. Morgan (1992) 'Age and Gender Differences in the Value of Productive Activities. Four Different Approaches', *Research on Aging* 14, 169-198.

Herzog, R.A., R.L. Kahn, J.N. Morgan, J.S. Jackson and T.C. Antonucci (1989) 'Age Differences in Productive Activities', *Journal of Gerontology* 44, S129-138.

Hochschild, A.R. (1975) 'Disengagement Theory: A Critique and Proposal', *American Sociological Review* 40, 553-569.

Höpflinger, F. and D. Baumgärtner (1999) '„Sandwich-Generation": Metapher oder soziale Realität?' *Zeitschrift für Familienforschung* 11, Heft 3, 102-111.

Hörl, J. and J. Kytir (1998) 'Die „Sandwich-Generation": Soziale Realität oder gerontologischer Mythos? Basisdaten zur Generationenstruktur der Frauen mittleren Alters in Österreich., *Kölner Zeitschrift für Soziologie und Sozialpsychologie* 50, 730-741.

Johnson, R.W. and A.T. Lo Sasso (2000) *The Trade-off between Hours of Paid Employment and Time Assistance to Elderly Parents at Midlife* (Washington: The Urban Institute) (Ms.).

Johnson, R.W., D. Toohey and J.M. Wiener (2007) *Meeting the Long-term Care Needs of the Baby Boomers: How Changing Families Will Affect Paid Helpers and Institutions* (Washington: The Urban Institute) (Ms.).

Kahn, R.L. and T.C. Antonucci (1980) 'Convoys over the Life Course: Attachment, Roles, and Social Support' in P.B. Baltes and O.G. Brim (eds) *Life-span Development and Behavior* (New York: Academic Press), 383-405.

Kalmijn, M. and C. Saraceno (2008) 'A Comparative Perspective on Intergenerational Support: Responsiveness to Parental Needs in Individualistic and Familialistic Countries', *European Societies* 10, 479-508.

King, R., T. Warnes and A. Williams (2000) *Sunset Lives* (Oxford: Berg).

Kinney, J.M. (1996) 'Home Care and Caregiving'. in J.E. Birren (ed.) *Encyclopedia of Gerontology* (San Diego: Academic Press), vol. I, 667-678.

Knijn, T. (2004) 'Family Solidarity and Social Solidarity: Substitutes or Complements?' in Knijn, T. and A. Komter (eds) *Solidarity between the Sexes and the Generations* (Cheltenham: Edward Elgar), 18-33.

Kogovšek, T. and V. Hlebec (2008) 'Measuring Ego-centered Social Networks: Do Cheaper Methods with Low Respondent Burden Provide Good Estimates of Network Composition?' *Metodološki zvezki* 5, 127-143.

Kohli, M. (1999) 'Private and Public Transfers between Generations: Linking the Family and the State', *European Societies* 1, 81-104.

Kohli, M. (2004) 'Intergenerational Transfers and Inheritance: A Comparative View' in M. Silverstein (ed.) *Intergenerational Relations across Time and Place* (*Annual Review of Gerontology and Geriatrics*, vol. 24). (New York: Springer), 266-289.

Kohli, M. (2006) 'Aging and Justice' in R. Binstock and L. George (eds) *Handbook of Aging and the Social Sciences, 6th ed.* (San Diego: Elsevier), 456-478.

Kohli, M., H.-J. Freter, M. Langehennig, S. Roth, G. Simoneit and S. Tregel (1993) *Engagement im Ruhestand. Rentner zwischen Erwerb, Ehrenamt und Hobby.* (Opladen: Leske + Budrich).

Kohli, M. and H. Künemund (1996) *Nachberufliche Tätigkeitsfelder – Konzepte, Forschungslage, Empirie* (Stuttgart: Kohlhammer).

Kohli, M., H. Künemund, A. Motel and M. Szydlik (2000) 'Families Apart? Intergenerational Transfers in East and West Germany'. in S. Arber and C. Attias-Donfut (eds) *The Myth of Generational Conflict: Family and State in Ageing Societies* (London: Routledge), 88-99.

Kohli, M., H. Künemund and C. Vogel (2008) 'Shrinking Families? Marital Status, Childlessness, and Intergenerational Relationships'. in A. Börsch-Supan, A. Brugiavini, H. Jürges, J. Mackenbach, J. Siegrist and G. Weber (eds) *Health, Ageing and Retirement in Europe (2004-2007) Starting the Longitudinal Dimension* (Mannheim: Mannheim Research Institute for the Economics of Aging), 164-171.

Kohli, M., K. Hank and H. Künemund (2009) 'The Social Connectedness of Older Europeans: Patterns, Dynamics and Contexts', *Journal of European Social Policy* 19, 327-240.

Kohli, M., M. Albertini and H. Künemund (2010) 'Linkages among Adult Family Generations: Evidence from Comparative Survey Research' in P. Heady and M. Kohli (eds) *Family, Kin-*

ship and State in Contemporary Europe. vol. 3: Perspectives on Theory and Policy (Frankfurt/M: Campus), 195-220.

Konrad, K.A., H. Künemund, K.E. Lommerud and J.R. Robledo (2002) 'Geography of the Family', *American Economic Review* 92, 981-998.

Künemund, H. (1999) 'Entpflichtung und Produktivität des Alters', *WSI-Mitteilungen* 52, 26-31.

Künemund, H. (2000) 'Pflegetätigkeiten in der zweiten Lebenshälfte - Verbreitung und Perspektiven' in W. Clemens and G. Backes (eds) *Lebenslagen im Alter. Gesellschaftliche Bedingungen und Grenzen* (Opladen: Leske + Budrich), 215-229.

Künemund, H. (2001) *Gesellschaftliche Partizipation und Engagement in der zweiten Lebenshälfte. Empirische Befunde zu Tätigkeitsformen im Alter und Prognosen ihrer zukünftigen Entwicklung* (Berlin: Weißensee Verlag).

Künemund, H. (2006) 'Changing Welfare States and the „Sandwich Generation" – Increasing Burden for the Next Generation?' *International Journal of Ageing and Later Life* 1, 11-30.

Künemund, H. (2008) 'Intergenerational Relations within the Family and the State' in C. Saraceno (ed.) *Families, Ageing and Social Policy: Intergenerational Solidarity in European Welfare States* (Cheltenham: Edward Elgar), 105-122.

Künemund, H. and M. Rein (1999) 'There is More to Receiving than Needing: Theoretical Arguments and Empirical Explorations of Crowding in and Crowding out', *Ageing and Society* 19, 93-121.

Künemund, H. and B. Hollstein (2000) 'Soziale Beziehungen und Unter-stützungsnetzwerke. in M. Kohli and H. Künemund (eds) *Die zweite Lebenshälfte – Gesellschaftliche Lage und Partizipation im Spiegel des Alters-Survey* (Opladen: Leske + Budrich), 212-276.

Lang, F.R. (1994) *Die Gestaltung Informeller Hilfebeziehungen im hohen Alter – die Rolle von Elternschaft und Kinderlosigkeit. Eine empirische Studie zur sozialen Unterstützung und deren Effekt auf die erlebte soziale Einbindung* (Berlin: Max-Planck-Institut für Bildungsforschung).

Levy, B.R., M.D. Slade, S.R. Kunkel and S.V. Kasl. (2002) 'Longevity Increased by Positive Self-perceptions of Aging', *Journal of Personality and Social Psychology* 83, 261-270.

Lewinter, M. (2003), 'Reciprocities in Caregiving Relationships in Danish Elder Care', *Journal of Aging Studies* 17, 357-377.

Li, Y. (2007) 'Recovering from Spousal Bereavement in Later Life: Does Volunteer Participation Play a Role?' *Journal of Gerontology* 62B, S257-S266.

Litwak, E. (1985) *Helping the Elderly. The Complementary Roles of Informal Networks and Formal Systems* (New York: Guilford Press).

Litwin, H. (ed.) (1996) *The Social Networks of Older People: A Cross-national Analysis* (Westport: Praeger).

Litwin, H. (2004) 'Social Networks, Ethnicity and Public Home-care Utilisation', *Ageing and Society* 24, 921-939.

Litwin, H. and S. Shiovitz-Ezra (2006) 'Network Type and Mortality Risk in Later Life', *The Gerontologist* 46, 735-743.

Litwin, H., C. Vogel, H. Künemund and M. Kohli (2008) 'The Balance of Intergenerational Exchange: Correlates of Net Transfers in Germany and Israel', *European Journal of Ageing* 5, 92-102.

Loomis, L.S. and A. Booth (1995) 'Multigenerational Caregiving and Well-being: The Myth of the Beleaguered Sandwich Generation', *Journal of Family Issues* 16, 131-148.

Lowenstein, A. and J. Ogg (eds) (2003) *OASIS – Old Age and Autonomy: The Role of Service Systems and Intergenerational Family Solidarity*, final report. Haifa (Ms.).

Lum, T.Y. and E. Lightfoot (2005) 'The Effects of Volunteering on the Physical and Mental Health of older People', *Research on Aging* 27, 31-55.

Lund, F. (2002) 'Crowding in' Care, Security and Micro-enterprise Formation: Revisiting the Role of the State in Poverty Reduction and in Development', *Journal of International Development* 14, 681-694.

Lyberaki, A. and P. Tinios (2005) 'Poverty and Social Exclusion: A New Approach to an Old Issue'. in A. Börsch-Supan, A. Brugiavini, H. Jürges, J. Mackenbach, J. Siegrist and G. Weber (eds) *Health, Ageing and Retirement in Europe: First Results from the Survey of Health, Ageing and Retirement in Europe* (Mannheim: MEA), 302-309.

Messeri, P., M. Silverstein and E. Litwak (1993) 'Choosing Optimal Support Groups: A Review and Reformulation', *Journal of Health and Social Behavior* 34, 122-137.

Moen, P., D. Dempster-McClain and R.M. Williams (1989) 'Social Integration and Longevity: An Event History Analysis of Women's Roles and Resilience', *American Sociological Review* 54, 635-647.

Motel-Klingebiel, A., C. Tesch-Römer and H.-J. von Kondratowitz (2005) 'Welfare States Do Not Crowd out the Family: Evidence for Mixed Responsibility from Comparative Analyses', *Ageing and Society* 25, 863-882.

Musick, M.A., J.S. House and D.R. Williams (2004) 'Attendance at Religious Services and Mortality in a National Sample', *Journal of Health and Social Behavior* 45, 198-213.

Neyer, F.J. and F.R. Lang (2003) 'Blood is Thicker than Water: Kinship Orientation across Adulthood', *Journal of Personality and Social Psychology* 84, 310–321.

Nichols, L.S. and V.W. Junk (1997) 'The Sandwich Generation: Dependency, Proximity, and Task Assistance Needs of Parents', *Journal of Family and Economic Issues* 18, 299-326.

OECD Health Project (2005) *Long-term Care for Older People* (Paris: OECD).

O'Reilly, P. and F.G. Caro (1994) 'Productive Aging: An Overview of the Literature', *Journal of Aging and Social Policy* 3, No. 6, 39-71.

Peek, C.W., B.A. Zsembik and R.T. Coward (1997) 'The Changing Caregiving Networks of Older Adults', *Research on Aging* 19, 333-361.

Penning, M.J. (1998) 'In the Middle: Parental Caregiving in the Context of Other Roles', *Journal of Gerontology* 53B, S188- S197.

Pezzin, L.E., R.A. Pollak and B.S. Schone (2008) 'Parental Marital Disruption, Family Type, and Transfers to Disabled Elderly Parents', *Journal of Gerontology* 63B, S349-S358.

Pichler, F. and C. Wallace (2007) 'Patterns of Formal and Informal Social Capital in Europe', *European Sociological Review* 23, 423-435.

Pichler, F. and C. Wallace (2009) 'Social Capital and Social Class in Europe: The Role of Social Networks in Stratification', *European Sociological Review* 25, 319-332.

Pillemer, K. and K. Lüscher (eds) (2004) *Intergenerational Ambivalences: New Perspectives on Parent-child Relations in Later Life* (Boston: Elsevier).

Putnam, R.D. (2000) *Bowling Alone: The Collapse and Revival of American Community* (New York: Simon and Schuster).

Putney, N.M. and V.L. Bengtson (2001) 'Families, Intergenerational Relations, and Kinkeeping in Midlife' in M.E. Lachman (ed.) *Handbook of Midlife Development* (New York: Wiley), 528-570.

Rasulo, D., K. Christensen and C. Tomassini (2005) 'The Influence of Social Relations on Mortality in Later Life: A Study on Elderly Danish Twins', *The Gerontologist* 45, 601-608.

Reil-Held, A. (2006) Crowding out or Crowding in? Public and Private Transfers in Germany', *European Journal of Population* 22, 263-280.

Roots, C.R. (1998) '*The Sandwich Generation: Adult Children Caring for Aging Parents* (New York: Garland Publications).

Rosenthal, C.J., A.M. Matthews and S.H. Matthews (1996) 'Caught in the Middle? Occupancy in Multiple Roles and Help to Parents in a National Probability Sample of Canadian Adults', *Journal of Gerontology* 51B, S274-S283.

Rosow, I. (1974) *Socialization to Old Age.* (Berkeley: University of California Press).

Rowe, J.W. and R.L. Kahn (1998) *Successful Aging* (New York: Pantheon).

Sirven, N. and T. Debrand (2008) *Promoting Social Participation for Healthy Ageing. A Counterfactual Analysis from SHARE*. Paris: IRDES working paper 7.

Spitze, G., J.R. Logan, G. Joseph and E. Lee (1994) 'Middle Generation Roles and the Well-being of Men and Women', *Journal of Gerontology* 49, S107-S116.

Staudinger, U.M. (1996) 'Psychologische Produktivität und Selbstentfaltung im Alter in M.M. Baltes and L. Montada (eds) *Produktives Leben im Alter*. (Frankfurt/M.: Campus), 344-373.

Sundström, G.L., L. Johansson and L.B. Hassing (2002) 'The Shifting Balance of Long-term Care in Sweden', *The Gerontologist* 42, 350-355.

Szydlik, M. (2000) *Lebenslange Solidarität? Generationenbeziehungen zwischen erwachsenen Kindern und Eltern* (Opladen: Leske + Budrich).

Szydlik, M. (2002) 'Wenn sich Generationen auseinanderleben, *Zeitschrift für Soziologie der Erziehung und Sozialisation* 22, 362-373.

van der Pas, S., T. van Tilburg and K. Knipscheer (2007) 'Changes in Contact and Support within Intergenerational Relationships in the Netherlands: A Cohort and Time-sequential Perspective' in T. Owens and J.J. Suitor (eds) *Advances in Life Course Research: Interpersonal Relations Across the Life Course (vol 12)* (London: Elsevier), 243-274.

Wagner, M. and C. Wolf (2001) 'Altern, Familie und soziales Netzwerk', *Zeitschrift für Erziehungswissenschaft* 4, 529-554.

Wall, K., S. Aboim, V. Cunha and P. Vasconcelos (2001) 'Families and Informal Support Networks in Portugal: The Reproduction of Inequality', *Journal of European Social Policy* 11, 213–33.

Ward, R.A. and G. Spitze (1998) 'Sandwiched Marriages: The Implications of Child and Parent Relations for Marital Quality at Midlife', *Social Forces* 77, 647-666.

Wenger, C.G. (1997) 'Review of Findings on Support Networks of Older Europeans', *Journal of Cross-Cultural Gerontology* 12, 1-21.

Zal, M.H. (1992) *'The Sandwich Generation: Caught between Growing Children and Aging Parents* (New York: Plenum Press).

Social Networks

Comments by António M. Fonseca

Demographic data indicate that the group of the oldest old will show a large increase in the next decades. This means that the older population is itself ageing! This is important, because if there are more older people now – and more oldest-old people in the near future – then the risk of dependency in a larger group of population will also increase, confronting us with new demands of providing them with social and psychological support. Parallel to this, there are nowadays fewer family members available to take care of older relatives (essentially, due to the decreasing birth rate and to greater participation of women in the labour market). These social transformations have led to a decline in the traditional solidarity between generations that has been very common in southern and eastern European countries (a kind of social life insurance for the final days of life), leading frequently to older people living alone during their final years of life.

The chapter by Kohli and Künemund is definitely a good starting point in this field, and three main conclusions of the paper are noteworthy: (i) social networks are crucial for the well-being of older individuals; (ii) maintaining social connectedness is an important prerequisite for successful ageing; (iii) a new concept of network focuses on the social relationships that come with opportunities and demands for individuals to be socially productive. These three conclusions will be considered in turn.

A particularly important aspect of the ageing process is the relationship between one's environment and the ageing experience. The residence context (or simply, the place where one lives) plays an important role in the comprehension of the different ageing patterns, and helps to explain why some people experience 'successful' ageing (and others do not). The notion of 'ageing-in-place' is, therefore, central to the comprehension of the relationships between the residence context and successful ageing. In this respect, we can identify two situations related to these issues: (i) living at home versus living in an institution, (ii) living in a rural setting versus living in an urban setting.

Concerning the first situation (living at home versus living in an institution), it can generally be said that residents in old-age homes tend to feel lonelier and less satisfied, distant from their social networks, living a monotonous day-to-day existence, with meagre or no investment made at all in their remaining life. Apparently, the degree of discrepancy between individual competencies and the institutional environment, and the personality characteristics of each individual, have a great deal to do with the evaluation of the impact of the institutionalisation of people in old-age homes. Regarding the second situation (living in a rural environment versus living in an urban environment), it is important to point out from the outset that rural areas may become aged- and feebly populated zones that have been abandoned by the younger inhabitants, gone in search of better ways of life. In that case, it's impossible to say in advance in which setting the social networks are better preserved; individual aspects (such as being more or less active and autonomous) and the level of social participation are important variables to consider with regard to the role of social networks.

In fact, looking closely at the feelings of loneliness that are largely verified in different samples of older people living in either rural or urban settings, we are faced with the tremendous importance of social networks for older people. Apparently, the social support networks provide not only emotional support but also some kind of instrumental support during situations of slight dependency. However, as an older person's state of health deteriorates, feelings of resigned peace mixed with loneliness tend frequently to emerge, alleviated only by the day-to-day companionship of their partners. Many who are ageing far away from their offspring and grandchildren are in positions that do not encourage healthy psychological conditions; their low expectations regarding what the future holds disguise a state that even a widespread social network cannot remedy.

We may well ask, then, what social networks will be available for supporting older people and helping them avoid loneliness – but we may also question to what extent older people themselves will be responsible for creating their own social networks and also being social resources for others – even younger people. Clearly, demographic change will have significant implications for family and household structures, community behaviour, networks and social interaction, and intergenerational relationships. Simultaneously, the 'demographic burden' hypothesis is spreading: social services and even economies are predicted to collapse under multiple strains, and families will no longer be there to compensate for failing public provision. Kohli and Künemund take a critical stance with regard to this very popular position in public opinion, arguing that it is necessary to go further with this statement. For them, reality is far more complex than that, and ageing must above all be seen as a challenge: for individuals who need to reassess their life courses in light of new longevity probabilities, and for communities charged with developing appropriate policy frameworks with which to address the opportunities and risks of this ageing 'boom'.

This double challenge has implications for social networks: new longevity probabilities imply different network patterns across the lifespan, and developing appropriate policy frameworks should enhance the promotion of social and interpersonal relationships as crucial dimensions for successful ageing. Particular attention must be paid to the implications of this ageing boom for intergenerational solidarity at the family and community level.

Why intergenerational? Because we must refuse to see older people confined to social ghetto-networks (holidays only for older people, communities only for retired older people…). Why also community level and not only family level? Looking at families as the major social network for older people can be short-sighted, considering that many families are not and will not be able to perform support and care – which means that we must see family as one social network among others (including informal, educational, recreational, peer-based or neighbourhood-based social networks). Voluntary intergenerational programmes/projects in the wider community can definitely play an important role in maintaining intergenerational interaction and support in light of the weakness or even absence of family networks.

Finally, a policy for the promotion of social networks for older people should be based on adequate knowledge of psychological, social, economic, health- and contextual conditions of the ageing process. These conditions have a differential value if we are reporting on a person living alone or with a spouse, in a big city or in small village, with good or poor health. Therefore, the answer to the question of how much need there will be of social support will always have an ecological validity that obliges us to observe and understand people within their settings and to respect their subjectivities as individuals.

Social Networks

Comments by John Doling

This chapter, in bringing together a substantial amount of literature and identifying broadly what we know, don't know and (most importantly) should know, provides an informed account of the state of the art with an agenda for the future.

My focus here in this discussion is on an as-yet small and quite specific aspect of this more general field: e-networking. Most of the current focus on social networking, and indeed most of the literature, is based on an assumption (often implicit) that social networking involves face-to-face contact. Kohli and Künemund's chapter suggests that while e-networking is not, at the present time, significantly widespread, its potential to affect the lives of older people is great, indicating that 'serious research in this new field is still rare and should be expanded' (see p. 159). I would like to offer a few more thoughts on the potential significance of e-networking, as well as an indication of some of the gaps in our knowledge.

One of the specific advantages of e-networking and information- and communication technologies (ICT) more generally to older generations arises because they are more likely (compared with the general population) to experience mobility difficulties. ICT neutralises geography. In doing so, it may meet some practical needs that can be met by other forms of networking (such as shopping, banking and acquiring information). But ICT can also facilitate interaction with family and friends, and open up many new possible contacts and activities (such as playing games, chatting and exchanging photographs). Some evidence (see Morris, 2007) suggests that ICT use increases confidence, allowing people to explore new horizons and develop new interests, to be better informed (about health, for example) and to take a more active role in the community (participating in debating local issues and voting, for example).

In short, ICT may thus offer older persons the possibility of enjoying many of the benefits that are generally attributed to social networking.

It is clear that across populations as a whole these possibilities are being increasingly used – evidenced by the interest in social networking sites, blogging and tweeting. But, while the term 'silver surfing' has entered the (English) vocabulary, and a number of social networking sites aimed specifically at older people have been set up, we know relatively little about what the Internet is being used for (and thus to what extent its potential is being realised). The indication from at least some countries is that the 'digital divide' can be equated with the 'generational' or 'grey' divide. In the UK, elderly households are only about a quarter as likely as younger households to have home access to Internet (Morris, 2007). Is this the same in all countries?

Further research could observe divisions within groups of older people – finding out, for example, what the significance of education and income is in determining who uses Internet and who does not (see Selwyn et al., 2003). It would also be useful to understand what interventions (for example, subsidies and training courses) help to increase usage. Again, would the same interventions work in all countries?

These questions of 'generalisability' across countries are important. Kohli and Künemund report on a wide body of literature covering quite a large number of countries, but with a marked bias toward literature about a few – with Germany, France and the US figuring quite prominently. The danger in this, implicitly or explicitly, is overlooking the large differences that may exist between the older and the newer member states, for example, thereby sliding into generalisations about all countries.

Two further questions are of particular importance. The first is whether any, or all, of the identified relationships concerning who uses e-networking (and who does not) constitute a generational- or age-effect. The second is whether any use of e-networking constitutes a substitute or a complement for face-to-face networking. The answers to both questions have important consequences for the development of appropriate policy interventions.

References

Morris, A. (2007) 'E-literacy and the Grey Digital Divide: A Review of Recommendations', *Journal of Information Literacy*, 2(3).

Selwyn N., S. Godard, J. Furlong and L. Madden (2003) 'Older Adults' Use of Information and Communications Technology in Everyday Life', *Ageing & Society*, 23: 561-582.

7

Subjective Well-being in Older Adults: Current State and Gaps of Research

Dieter Ferring / Thomas Boll

1   Policy questions

Subjective well-being (SWB) refers to an evaluation of an individual's life from his or her *own perspective*. It contrasts sharply with evaluations made from the point of view of *external* observers (researchers or policymakers), which are based on objective criteria related to health, education, income or other aspects (Diener, 2006). Recently, prominent SWB researchers have argued forcefully in favour of supplementing traditional objective indicators of well-being or quality of life (such as economic indicators) with indicators of SWB (thus, people's evaluations and feelings about their lives). This should provide the public and politicians with more complete and relevant information for public discussion and political decision-making (see Diener, Kesebir and Lucas, 2008). In our opinion, these arguments apply with equal measure to the quality of life and SWB of older adults. The following list is an overview of policy questions about these issues. The list comprises five sets of questions that will in part be answered in later sections of the chapter:

- *Level, variation and developmental course of SWB*: What is the average level of SWB in older adults (compared to other age groups), and how does SWB vary between and within older adults in the course of their life?
- *Sources of SWB*: Which life circumstances contribute to SWB of older adults? Is there a demand to remove SWB-decreasing living conditions of older people by policy measures? (Such measures could involve local, regional, national or supranational levels of policymaking).

- *Effects of SWB*: To what extent does high SWB of older adults produce positive individual and societal outcomes (such as good health, high work performance, high community involvement)?
- *Value issues about SWB*: Should policymaking aim to reach or keep a certain level of SWB, or to increase SWB, in older adults?
- *Promoting SWB*: What are the possibilities, the limits, and the effects of policy measures (referring to pensions, support services, opportunities for lifelong learning) in influencing SWB of older citizens?

The next section presents some answers to the policy questions on the basis of existing research, and highlights where further research is needed to give satisfactory answers.

2   Major progress in understanding

This section is comprised of five subsections: the first three provide a conceptual clarification of the term SWB, followed by a discussion of measurement issues. Subsequently, we highlight the structure of SWB and the theoretical status of the concept in causal networks. On the basis of these explanations, the fourth subsection presents the main findings on the developmental course, on the causes and consequences of SWB in older adults and on what relevant interventions might involve. This is followed by a fifth subsection discussing models of SWB regulation in older adults.

2.1   Conceptual clarification and measurement issues

As mentioned above, SWB refers to an evaluation of an individual's life from his or her *own perspective,* which contrasts with evaluations made from the point of view of external observers (including researchers and policymakers), which are based on objective criteria. Leading researchers agree that SWB has a *cognitive* component (namely, the evaluation of one's life) and an affective component (namely, the presence of positive- and absence of negative feelings) (Diener, 2006). This consensus is also shared by prominent researchers of SWB in late life (Diener and Suh, 1998; Ferring et al., 2004; Pinquart and Sörensen, 2000; Staudinger, 2000). SWB can refer broadly to one's *life as a whole* or more narrowly to *specific life domains* (that is, health, material wealth, social relationships). This difference in scope has been considered particularly with respect to the cognitive component of SWB by distinguishing between global 'life satisfaction' (LS) and 'domain satisfaction' (DS).

Most empirical studies on SWB of older adults have relied on *self-reports* (obtained in questionnaires or interviews) as measures of SWB. The cognitive components of SWB (life satisfaction) have been assessed by

(1) *single-item measures* ('How do you feel about your life as a whole?' followed by a rating scale of satisfaction/dissatisfaction; see Andrews and Withey, 1976),

(2) direct *multiple-item measures* of global life satisfaction (Satisfaction with Life Scale by Diener, Emmons, Larsen and Griffin, 1985), or

(3) *global life-satisfaction indices,* as aggregated scores over several domain-specific satisfaction ratings (Personal Well-Being Index; Cummins et al., 2003).

The affective components of SWB have been assessed by either single-item measures (see Andrews and Withey, 1976) or multiple-item scales of affect (see Affect Balance Scale by Bradburn, 1969; Positive And Negative Affect Scale, PANAS by Watson, Clark and Tellegen, 1988). Table 7.1 provides a classification of the SWB measures. Even though SWB has a cognitive *and* an affective component, the measurement in many existing studies has focused on just one of them – that is, life satisfaction (often global life satisfaction assessed by just one item) *or* positive/negative affect. A domain-specific measure of positive or negative affect has – to the best of our knowledge – not been realised yet, and this represents an important task for further research.

Table 7.1: Classification of common SWB measures

Type of variable	Focus	
	One's life as a whole	Specific areas of one's life
Cognitive evaluation	Global life satisfaction (e.g., Life Satisfaction Index)	Domain satisfaction (e.g., ratings of satisfaction with one's health)
Positive affect	Global positive affect (e.g., positive subscale of Affect Balance Scale)	Domain-specific positive affect
Negative affect	Global negative affect (e.g., negative subscale of Affect Balance Scale)	Domain-specific negative affect

Kahneman and colleagues have proposed *alternative approaches* to replace global self-reports of life satisfaction: The *Experience Sampling Method (ESM)* and *Daily Reconstruction Method (DRM;* Kahneman et al., 2004). The ESM requires that subjects use an electronic diary that beeps at random times and asks respondents to indicate the intensity of positive and negative feelings in everyday life. DRM requires that respondents fill out a diary about events of their previous day and ascertain how they felt during each event on selected affect dimensions. To the best of our knowledge, these methods have not yet been applied in studies on SWB in later life.

Only a few studies have examined the *equivalence of SWB measures* across age groups, which is a precondition for a meaningful interpretation of possible mean differences between such groups. Here, measurement equivalence refers to the extent to which a given measure of SWB has identical relations to the theoretical construct of SWB in different age groups. Existing studies have examined this for the cognitive or the affective component of SWB, or for both components. Results have been mixed, indicating that measures were (see Lawrence and Liang, 1988) or were not (see Pons et al., 2000) equivalent.

Self-report measures pose specific challenges for old and very old individuals, if they suffer from *cognitive impairments and/or language disturbances*: To what extent are they able to comprehend the survey questions appropriately and to give meaningful answers? Moreover, at what degree of cognitive impairment do they lose their ability to do so? In that case, what other avenues exist for assessing *subjective* – not objective – well-being? Several instruments have been developed for persons suffering from dementia, in which data had to be provided by the individuals themselves and/or by other informants (thus, family members or professional caregivers). Problems with using such instruments have been extensively discussed. The current state of the art can be described as follows (Roick et al., 2007):

(1) From a gerontological and ethical point of view, one should not abstain from collecting self-reports of SWB.

(2) Measurement of SWB through self-report is possible in older adults with mild- and even moderate degrees of dementia.

(3) Measures can be designed in a way that facilitates participation despite cognitive impairments (thus, face-to-face administration by trained interviewers, simplified item wording and response formats; see Logsdon et al., 2002).

(4) Consensus between self-reports of SWB and reports provided by other persons (including family members, caregivers) is rather low. Other persons systematically report a lower level of SWB than older adults report themselves.

(5) Reports of SWB by family members or caregivers are systematically influenced by attributes of the person reporting (for example, caregiver burden, depression, attitudes toward dementia).

(6)   Less biased ratings may be provided by independent observers who – in contrast to family members or caregivers – should not be motivated to distort their judgements in a positive or negative direction.

The prevalence prognosis of dementia shows a clear rise with age.[1] The development of suitable measurement tools for the subjective- and the observer perspective represents an important task for the future – in order to develop a broad understanding of the psychosocial aspects of these diseases (see section 3.1).

2.2   Structure of SWB

Psychologists and economists have conducted several studies on the structure of SWB. In so doing, they refer mostly to the relations between the cognitive subcomponents of SWB: the relation between domain satisfaction (DS) and global life satisfaction (LS). These links have been theoretically conceived in two different ways (see Schimmack, 2008). The 'bottom-up' approach assumes that global LS depends on DS (thus, satisfaction with a limited set of life domains, including health, job, financial situation, family life, friendships). In contrast, the 'top-down' approach assumes that the various measures of DS depend on global LS. These two models have different implications: 'Bottom-up' approaches predict that changes in DS (such as financial satisfaction) produce changes in global LS, whereas 'top-down' approaches predict that changes in DS will *not* produce changes in LS. The current evidence from the psychological literature favours more the 'bottom-up' approach (Schimmack, 2008).

Recent studies by economists based on data from very large and approximately representative samples across all adult age groups (*German Socio-Economic Panel*; *United States General Social Survey*) have provided further evidence for the 'bottom-up' approach. First, LS can be successfully predicted from a set of DS at a certain point of time: LS turned out to be most strongly related to financial satisfaction, health satisfaction and job satisfaction, but most weakly related to house satisfaction and environmental satisfaction (van Praag et al., 2003). Second, the average trend in LS across the life cycle can be successfully predicted from average life-cycle trends of four DS (satisfaction with family life, financial situation, job and health; see Easterlin, 2006). Third, differences between men and women in the life-cycle trend of LS correspond to the male-female differences in the life-cycle trends in satisfaction with two

---

1   According to estimations by Ferri et al. (2005), the number of people with dementia over the age of 60 in the Western European region will increase from 4.9 million in 2001 to 9.9 million in 2040. The increase will be from 1 to 3.2 million for the Eastern European regions, respectively.

domains: *'Early in adult life, women are more satisfied than men with their family life and finances and correspondingly happier …. In late life men feel better about their family and financial circumstances and are the happier of the two'* (Plagnol and Easterlin, 2008: 609). Fourth, the functional relation of LS to educational level (positive curve) and to year of assessment (fairly flat curve) can be predicted clearly from the relationship of four DS (family life, financial, job and health satisfaction) to education and to year of assessment (Easterlin and Sawangfa, 2007).

An additional advantage of this domain approach is that it can be easily supplemented with assumptions about objective characteristics of the domains that may produce changes in DS (an increase in income produces an increase of financial satisfaction, decreasing health accompanies diminishing health satisfaction, and so forth). Following this approach, van Praag et al. (2003) found significant and in part specific relations between socioeconomic and socio-demographic variables and the various DS. However, such issues have not been analysed specifically for older adults.

Other work within the 'bottom-up' approach has assumed that the subjective importance of a life domain moderates the effect of DS on LS, such that satisfactions with subjectively more important domains carry more weight in determining overall LS than satisfactions with subjectively less important domains. Hsieh (2003) provides evidence that considering *the relative subjective importance weights* of life domains improves the prediction of global LS from DS, compared to treating every life domain as having equal weight. This leads to the question whether the importance of certain life domains and the size of the effect of DS on LS might differ between older and younger adults. A likely hypothesis is that health and housing are more important in the very old, compared to the younger age groups – but this has yet to be tested.

2.3   *Theoretical status of SWB in causal networks*

The *theoretical status of SWB* in causal networks has been conceived differently in the literature, and these differences can be represented in four basic models (see Figure 7.1):

(1)   *Effect model.* Most of the literature regards SWB *as an effect:* as the result of personal and/or environmental variables. Environmental factors can be described on different levels of analysis (micro, meso, macro, as described by Bronfenbrenner, 1979).

(2) *Predictor model.* A smaller part of the literature sees SWB *as a cause of individual and social outcomes* (such as health, longevity, voluntary engagements; see Lyubomirsky, King and Diener, 2005, for a review).

(3) *Mediator model.* Very few studies consider SWB *as a mediating variable that is caused by personal and/or environmental factors* – and which, in turn, has an effect on individual and/or social outcomes (see Murrell and Meeks, 2002).

(4) *Moderator model.* SWB could be regarded *as a moderator* (a third variable that may modify the effect of certain causes on certain outcomes). We did not find any study that analysed this role of SWB in older adults.

Figure 7.1: Theoretical status of SWB in causal networks.

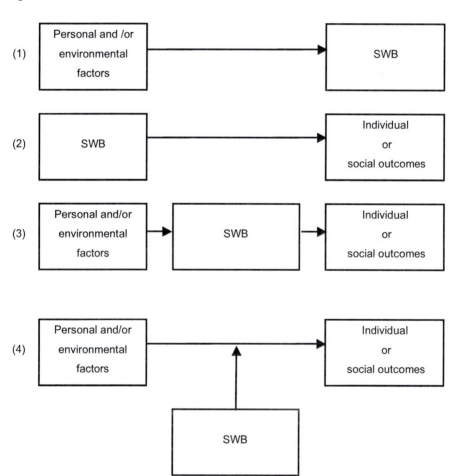

The aforementioned models provide, however, a somewhat simplified linear picture of the causal relations. Effects of personal or environmental factors (such as critical life events) on SWB may also be moderated by other relevant variables (such as social support). Moreover (and as will become evident later), *regulative models* seem to be more appropriate for describing the theoretical status of the construct. These models realise an integrative view which incorporates all models depicted above: SWB is caused by personal and/or environmental factors; SWB has individual and/or social consequences; the level of SWB is perceived by the individual and evokes regulative responses on the part of him or her; regulative activities of the individual may lead to a change in SWB level (see section 2.5).

2.4   Main findings about developmental course, causes and consequences of SWB in older adults

2.4.1 SWB as a function of age

To what extent does SWB level change with age? Competing hypotheses have been proposed.

- H1: A *marked decrease of SWB occurs in old age.* This hypothesis emerged from the common-sense assumption, and from gerontological evidence, that older persons face several losses (regarding health, autonomy, income, social relations…) that may interfere with their ability to get their needs met; frustration of needs, in turn, is assumed to decrease SWB.

- H2: *Little or no change of SWB occurs in old age.* This hypothesis was derived from research on personality factors in SWB: Because SWB is strongly related to personality traits (such as neuroticism or extraversion), which are very stable across adulthood (see DeNeve and Cooper, 1998; Diener et al., 2003; Lucas, 2008), little or no change in SWB should occur during that period of life. The little-or-no-change hypothesis can also be derived from three other lines of reasoning: From adaptation-theoretical considerations about SWB (Frederick and Loewenstein, 1999), from discrepancy-theoretical frameworks of satisfaction (see Thomae, 1970; Michalos, 1986), and action-theoretical models of regulation of SWB across the lifespan (see Baltes and Baltes, 1990; Brandtstädter and Rothermund, 2002).

Different empirical designs have been used to describe the actual developmental course of SWB in older adults. Numerous *cross-sectional* studies have compared groups of persons of different age at one point of measurement (for example, middle-aged- vs. old individuals). Narrative literature reviews (see Diener and Suh, 1998) and meta-analyses (see Pinquart, 1997, 2001) suggest that – despite losses – there is no pronounced decline in global life satisfaction. This phenomenon has been labelled as the 'paradox of subjective well-being' (see Staudinger, 2000) or as the 'stability-despite-loss paradox of subjective well-being' (see Schilling, 2006; Gerstorf et al., 2008a).

However, cross-sectional studies cannot disentangle age effects from cohort effects (thus, whether the different SWB means found, for instance, between groups of 60-79- versus 80-99 year-old individuals are due to differences in age or in birth cohort). In order to specifically detect age effects, several *longi-tudinal* studies have been performed; these studies measure SWB of the same individuals at two or more consecutive points of time. Most of them referred to the cognitive component of SWB (life satisfaction) as dependent variable (see Gerstorf et al., 2008a; Gerstorf et al., 2008b; Mroczek and Spiro, 2005; Schilling, 2006), whereas only a few focused on the affective component of SWB (see Charles et al., 2001; Costa and McCrae, 1993; Kunzmann et al., 2000).

Regarding the *cognitive* component (life satisfaction), most longitudinal studies found a rather slight, sometimes even non-significant average decline in life satisfaction from middle- to old age (up to about 70 or 85 years). Studies that included also individuals beyond 85 years old up to 100+ years found a more pronounced decline in very old age, compared to the earlier phases of old age (Gerstorf et al., 2008a; Gerstorf et al., 2008b). Moreover, these studies also introduced a new perspective by analysing life satisfaction not merely as a function of chronological age (= distance from birth), but also as a function of 'distance from death', which is a variable presumably related to mortality processes. Distance from death explained more variance of change in life satisfaction than did chronological age (see also Palgi et al., 2010). Analyses with distance from death as independent variable also revealed two different stages of change in life satisfaction: A phase of slight decline occurring in a larger temporal distance from death ('preterminal decline') and a phase of a much steeper decline (thus, at least by factor 2) about four years before death ('terminal decline'). This decline was most pronounced in the oldest-old (85+) individuals. The decline of SWB in the very old may be even more pronounced than the empirical findings suggest: Dropout of subjects between successive points of measurement may lead to an underestimation of the real decrease of SWB, because older individuals with decreasing SWB can be expected to be less motivated and/or able to participate in the following measurement waves.

Notable inter-individual differences concerning the transition point between the two phases were found: Some people reached the terminal phase many years before they died, whereas others reached it just a few years before their death – and still others never entered this phase at all. However, these studies are still descriptive; the direction of causality remains open: Do mortality-related processes or do cognitions of impending death reduce life satisfaction? Or, does reduced life satisfaction promote mortality?

Two positions may be mentioned here: Authors of the working group around Paul Baltes have argued that 'psychological mortality', described by a decrease in SWB and reduced cognitive functioning, will precede physical mortality (Smith, 2002). On the other side is evidence from the Heidelberg centenarian study that there is – even in old age – variability in SWB, underlining the plasticity of SWB (Jopp and Rott, 2006).

With respect to the *affective* components of SWB, the longitudinal studies revealed a different developmental pattern of findings for positive vs. negative affect. In most studies, *positive* affect remained stable for the middle-aged to the young-old groups – but for the oldest-old groups, some decline of positive affect was documented (see Charles et al., 2001; Kunzmann et al., 2000). However, the evidence for age-related change in *negative* affect is mixed. Whereas some studies showed a notable decline of negative affect in the younger groups – albeit not in the oldest age groups (see Charles et al., 2001) – another study provided no longitudinal evidence for age-related change of negative affect (Kunzmann et al., 2000). Again, most longitudinal studies reported considerable inter-individual differences in the intra-individual change patterns (thus, positive affect of some older adults declined, whereas that of others increased or remained constant).

Isaacson and Smith (2003) show that a unique age effect on positive and negative affect diminishes if one controls for demographic-, personality-, health- and cognitive-functioning variables. Using data from the *Berlin Aging Study*, the authors report that personality and general intelligence emerged as the strongest predictors of positive and negative affect. These findings reflect the fact that SWB is related to personality variables, but also to differing life conditions (thus, health, functional status). Health and functional problems may thus be more reliable predictors of positive and negative affect in old age than in young or middle adulthood, where job and social relations constitute the main issues.

2.4.2 SWB as a function of current life circumstances

The following discussion relies on a meta-analysis of 286 empirical studies provided by Pinquart and Sörensen (2000). The authors analysed the association of socioeconomic status, social network and competence with SWB in old people; SWB was covered by life satisfaction, happiness and self-esteem measures. Four main findings may be summarised as follows.

First, the authors report significantly higher correlations of all three measures of SWB with income than with education (both being indicators of SES); this finding highlights the importance of the *financial situation* for SWB in old age. Second, *competence* (indicated by activities and instrumental activities of daily living) was related to SWB – the higher the competence, the higher the reported life satisfaction, happiness and self-esteem. Third, the *quality* of social contacts was more strongly related to all three SWB indicators than the *quantity* of social contacts. A specific finding was that the quantity of contact with friends was more important for all indicators of SWB in old age than quantity of contact with family and adult children.[2] Fourth, age and gender moderated the relations between income and social contacts on the one hand and SWB indicators on the other.

Two gender effects were reported; first, the analysis showed that the relationship between SES and life satisfaction or happiness was stronger in the male samples; there was no significant influence of gender on the association between SES and self-esteem. Second, social contacts were more closely related to life satisfaction and happiness for women than for men; again, no effect emerged for self-esteem. Moderating effects of age were tested by contrasting a group with an age 70 years or less with a group above 70 years of age. Here, three effects were observed: (1) Life satisfaction and happiness showed a lower correlation with SES in the older group than in the younger group; (2) the correlation of SWB with social integration was larger for older participants than for younger participants; in the older samples, correlations between the quantity and the quality of social contacts with SWB were significantly higher than in younger samples; (3) a stronger relation between competence and happiness was reported for older compared with younger samples; no relations were found for life satisfaction and self-esteem.

---

2   Kohli and Künemund (chapter 6 of this book) emphasise the importance of social networks for the well-being of older adults. The authors speak of well-being in general. Their propositions may thus refer also to *objective* well-being, and not exclusively to *subjective* well-being, which is the primary focus of this chapter.

All in all, the results highlight the possible importance of life circumstances for SWB in old adults: Low income, low educational status, low quality of social contacts, and functional deficits accompany reduced life satisfaction, low happiness and reduced self-esteem. However, the findings are only correlational ones; they do not allow for definitive causal inferences. More longitudinal and intervention studies are needed in order to justify such conclusions.

2.4.3 SWB as a function of critical life events

A few *prospective longitudinal studies* have been conducted with large and approximately representative samples in which SWB was assessed several years *before* and *after* the occurrence of age-related life events such as retirement, disability or widowhood, which thus provide additional data on the developmental course of SWB. The general result pattern here indicates that long-term levels of SWB do change in response to such events, and that the pattern of adaptation varies with the kinds of life events. Depending on the type of event, the adaptation of SWB (the return to the pre-event level) may be rather slow and may be incomplete. For instance, adaptation to widowhood took about seven years on average (see Lucas et al., 2003). SWB decline in response to severe disability turned out to be even more pronounced, and adaptation took either longer or never happened at all (Lucas, 2007a,b). Such studies also provide evidence for large individual differences in the amount of adaptation that may result from variation in concrete circumstances of the event type under consideration and/or from variability in the response of different persons to the same type of event.

Recent studies have also made preliminary attempts to analyse more deeply inter-individual differences in event-related intra-individual change of SWB. In a latent-class analysis, Pinquart and Schindler (2007) found three groups of persons with different patterns of response to retirement (SWB decline followed by increase; initial increase but subsequent decrease; temporary small increase), and identified correlates of the different patterns of response. For instance, the group characterised by a decline in life satisfaction some time after retirement showed worse physical health, below-average age before retirement, unemployment before retirement, low socioeconomic state, and they lacked a spouse.

All in all, the notion holds that critical life events have an impact on SWB. As research indicates, the effect of critical life events is shaped by individual regulation. These processes have turned out to be a function of event characteristics (controllability, predictability, irreversibility), personal factors (habits, coping styles, competences) and context factors (such as social support); their specific interplay explains inter-individual and intra-individual differences in SWB across time (Filipp and Aymanns, 2010). It should be emphasised, as a specific point concerning the situation of the elderly person, that there is a heightened probability of critical life events (especially losses of persons and losses concerning physical and functional performance) with ongoing age. This qualifies especially old age as a period that sets heightened demands to individual regulation of SWB.

2.4.4 SWB as a function of intentional activities

Other potential determinants of SWB are *intentional activities:* those activities that people can choose to engage in and which require some degree of effort to enact. Whereas several empirical studies have examined their relation to SWB in samples of younger adults or in mixed-aged adult samples (for an overview, see Lyubomirsky, Sheldon, and Schkade, 2005), such studies specifically devoted to older adults are rare. A recent exception (Pushkar et al., 2010) investigated longitudinally how continuity and change of various intentional activities affected changes in positive and negative affect of retired men and women. Increased activity frequency as well as increased feelings of ability, ease and future intentions to perform such activities predicted higher positive affect; increases in felt ability and ease of activity predicted decreasing negative affect.

Links between a special subset of intentional activities (namely, *socially-productive activities* such as volunteering, informal helping, caring) and SWB have received special interest in research – but only a few studies have covered a broader range of such activities performed specifically by older adults. For instance, McMunn et al. (2009) examined links between various social activities (including paid work, voluntary work, caring for someone) and life satisfaction of participants of state-pension age or older. Participants in paid- or voluntary work showed more life satisfaction than non-participants, whereas carers and non-carers did not differ in life satisfaction. Such inconsistent findings suggest that one should differentiate carefully between the various kinds of such activities when analysing their relation to SWB. In contrast to the results of the aforementioned study, a meta-analysis of 84 studies comparing mainly older adult caregivers and non-caregivers found a significantly lower SWB among caregivers, and a substantial heterogeneity of effect sizes across studies (Pinquart and Sörensen, 2003). The extended literature on caregiving elucidates that caregiving is a rather heterogeneous category, which complicates the analysis of links to SWB. Caregiving by older adults may be addressed to different receivers (physically frail vs. demented spouses), may require various amounts of time and effort, may interfere differently with other goals of the caregiver, and may be associated with different amounts of strains and gains. Moreover, stress and coping models of caregiving (see for example Pinquart and Sörensen, 2005) suggest that the effect of caregiving on SWB may depend on the mix of the aforementioned factors and on the subjective interpretations of the caregiving situation – as well as on the availability of personal and social resources for coping with the given demands.

Voluntary work is another category of socially productive behaviour of older adults that has received the exclusive focus of some studies. In a meta-analysis based on 29 studies, Wheeler et al. (1998) found that older volunteers involved in direct (face-to-face) or indirect helping roles were characterised by a significantly greater quality of life (mostly indicated by measures of life satisfaction) compared to non-volunteers – even after controlling for health- and socioeconomic status of the participants. The effect was found to be more pronounced in studies involving direct (compared to indirect) helping relationships. More recent studies have found evidence for a positive relationship between the amount of volunteering and measures of SWB – either life satisfaction (van Willigen, 2000) or positive affect (Greenfield and Marks, 2004). However, the cross-sectional design of most studies precludes definitive causal inferences (thus, whether socially productive activities have an effect on SWB or vice versa, or whether some third variable – such as living conditions or personality traits – has an effect on both).

Siegrist and Wahrendorf (chapter 5; this volume) review the literature and present data on links between *socially productive activities* (paid work, volunteering, informal helping) and *well-being* as outcome variable. However, *well-being (WB)* as conceived by these authors cannot be equated with *subjective well-being (SWB)*. Although both constructs have similar names and refer to the individual's point of view, their precise content differs substantially. Whereas SWB includes life satisfaction, domain satisfaction, and positive and negative affect as central elements (see section 2.1 of this chapter), WB is defined with reference to self-assessed health and to quality of life as indicated by the individual's sense of control, autonomy, self-realisation and pleasure. Thus, Siegrist and Wahrendorf provide valuable information about the effects of socially productive activities of (early) old adults in areas beyond SWB.

2.4.5 Interventions to promote SWB in older adults

Interventions to promote SWB in older adults may refer to different levels of the person-environment relation. On the *micro-level*, certain psycho-educational and psychotherapeutic approaches focus directly on the individual. Pinquart and Sörensen (2001) analysed 122 studies, and their findings in general support the assertion that psychosocial and psychotherapeutic interventions significantly reduce depression and improve psychosocial well-being in older adults. The following factors were associated with improvement of life satisfaction and reduction of depression in older adults through specific interventions: (1) psychotherapeutic interventions (cognitive-behavioural therapy, relaxation, reminiscence) were more effective than psycho-educational ones (such as social activity programmes); additionally, cognitive-behavioural- as well as control-enhancing strategies proved to be quite effective for nursing home residents; (2) interventions targeting already depressed persons before treatment were more effective than those addressing at-risk groups; (3) more than nine sessions produced better results than sessions with a lower number of contacts; (4) qualification of the therapist was a crucial factor in predicting reduction of depression and improvement of psychosocial well-being. Moreover, nursing home patients reported greater reduction in depression, but not in life satisfaction compared to community resident older adults. Furthermore, the younger old (below the age of 77) showed a greater improvement in depression, but not in LS, compared to older persons (above the age of 77). All in all, one may conclude with Knight (2004) that the effectiveness of psychotherapy and mental health services in general has been established regarding older adults. Findings thus strongly support the development and elaboration of structures offering psychosocial intervention and psychotherapeutic interventions for these people.

Further approaches for promoting SWB of older adults are conceivable that refer to the meso- or macro level, such as local, regional or national policy measures. Social and transport policy initiatives are discussed as possible measures that could improve SWB by enhancing the provision of care (see Vaarama, 2009) or the out-of-home mobility for older adults (see Mollenkopf et al., 2004). To the best of our knowledge, their effectiveness has not yet been *empirically* evaluated.

Whereas empirical findings are still lacking, three different *theoretical* approaches already exist that allow for generating hypotheses concerning whether and how policy interventions may have an effect. These approaches, referring to the promotion of SWB in general, can easily be applied to the issue of enhancing SWB in older adults. First, mainstream economists (see Lewin, 1996) hold that SWB depends predominantly on one's objective living conditions (income, for example), and this leads one to expect that policy measures improving the objective living conditions of older adults will also increase their SWB. Second, some psychologists maintain that the effect of objective living conditions on SWB depends on internal processes. For instance, the 'hedonic treadmill model' assumes that individuals are characterised by set point levels of happiness, which in turn are determined by personality and genetic factors. A change of living conditions is expected to change a person's happiness above or below this set point, and adaptive processes are assumed to return the individual rather rapidly and completely to this point (see Brickman and Campbell, 1971). Viewed from this perspective, policy measures changing economic or social conditions could likely be regarded either as unnecessary or as having at least no permanent effects on SWB. However, this theoretical perspective (and the implied pessimism regarding interventions for improving well-being) have been criticised in several respects, and revisions have been proposed to the 'hedonic treadmill model' that provide more hope for policymakers that sustainable change of SWB can be reached (Diener et al., 2006). Furthermore, cross-national research – especially findings comparing countries with differing socioeconomic and social welfare systems – makes it very likely that objective living conditions have an impact on SWB.

A third theoretical approach that has emerged among some economists and psychologists favours a moderate adaptation model: SWB is conceived as the outcome of both objective factors and subjective factors. For example, satisfaction with a certain life domain is assumed to result from the extent to which one's attainments in that life domain match one's aspirations (see Easterlin, 2006; Plagnol and Easterlin, 2008). Overall satisfaction, in turn, is conceived as the net outcome of the satisfaction with various life domains (including material life conditions, family life, health). This model can inspire two principle categories of policy measures for improving SWB in older adults. The first refers to measures for changing objective living conditions under the assumption that the aspirations regarding these conditions remain constant. The second one includes measures that aim at lowering the aspirations of older adults; this option is particularly relevant if objective living conditions cannot be changed. Obviously, much more needs to be known about the content, stability and modifiability of the aspirations of older adults' in order to anticipate the effectiveness of these two categories of policy measures for promoting SWB.

2.4.6 Effects of SWB in older adults

Based on prior analyses of adaptive effects of positive emotions, Lyubomirsky, King and Diener (2005) developed a conceptual model that proposes that positive affect (a central component of SWB) leads people to pursue positive goals and thus to be more 'successful' in various life domains. Moreover, these authors provide an extensive meta-analysis that tests their model against evidence from a total of 225 cross-sectional, longitudinal, and experimental studies on the link between positive affect and positive outcomes in the areas of social behaviour, social relationships, work performance, income, health, creativity and problem solving. They concluded that people who experience more positive affect are more likely to have good social relationships, higher incomes, superior work performance, more community involvement, and better health compared to people with less positive affect. Although the meta-analysis was not specifically devoted to older adults, about a dozen studies focussing on this age group were included in the meta-analysis, most of them referring to health-related outcomes of SWB: Positive affect turned out to be positively related to health or longevity, or to be negatively related to mortality. Future research should provide stronger evidence from intervention studies that SWB causes such outcomes. Moreover, the effects of positive affect and other components of SWB on positive outcomes beyond health should be investigated further.

2.5   Theoretical models about SWB regulation in older adults

The finding that despite age-linked losses there is no strong decline in at least some measures of SWB (such as life satisfaction) across age (with the exception of the terminal phases of life) has raised the question which regulative processes might account for that phenomenon. Several approaches have been proposed here, and the historically more important and / or scientifically more influential ones all stem from lifespan developmental psychology or psychological gerontology. One may list here Thomae's (1970) *cognitive theory of ageing*, the *model of selective optimisation with compensation* proposed by Baltes and Baltes (1990), the *dual-process model of assimilative and accommodative coping* by Brandstädter and colleagues (see Brandtstädter and Rothermund, 2002), as well as the *motivational theory of lifespan development* elaborated by Heckhausen et al. (2010). The primary focus of these models is on *self*-regulation of SWB through intentional or non-intentional processes of the individual. All of these theories claim that SWB in old age is open to regulation processes in which an objective life situation that may be 'good' or 'bad' is evaluated by the individual as either 'good' or 'bad'; a cross-classification of objective life situations and their subjective evaluation results in four constellations, depicted in Table 7.2.

Table 7.2:   Cross-tabulation of objective life situations and their subjective evaluations

Subjective evaluation	Objective life situation	
	Good	Bad
Good	Well-being	Satisfaction paradox
Bad	Dilemma of discontentment	Deprivation

Of particular interest are the two constellations in which subjective evaluation and objective life situation diverge (satisfaction paradox, dilemma of discontentment), pointing to the plasticity and the regulation of SWB (see Ferring and Filipp, 1997). When discussing the objective life situation in age, one should therefore keep in mind its transformation into subjective reality. Such a view should by no means foster a kind of 'pessimistic' view on the possible effects that an amelioration and elaboration of objective living conditions may have on SWB (see below).

3   Remaining gaps in knowledge: main challenges

This section uses insights and arguments from the preceding sections to highlight knowledge gaps and indicate potential research directions on SWB in older adults. Theoretical, methodological and empirical issues will be covered.

3.1   Measurement of SWB in older adults

With respect to measurement of SWB, three points are deserving of further development.

SUFFICIENTLY BROAD MEASUREMENT OF SWB
SWB is an umbrella concept covering four constructs (namely, general life satisfaction, domain satisfaction, positive affect, and negative affect). In order to realise a comprehensive and sound measurement, not just one indicator of SWB should be used, but rather all four. To ensure psychometric quality, one should also use multiple indicators for each construct.

MEASUREMENT EQUIVALENCE OF SELF-REPORT MEASURES OF SWB

Until now, little has been known about the extent to which frequently applied measures of SWB are equivalent for people of different age groups. Existing studies have been confined to comparisons between adolescents vs. older adults (Pons et al., 2000) or between young-old vs. oldest-old groups (Lawrence and Liang, 1988). Moreover, studies have considered just global measures of life satisfaction without taking into account measures of domain satisfaction. Given the importance of measurement equivalence for an adequate interpretation of comparisons of SWB means across age groups, more research is needed in this context. First, measurement equivalence should be examined across a broader range of age groups that cover all phases of adult life. Second, it should be clarified whether people of a different age rely on the same or on a different kind of information when they rate how satisfied they are with their life as a whole. People of a different age may rely on information about different domains of life when they rate global life satisfaction, given that the relative importance of health, financial situation and work decreases with age (Hsieh, 2005). Structural equation modelling is a powerful tool for testing the measurement equivalence of various domain satisfactions in relation to global life satisfaction as a latent variable.

ASSESSMENT OF SWB IN OLDER ADULTS WITH SEVERE COGNITIVE IMPAIRMENT: BEYOND SELF-REPORT

More emphasis should be put on the further development of assessment approaches to SWB of older individuals with severe cognitive impairments who are unable to provide reliable and valid self-reports of their SWB. Even though SWB is by definition *subjective*, SWB can manifest itself not just in self-reports, but also in other verbal and non-verbal behaviour.[3] Table 7.3 contains a classification of indicators of SWB on which future research may concentrate.

To circumvent the cognitive limitations of the person in question in the assessment of such indicators, two kinds of informants can be referred to: (1) People having frequent contact with the target person (family members, professional caregivers) and (2) People not related to the target person in daily life (independent observers).[4]

---

3   However, if there is no evidence that an older person has at least a minimally differentiated mental representation of his or her life, the concept of 'life satisfaction' cannot be meaningfully used. Such an individual may still be 'satisfied' or 'dissatisfied' in an unspecified sense, but speaking of '*life satisfaction*' would be misleading here.

4   Past research has also relied on these kinds of informants to get information about *objective/ external* life circumstances or about the competences of elderly individuals.

Future research will have to work out the details of the aforementioned behavioural indicators in guidelines and instruments for their assessment. Because the presence of positive and absence of negative affect are important components of SWB, it should be fruitful to use theoretical and methodological knowledge from research on affects (see Coan and Allen, 2007) in projects developing SWB measures for older adults with severe cognitive impairments. Some of these studies should also provide a further analysis of biases that external observers (such as family members or professional caregivers) manifest in rating SWB of older persons.

Table 7.3: Indicators of SWB beyond self-report

Component of SWB	Indicator	
	Verbal Behaviour	Non-verbal Behaviour
Life satisfaction	Scores aggregated across domain-specific measures, as described below	Scores aggregated across domain-specific measures, as described below
Domain satisfaction	- verbal reactions - para-verbal reactions referring to real or symbolically presented 'life domains'	- mimic and glance - gesture - posture - gross-bodily movement referring to real or symbolically presented 'life domains'
Positive / negative affect	- verbal reactions - para-verbal reactions referring to real or symbolically presented 'life domains'	- mimic and glance - gesture - posture - gross-bodily movement referring to real or symbolically presented 'life domains'

3.2   Limited age range and age segmentation in studies on SWB in older adults

The upper end of the age scale has been rather low in many gerontological studies; in particular, there is a scarcity of very old people (85+) in most studies, which may be explained by their comparatively poor functional state. Moreover, the oldest age group is defined rather broadly and differently in many studies (50+, 65+ and so forth). It is thus impossible to get specific information of SWB in more narrowly defined age groups (80-85, 85-90, …) and to describe the development of SWB within the category of older adults aged 65+. A few studies offer a more differentiated/fine-grained scaling of the age variable up to 100+ years, which points in the right direction; we need more studies of that kind in future.

3.3   Lack of representative and longitudinal studies on SWB in older adults

There are just a few European studies on SWB in older adults based on sufficiently representative samples. Only some of these studies have a longitudinal
design (see also below): The *German Socio-Economic Panel*, the *German Aging
Survey*[5], the *British Household Panel Study*, the *Eurobarometer*, and the *Survey on
Health, Ageing and Retirement*. All of these studies cover extended periods of
time and include several points of measurement.

However, studies can be representative only to a certain degree. Because
in democracies participation in empirical studies is voluntary, self-selection
may occur. This problem is more serious in gerontological studies. Overestima-
tions of SWB may result already in cross-sectional studies, if the self-selection
of older participants is correlated with the level of SWB, such that individuals with higher SWB are more likely to participate, whereas individuals with
lower SWB will more likely refuse participation. A biased estimation of the
developmental course of SWB is likely to occur in longitudinal studies, too,
if the dropout in the following waves of measurement is correlated with the
change (the decline) of SWB. The actual decrease of SWB in very old age may
be underestimated if individuals with decreasing SWB become less motivated
or able to participate in subsequent measurement waves. Moreover, given that
13% of older adults between 80-85, and 21% of older adults between 85 and
90 are (severely) demented in Europe[6] (and thus may be unable and/or not
sufficiently motivated to participate in a normal survey), the dropout rate is
considerable, and results may be rather biased. So, a check of representativeness and a dropout analysis should be integrated in these studies, and findings
should be interpreted in light of systematic selection biases.

3.4   Ignorance about the causes of SWB in older adults

SWB is not the product of a monocausal process; it is caused and initiated by
several interlinked factors on the societal and individual level. With respect to
its causal links, equifinality (that is, differing causes and conditions leading to
similar results) and multifinality (that is, similar causes and conditions leading
to differing results) may be considered, and causes have to be analysed in their
linear as well as non-linear interrelation.

---

5   *German Ageing Survey*: Nationwide representative cross-sectional and longitudinal survey
    of the German population aged over 40. Up to now, the survey has had three waves of as-
    sessment (1996, 2002, 2008) that cover material living conditions, critical life events, health,
    social relationships, activities, self-concept and life goals, quality of life and SWB.

6   European Community Concerted Action on the Epidemiology and Prevention of Dementia
    Group (2009). Retrieved at: http://www.dementia-in-europe.eu/?lm2=HIXPJGBKGFTQ

The exploration of causal relations requires more appropriate research designs (such as prospective longitudinal studies, experimental/intervention studies and adequate causal modelling data analyses) – and here, there is also a considerable research gap; studies should also examine effects not just of *inter-individual variation*, but also of *intra-individual change* in external life circumstances on SWB (see Lucas et al., 2003, 2007a,b); studies should also look for differential trajectories of life circumstances (see Pinquart and Schindler, 2007, for an innovative approach).

3.5 Lack of knowledge about the regulation of SWB in older adults

Existing models of SWB regulation focus predominantly on processes related to the *individual* him- or herself (see section 2.5). However, the regulation of older adults' SWB could also involve their interaction partners. To distinguish this phenomenon from an individual's *self*-regulation, we call it *cooperative* regulation of SWB: this kind of regulation has been widely neglected in previous research. The following processes seem to be involved in the cooperative regulation of SWB in a dyad consisting of A (the older person) and B (a family member, a friend, a neighbour, a co-resident in a nursing home or a professional caregiver):

(1)   A communicates his or her SWB to B,
(2)   B perceives, interprets and evaluates this message,
(3)   B acts toward A with the intention to influence (increase) A's SWB.

Researchers may import concepts and methods from other relevant fields of inquiry, such as communication of emotions, self-disclosure, perception and recognition of emotion, emotional intelligence and mobilisation of support. Moving from dyads to larger social system research should also focus on (a) how information about SWB level of a group of individuals is conveyed (through expert hearings or through the media, for example) to responsible agents in care institutions or in political systems and (b) how this information is used (if at all) in the planning and implementation of policy measures to raise SWB. The notion of 'cooperative regulation' will certainly allow better representation of the complexity of SWB regulation by older adults in their social context. This will also challenge methodology, because dyadic or 'polyadic' assessment strategies will have to be further developed.

3.6   Lack of knowledge about consequences of SWB in older adults

Although there is growing research within 'positive psychology' focussing on SWB as a resource for optimal functioning of individuals (see Gable and Haidt, 2005), few studies adequately explore the (positive) effects that SWB of older adults may have on individual and social outcomes.

3.7   Lack of knowledge about effective policy measures to promote SWB in older adults

As already mentioned in subsection 2.4.5, empirical studies of the outcomes of policy measures for promoting SWB in older adults are still missing. This subsection considers the major determinants of SWB that might be influenced by policy interventions: Living conditions, critical life events and intentional activities. In considering these factors, one must, of course, bear in mind that for some of them further research is already needed to establish their causal relevance. The subsequent arguments are under the proviso that such evidence can be provided.

Living conditions. Global life satisfaction is systematically related to domain satisfaction regarding one's family life, financial situation, health condition, and job situation (the latter will be relevant only for older adults still working). Policy measures aiming to improve SWB should focus on the living conditions that are likely to contribute to the domain satisfaction in question. Relevant policy measures will thus refer to an improvement of the financial situation (such as public transfers to the poor aged, guaranteed minimum pensions), of health (such as community health and care services, prevention and curing programmes, health and dependency insurances) and of family life or social contact more generally (such as programmes for facilitation of out-of-home mobility).

Age-related life events. Although severe illness, functional disability and widowhood all have pronounced effects on SWB, substantial inter-individual differences in response to these events have been documented. Policy measures may help to prevent or postpone these life events (such as disease prevention, health education) and may facilitate coping with them if they occur (through promotion of healthcare, rehabilitation measures, psychotherapeutic counselling).

Intentional activities. Policy measures can help to expand the amount, range, quality and flexibility of opportunities for intentional activities, in general, and for socially productive activities, in particular. They may encourage older adults in performing them not just for the sake of their own SWB, but also for the benefit of communities.

4   Current state of play of European research infrastructures and networks

This section outlines the existing longitudinal studies and existing research infrastructures in Europe. In so doing, we make no claim to be comprehensive, given that several new programmes may already be under consideration by the European Commission.

Longitudinal studies. The following table lists ongoing or already finished longitudinal studies in Europe based on information provided by the US NIH *National Institute on Aging.* Already at the level of this crude tabulation is it evident that existing studies differ in age-group composition of their samples and in the covered time spans. Furthermore, only the *Survey on Health, Aging and Retirement (SHARE)* is a 'truly' European study – although not all European countries are involved (11 countries were included in the first wave, 16 in the third wave).

Table 7.4:   Longitudinal studies on ageing in Europe

Acronym	Title	Age groups/ Age range	Start/end
BOLSA	Bonn Longitudinal Study of Aging	60-75	1965-1984
BASE	Berlin Aging Study	70-105	1990-2000
ELSA	English Longitudinal Study of Aging	50+	2002-
Excelsa	Cross-European Longitudinal Study of Ageing	30-85	2002-
H70	Gothenburg Study (Göteborg)	70	1971-2001
ILSE	Interdisciplinary Longitudinal Study of Adult Development	41-43, 61-63	1994-1998
ILSA	Italian Longitudinal Study on Aging	65-84	1992-2004
LASA	Longitudinal Aging Study Amsterdam	55-85	1992-2002
Lund 80+	Lund 80+	80	1988
MAAS	Maastricht Aging Study	24-81	1992
NLSAA	Nottingham Longitudinal Study of Activity and Ageing	65+	1983-193
Rotterdam	Rotterdam Study	55+	1990-1999
SAP	Southampton Ageing Project	65+	1977/78-1998
SHARE	Survey on Health, Ageing, and Retirement in Europe	50+	2004-
TamELSA	Tampere Longitudinal Study on Aging	60-89	1979-1999

Source: National Institute on Aging, NIH (http://www.nia.nih.gov/ResearchInformation/ScientificResources).

Although the list may at first glance seem impressive, a second look shows that obtaining sound information about the ageing process, in general, and regulating SWB in old people, in particular, requires further longitudinal research as a collaborative European effort.

Research infrastructures and networks related to SWB in older adults. Several research networks and studies have been implemented (especially starting with the Fifth Framework Programme); in addition, regular representative surveys of the general and the older population have been initiated, in the meantime. Networks, work groups and surveys are listed in the following table. Again, we do not claim to be exhaustive here.

Table 7.5: European research infrastructures, networks, and surveys related to SWB in older adults

International research co-operations, networks and studies
• ENABLE-AGE: Enabling Autonomy, Participation and Well-Being in Old Age: The Home Environment as a Determinant for Healthy Aging (2002-2004). http://www.enableage.arb.lu.se/index.html
• ESAW: European Study on Adult Well-Being (2002-2003). http://esaw.bangor.ac.uk//
• OASIS: Old Age and Autonomy: The Role of Service Systems and Intergenerational Family Solidarity; non-European participant: Israel (2000-2004). http://oasis.haifa.ac.il/
• EUROFAMCARE: Family care for older people in Europe (2003-2005). http://www.uke.de/extern/eurofamcare/
• MOBILATE: Enhancing outdoor activities in later life. Personal Cooping, environmental resources and technical support (2000-2002). http://www.ist-world.org/ProjectDetails.aspx?ProjectId=a5707e88509e4a5abe798c c27b86e268
• ERA-AGE - European Research Area in Ageing (2004-2009). http://era-age.group.shef.ac.uk/
• ERA-AGE2 European Research Area in Ageing (2009-) http://era-age.group.shef.ac.uk/
• FLARE Future Leaders in Aging Research (2007-). http://era-age.group.shef.ac.uk/content/188/
• FUTURAGE: A Road Map for Aging Research (2009-). http://futurage.group.shef.ac.uk/
Regular representative surveys: general and older adult population
• Eurobarometer (twice a year). http://ec.europa.eu/public_opinion/index_en.htm
• EVS: European Values Study (1981, 1990, 1999, 2008). http://www.europeanvaluesstudy.eu/
• ECHP: European Community Household Panel (1994-2001: every year). http://epp.eurostat.ec.europa.eu/portal/page/portal/microdata/echp
• ESS: European Social Survey (twice a year). http://ess.nsd.uib.no/

5   Required research infrastructures, methodological innovations, data, networks and so forth – and consequences for research policy

This section highlights some potential directions of SWB research, beginning with a focus on SWB measurement. First, additional research programmes are needed that coordinate existing research and synthesise existing evidence in order to refine measures of SWB in older adults. Subsequently, a research programme is needed that develops and adapts measures of SWB for older adults with advanced neurodegenerative diseases, since these individuals are unable to fill out questionnaires, let alone handle experience samplers.

A further avenue for research has to do with the implementation of a research infrastructure for a *'SWB Monitor in Older Adults'*. Such a monitor should include an ongoing measurement of SWB in representative samples using multiple indicators of different type and specificity, assessed at several times of measurement. This would best be realised by a prospective longitudinal study of long duration. Existing studies – such as the *German Socio-Economic Panel* or the *British Household Panel Study* – could be refined in such a direction by including more-elaborate SWB measures as well as measures of additional causes and outcomes of SWB in old age. Besides (and complementary to) this, a cooperative trans-disciplinary network should be established. This network should include SWB researchers of the general population, gerontologists, economists, and representatives from other relevant disciplinary groups (such as medical doctors, sociologists).

6   What (and when) can we deliver on policy questions?

SWB in older adults constitutes a large research domain that has already produced an impressive body of knowledge. Only the major essentials will be summarised here and combined with reflections about their significance for policy questions in this field. As a general point we would like to emphasise that strategies for communicating research findings to policymakers have to be further elaborated in the future so that these may have a greater impact on political decision-making.

(1) Knowledge about the concept of SWB and its assessment
Researchers of SWB in older adults have broadly agreed upon a definition of SWB that emphasises the *individual's point of view* in evaluating his or her life. This reflects a distinct component of older adults' quality of life beyond objective dimensions that deserves more attention by policymakers. The internal structure of SWB has been clarified (such as cognitive and affective components; general and domain-specific facets and their interrelation), and several instruments for assessing the various components of SWB of older adults are available. The reliability of the instruments has been established – and also some necessary conditions for their validity. Future research may provide further refinement – for instance, with respect to measurement devices for older adults with cognitive impairments.

(2) Descriptive knowledge
Longitudinal studies have revealed the course of SWB in later life. A rather small decline from middle- to old age and a pronounced decline in very old age have been found (especially large in the final years before death). However, notable inter-individual deviations from this average course have been documented. Further research is needed in order to deliver more differentiated information about the amount and kind of inter-individual differences and intra-individual change in SWB.

(3) Knowledge about effects
Although one cannot assume a direct causal relationship, the present evidence indicates that SWB may promote health and longevity in older adults. Future research may find additional positive outcomes beyond health that have already been documented for younger age groups: Positive social relationships, socially productive behaviour and creativity. Thus, SWB of older adults isn't just relevant because it reflects a pleasant experience, but also because it has positive individual and social consequences which then reinforce a positive SWB.

(4) Knowledge about potential causes
Important potential causes of SWB in older adults have been identified in the area of current life circumstances (regarding financial situation, social relationships). Future research should provide stronger evidence for their causal role through longitudinal and intervention studies. Age-related losses (such as death of spouse, or becoming severely disabled) have also been described as potential causes of SWB changes. Prospective longitudinal studies have shown

here a large decline of SWB after the event, which shows a rather slow return (and sometimes even no return) to the pre-event level over the years.

Various kinds of intentional activities (such as socially productive behaviour) have also been identified as systematically related to SWB in older adults. The amount, type and particular circumstances appear to moderate the relation between these activities and SWB.

Theoretical models of SWB in older adults can provide a deeper explanation of well-documented phenomena regarding SWB (for example, having to do with the stability-despite-loss-paradox; and individual and temporal differences in response to similar circumstances). The best available models are cognitively oriented and assume that not reality per se (such as living conditions, critical life events) determines SWB, but rather reality as subjectively perceived and evaluated by the individual in relation to his or her desires (needs, aspirations, goals and so forth). Fulfilment of desires is assumed to promote, and frustration of desires is assumed to deteriorate, SWB. Given the central role of reality perceptions and desires for SWB, more knowledge is needed about the content and the dynamic of these mental states of older adults.

(5) Knowledge relevant for interventions

Effective psychotherapeutic interventions for improving SWB in older adults are available; such interventions operate on a micro level and focus directly on the individual. Interventions operating on the meso- or macro level (and which may be instantiated by policy measures) can be further developed in future. Theoretical frameworks (for example, models about SWB and its regulation) and some robust empirical findings are already available; these frameworks can guide the construction and implementation of such interventions. This knowledge will also help in getting a realistic view of the chances and limits of policy measures to influence SWB. According to well-founded models of SWB, the effects of policy measures (intended, for example, to improve financial, social and health conditions) ultimately depend not primarily on changed living conditions *per se*, but rather on how they are perceived and evaluated by older adults in relation to their desires. Thus, any planning, implementation and evaluation of policy measures should pay attention to how living conditions are actually appraised. Based on such studies, public communication strategies can be used to influence SWB by changing older adult's perceptions and desires.

A short outlook

In closing, we would like to emphasise that in a European society in which the proportion of older persons is continuously increasing, while the proportion of children, adolescents and younger adults is continuously decreasing, research on conditions that guarantee and enhance SWB in old age (and other age groups) will become increasingly important. This research is and will be multidisciplinary, as is reflected by the various disciplines that are already engaged here. Still needed is better coordination of these disciplinary and complementary approaches; last but not least, this may also improve the impact of specific research findings on social policy measures for older adults.

References

Andrews, F.M. and S.B. Withey (1976) *Social Indictors of Well-being* (New York: Plenum Press).

Baltes, P.B. and M.M. Baltes (1990) 'Psychological Perspectives on Successful Aging: The Model of Selective Optimization with Compensation' in P.B. Baltes and M.M. Baltes (eds) *Successful Aging: Perspectives from the Behavioral Sciences* (New York: Cambridge University Press), pp. 1-34.

Bradburn, N.M. (1969) *The Structure of Psychological Well-being* (Chicago: Aldine).

Brandtstädter, J. and K. Rothermund (2002) 'The Life-course Dynamics of Goal Pursuit and Goal Adjustment: A Two-process Framework'. *Developmental Review*, 22, 117-150.

Brickman, P. and D.T. Campbell (1971) 'Hedonic Relativism and Planning the Good Society' in M.H. Appley (ed.) *Adaptation Level Theory: A Symposium* (New York: Academic Press), pp. 287-302.

Bronfenbrenner, U. (1979) *The Ecology of Human Development: Experiments by Nature and Design* (Cambridge: Harvard University Press).

Charles, S.T., C.A. Reynolds and M. Gatz (2001) 'Age-related Differences and Change in Positive and Negative Affect over 23 Years'. *Journal of Personality Assessment*, 80, 136-151.

Coan, J.A. and J.J.B. Allen (eds) (2007) *Handbook of Emotion Elicitation and Assessment* (New York: Oxford University Press).

Costa, P.T. and R.P. McCrae (1993) 'Psychological Research in the Baltimore Longitudinal Study of Aging'. *Zeitschrift für Gerontologie*, 26, 138–141.

Cummins, R.A., R. Eckersley, J. Pallant, J. van Vugt and R. Misajon (2003) 'Developing a National Index of Subjective Wellbeing: The Australian Unity Wellbeing Index'. *Social Indicators Research*, 64, 159-190.

DeNeve, K.M. and H. Cooper (1998) 'The Happy Personality: A Meta-analysis of 137 Personality Traits and Subjective Well–being'. *Psychological Bulletin*, 124, 197–229.

Diener, E. (2006) 'Guidelines for National Indicators of Subjective Well-being and Ill-being'. *Applied Research in Quality of Life*, 1, 151-157.

Diener, E., R.A. Emmons, R.J. Larsen and S. Griffin (1985) 'The Satisfaction with Life Scale', *Journal of Personality Assessment*, 49, 71-75.

Diener, E. and M.E. Suh (1998) 'Subjective Well-being and Age: An International Analysis', *Annual Review of Gerontology and Geriatrics*, 17, 304–324.

Diener, E., S. Oishi and R.E. Lucas (2003) 'Personality, Culture, and Subjective Well-being: Emotional and Cognitive Evaluations of Life', *Annual Review of Psychology*, 54, 403-425.

Diener, E., R.E. Lucas and C. Scollon (2006) 'Beyond the Hedonic Treadmill: Revising the Adaptation Theory of Well-being', *American Psychologist*, 61, 305-314.

Diener, E., P. Kesebir and R. Lucas (2008) 'Benefit of Accounts of Well-being - for Societies and for Psychological Science', *Applied Psychology: An International Review*, 57, 37-53.

Easterlin, R.A. (2006) 'Life Circle Happiness and its Sources: Intersections of Psychology, Economics and Demography', *Journal of Economic Psychology*, 27, 463-482.

Easterlin, R.A. and O. Sawangfa (2007) 'Happiness and Domain Satisfaction: Theory and Evidence', Bonn: IZA discussion paper 2584.

Ferri, C.P., M. Prince, C. Brayne, H. Brodaty, L. Fratiglioni, M. Ganguli, K. Hall, K. Hasegawa, H. Hendrie, Y. Huang, A. Jorm, C. Mathers, P. R Menezes, E. Rimmer, M. Scazufca, for Alzheimer's Disease International (2005) 'Global Prevalence of Dementia: A Delphi Consensus Study', *Lancet*, 366, 2112-2117.

Ferring, D. and S.-H. Filipp (1997) 'Subjektives Wohlbefinden im Alter: Struktur- und Stabilitätsanalysen' [Subjective Well-being in Old Age: Analyses of Structure and Stability], *Psychologische Beiträge*, 39, 236-258.

Ferring, D., C. Balducci, V. Burholt, C.G. Wenger, F. Thissen, G. Weber and I. Hallberg-Rahm (2004) 'Life Satisfaction of Older People in Six European Countries: Findings from the European Study on Adult Well-being', *European Journal of Ageing*, 1, 15-25.

Filipp, S.-H. and P. Aymanns (2010) *Kritische Lebensereignisse und Lebenskrisen [Critical Life Events and Life Crises]* (Stuttgart: Kohlhammer).

Frederick, S. and G. Loewenstein (1999) 'Hedonic Adaptation' in D. Kahneman, E. Diener and N. Schwarz (eds) *Well-being: The Foundations of Hedonic Psychology* (New York: Sage), pp. 302–329.

Gable, S.L. and J. Haidt (2005) 'What (and Why) is Positive Psychology?' *Review of General Psychology*, 9, 103-110.

Gerstorf, D., N. Ram, C. Röcke, U. Lindenberger and J. Smith (2008a) 'Decline in Life-satisfaction in Old Age: Longitudinal Evidence for Links to Distance to Death, *Psychology and Aging*, 23, 154-168.

Gerstorf, D., N. Ram, R. Estabrook, J. Schupp, G.G. Wagner and U. Lindenberger (2008b) 'Life Satisfaction Shows Terminal Decline in Old Age: Longitudinal Evidence from the German Socio-Economic Panel Study (SOEP)', *Developmental Psychology*, 44, 1148-1159.

Greenfield, E.A. and N.F. Marks (2004) 'Formal Volunteering as a Protective Factor for Older Adults' Psychological Well-being', *Journals of Gerontology: Series B: Psychological Sciences and Social Sciences*, 59B, S258-S264.

Heckhausen, J., C. Wrosch and R. Schulz (2010) 'A Motivational Theory of Life-span Development', *Psychological Review*, 117, 32-60.

Hsieh, C.-M. (2003) 'Counting Importance: The Case of Life Satisfaction and Relative Domain Importance', *Social Indicators Research*, 61, 227-240.

Hsieh, C.-M. (2005) 'Age and Relative Importance of Major Life Domains', *Journal of Aging Studies*, 19, 503-512.

Isaacson, D. and J. Smith (2003) 'Positive and Negative Affect in Very Old Age', *Journals of Gerontology Series B: Psychological Sciences and Social Sciences*, 58, P143-P152.

Jopp, D. and C. Rott (2006) 'Adaptation in Very Old Age: Exploring the Role of Resources, Beliefs, and Attitudes for Centenarians' Happiness', *Psychology and Aging*, 21, 266-280.

Kahneman, D., A.B. Krueger, D.A. Schkade, N. Schwarz and A.A. Stone (2004) 'A Survey Method for Characterizing Daily Life Experience: The Day Reconstruction Method', *Science*, 306, 1776-1780.

Knight, B. (2004) *Psychotherapy with Older Adults* (New York: Sage).

Kohli, M. and H. Künemund (2010) 'Social Networks' in L. Bovenberg, A. van Soest and A. Zaid (eds) *Aging, Health and Pensions in Europe* (Houndmills, Basingstoke, Hampshire: Palgrave Macmillan), pp. xxx-yyy.

Kunzmann, U., T.D. Little and J. Smith (2000) 'Is Age-related Stability of Subjective Well-being a Paradox? Cross-sectional and Longitudinal Evidence from the Berlin Aging Study', *Psychology and Aging*, 15, 511-526.

Lawrence, R.H. and J. Liang (1988) 'Structural Integration of the Affect Balance Scale and the Life Satisfaction Index A: Race, Sex and Age Differences', *Psychology and Aging*, 3, 375-384.

Lewin, S. (1996) 'Economics and Psychology: Lessons for our Own Day from the Early Twentieth Century', *Journal of Economic Literature*, 34, 1293-1323.

Logsdon, R.G., L.E. Gibbons, S.M. McCurry and L. Teri (2002) 'Assessing Quality of Life in Older Adults with Cognitive Impairment', *Psychosomatic Medicine*, 64, 510-519.

Lucas, R.E. (2007a) 'Adaptation and the Set-point Model of Subjective Well-being. Does Happiness Change after Major Life Events?', *Current Directions in Psychological Science*, 16, 75-79.

Lucas, R.E. (2007b) 'Long-term Disability is Associated with Lasting Changes in Subjective Well-being: Evidence from Two Nationally Representative Longitudinal Studies', *Journal of Personality and Social Psychology*, 92, 717-730.

Lucas, R.E. (2008) 'Personality and Subjective Well-being' in M. Eid and R.J. Larson (eds), *The Science of Subjective Well-being* (New York: Guilford Press), pp. 171-194.

Lucas, R.E., A.E. Clark, Y. Georgellis and E. Diener (2003) 'Reexamining Adaptation and the Set Point Model of Happiness: Reactions to Changes in Marital Status', *Journal of Personality and Social Psychology*, 84, 527-539.

Lyubomirsky, S., L. King and E. Diener (2005) 'The Benefits of Frequent Positive Affect: Does Happiness Lead to Success?' *Psychological Bulletin*, 131, 803-855.

Lyubomirsky, S., K.M. Sheldon and D. Schkade (2005) 'Pursuing Happiness: The Architecture of Sustainable Change', *Review of General Psychology*, 9, 111-131.

McMunn, A., J. Nazroo, M. Wahrendorf, E. Breeze and P. Zaninotto (2009) 'Participation in Socially-productive Activities, Reciprocity and Wellbeing in Later Life: Baseline Results in England', *Ageing and Society*, 29, 765-782.

Michalos, A.C. (1986) 'An Application of Multiple Discrepancies Theory (MDT) to Seniors', *Social Indicators Research*, 18, 349-373.

Mollenkopf, H., F. Marcellini, I. Ruoppila, Z. Széman, M. Tacken and H.-W. Wahl (2004) 'Social and Behavioural Science Perspectives on Out-of-home Mobility in Later Life: Findings from the European Project MOBILATE', *European Journal of Ageing*, 1, 45-53.

Mroczek, D.K. and A. Spiro III (2005) 'Change in Life Satisfaction During Adulthood: Findings from the Veterans Affairs Normative Aging Study', *Journal of Personality and Social Psychology*, 88, 189-202.

Murrell, S.A. and S. Meeks (2002) 'Psychological, Economic and Social Mediators of the Education-health Relationship in Older Adults', *Journal of Aging and Health*, 14, 527-550.

Palgi, Y., A. Shrira, M. Ben-Ezra, T. Spalter, D. Shmotkin and G. Kave (2010) 'Delineating Terminal Change in Subjective Well-being and Subjective Health: Brief Report', *Journals of Gerontology Series B: Psychological Sciences and Social Sciences*, 65B, 61-64

Pinquart, M. (1997) 'Selbstkonzept- und Befindensunterschiede im Erwachsenenalter: Ergebnisse von Metaanalysen', [Differences in Self-concept and Well-being across the Adult Life Span: Results of Meta-analyses]. *Zeitschrift für Gerontopsychologie und -Psychiatrie*, 10, 17-25.

Pinquart, M. (2001) 'Age Differences in Perceived Positive Affect, Negative Affect and Affect Balance in Middle and Old Age', *Journal of Happiness Studies*, 2, 375-405.

Pinquart, M. and S. Sörensen (2000) 'Influences of Socioeconomic Status, Social Network, and Competence on Subjective Well-being in Later Life: A Meta-analysis', *Psychology and Aging*, 15, 187-224.

Pinquart, M. and S. Sörensen (2001) 'How Effective are Psychotherapeutic and Other Psychosocial Interventions with Older Adults? A Meta-analysis', *Journal of Mental Health and Aging*, 7, 207-243.

Pinquart, M. and S. Sörensen (2003) 'Differences Between Caregivers and Noncaregivers in Psychological Health and Physical Health: A Meta-analysis', *Psychology and Aging*, 18, 250-267.

Pinquart, M. and S. Sörensen (2005) 'Caregiving Distress and Psychological Health of Caregivers' in K.V. Oxington (ed.) *Psychology of Stress* (New York: Nova Science Publishers), pp. 165-206.

Pinquart, M. and I. Schindler (2007) 'Changes of Life Satisfaction in the Transition to Retirement: A Latent-class Approach', *Psychology and Aging*, 22, 442-455.

Plagnol, A.C. and R.A. Easterlin (2008) 'Aspirations, Attainments and Satisfaction: Life Cycle Differences between American Women and Men', *Journal of Happiness Studies*, 9, 601-619.

Pons, D., F.L. Atienza, I. Balaguer and M.L. Garcia-Merita (2000) 'Satisfaction with Life Scale: Analysis of Factorial Invariance for Adolescents and Elderly Persons', *Perceptual and Motor Skills*, 91, 62-68.

Pushkar, D., J. Chaikelson, M. Conway, J. Etezadi, C. Giannopoulus, K. Li and C. Wrosch (2010) 'Testing Continuity and Activity Variables as Predictors of Positive and Negative Affect in

Retirement', *Journals of Gerontology Series B: Psychological Sciences and Social Sciences*, 65B, 42-49.

Roick, C., A. Hinz and H.-J. Gertz (2007) 'Kann Lebensqualität bei Demenzkranken Valide Bestimmt Werden? Eine Aktuelle Übersicht über Messinstrumente und Methodische Probleme' [Is Quality of Life in Dementia Patients Validly Estimable? A Current Review about Measuring Instruments and Methodological Problems], *Psychiatrische Praxis, 34*, 108-116.

Schilling, O. (2006) 'Development of Life Satisfaction in Old Age: Another View in the "Paradox"'. *Social Indicators Research, 75*, 241-271.

Schimmack, U. (2008) 'The Structure of Subjective Well-being' in M. Eid and R.J. Larson (eds) *The Science of Subjective Well-being* (New York: Guilford Press), pp. 97-123.

Sigrist, J. and M. Wahrendorf (2010) 'Social Productivity and Well-being' in L. Bovenberg, A. van Soest and A. Zaid (eds), *Aging, Health and Pensions in Europe* (Houndmills, Basingstoke, Hampshire: Palgrave Macmillan), pp. xxx-yyy.

Smith, J. (2002) 'The Fourth age: A Period of Psychological Mortality?' In Max-Planck-Gesellschaft zur Förderung der Wissenschaften and Ernst Schering Research Foundation (eds), *Biomolecular Aspects of Aging: The Social and Ethical Implications (Max-Planck-Forum No. 4)* (München: Max-Planck-Gesellschaft), pp. 75-88.

Staudinger, U. (2000) 'Viele Gründe Sprechen Dagegen, und Totzdem Geht es Vielen Menschen Gut: Das Paradox des Subjektiven Wohlbefindens' [*Many Reasons Speak Against it, Yet Many People Feel Good: The Paradox of Subjective Well-being*], *Psychologische Rundschau, 51*, 185-197.

Thomae, H. (1970) 'Theory of Aging and Cognitive Theory of Personality', *Human Development, 13*, 1-13.

Vaarama, M. (2009) 'Care-related Quality of Life in Old Age', *European Journal of Aging, 6*, 113-125.

van Praag, B.M.S., P. Frijters and A. Ferrer-i-Carbonell (2003) 'The Anatomy of Subjective Well-being', *Journal of Economic Behavior and Organization, 51*, 29-49.

van Willigen, M. (2000) 'Differential Benefits of Volunteering across the Life Course. *Journals of Gerontology, Series B: Psychological Sciences and Social Sciences*, 55B, S308-S318.

Watson, D., L.A. Clark and A. Tellegen (1988) 'Development and Validation of Brief Measures of Positive and Negative Affect: The PANAS Scales', *Journal of Personality and Social Psychology, 54*, 1063-1070.

Wheeler, J.A., K.M. Gorey and B. Greenblatt (1998) 'The Beneficial Effects of Volunteering for Older Volunteers and the People They Serve: A Meta-analysis', *International Journal of Aging and Human Development, 47*, 69-79.

Subjective Well-being in Older Adults: Current State and Gaps of Research

Comments by Michael Dewey

1   Introduction

Ferring and Boll provide us with an overview of the concept of subjective well-being (SWB) and its application in studies of older people. They distinguish the cognitive aspects from the affective, and refer to different measurement approaches that have been used. Clearly, as they point out, SWB relates to the studies of older people in many ways. Is it interesting in itself, interesting as a cause, or interesting as an effect? The relationship between SWB and various factors such as age, socioeconomic status, life events and personality is also examined.

2   The clash of cultures

One of the striking things for me, as an epidemiologist working in the area of mental health of older people, is just how dissimilar the language and concepts are in different fields of scientific enquiry – even when we claim to have the same goals. This represents both an opportunity and a challenge. It's an opportunity because we may be able to shed new light on the problem, and it's a challenge because we needlessly seem to repeat the work of others without realising it. An example is the extensive work that has gone into measuring the health-related quality of life in people with dementia-related illnesses – which seems to have drawn little from the work mentioned here on SWB in people with dementia. The intersection of the references cited here and those in the review by Banerjee et al. (2009) is very small.

3   *The research agenda*

3.1   European strengths

There is a strong tradition of epidemiological studies and social surveys in Europe, to which the list provided by Ferring and Boll testifies.

Ferring and Boll refer to the paucity of studies of measurement equivalence. This is an area where there is substantial European expertise in item-response theory and also in the use of vignettes. This expertise should be tapped for future research. Ferring and Boll rather underplay the advantages of experience sampling, which is becoming simpler and cheaper with technological advances.

3.2   European limitations

Too many studies have investigated only a limited age range, and coverage of the oldest old is limited. In part, this is due to interest in the period around retirement and also to the much larger number of people in the younger age range. As Ferring and Boll point out, SWB is subject to adaptive homeostatic mechanisms, which implies that one has to measure at the time when things are changing. In this respect, it is unfortunate that changes in health status and in functioning occur in the older old, whereas many cohorts have a focus much earlier. Figure 1 shows data from the SHARE project, and displays the cumulative proportion of people with various levels of self-perceived health and of limiting conditions. As can be seen, the prevalence of poor health does not really increase until age 70 (and that of severe limiting conditions not until 75), but much of the interest in social surveys has been the period around retirement.

Policy decisions about the health of older people are based on established but limited collaborations. The two most quoted mental health collaborations – EURODEM (Hofman and the EURODEM Prevalence Research Group, 1991) on dementia, and EURODEP (Copeland et al., 1999) on depression – both included centres in Finland, Germany, Italy, the Netherlands, Spain, Sweden and the United Kingdom. EURODEM also covered Norway, and EURODEP Belgium, France, Ireland and Iceland. This was not even complete coverage of the old EU.

There has been some failure to capitalise on best analysis methods. For instance, Ferring and Boll refer to dropout problems, but methods to handle this have been developed over the last two decades and should be more widely used.

Figure 1:   Cumulative proportion of people with various levels of self-perceived health and of limiting conditions

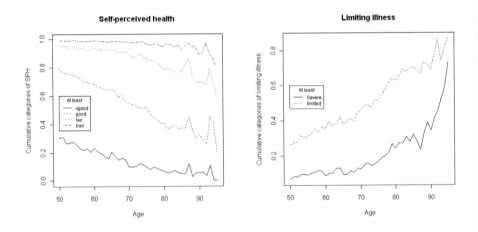

3.3   European opportunities

Existing cohorts of older people that started in the 1970s and 1980s offer an opportunity to study cohort changes. For instance, the group in Gothenburg has begun to study cohorts several decades on, and finds substantial differences in symptoms over the years. So far, only the results on sexual behaviour of older people have been published (Beckman et al., 2008) but the message is clear that we cannot rely on what older people said in the late twentieth century to plan for the current one.

3.4   European threats

- Response rates have been falling steadily in social surveys and epidemiological studies.
- The desire to have one study to answer everything seems to be on the rise.
- Regulatory issues have become increasingly important, as they influence the activities of researchers and the access to data sources.

4   *Summary*

Ferring and Boll provide us with much food for thought as we plan research into the well-being of older people in the EU.

References

Banerjee, S., K. Samsi, C.D. Petrie, J. Alvir, M. Treglia, E.M. Schwam and M. del Valle (2009) 'What Do We Know about Quality of Life in Dementia? A Review of the Emerging Evidence on the Predictive and Explanatory Value of Disease Specific Measures of Health-related Quality of Life in People with Dementia', *International Journal of Geriatric Psychiatry*, 24: 15-24.

Beckman, N., M. Waern, D. Gustafson and I. Skoog (2008) 'Secular Trends in Self-reported Sexual Activity and Satisfaction in Swedish 70-Year-Olds: Cross-sectional Survey of Four Populations, 1971–2001', *British Medical Journal*, 337:a279.

Copeland, J.R.M., A.T.F Beekman, M.E. Dewey, C. Hooijer, A. Jordan, B.A. Lawlor, A. Lobo, H. Magnússon, A. Mann, I. Meller, M.J. Prince, F. Reischies, C. Turrina, M.W. deVries and K.C.M. Wilson (1999) 'Depression in Europe. Geographical Distribution among Older People', *British Journal of Psychiatry*, 174: 312-321.

Hofman, A. and the EURODEM Prevalence Research Group (1991) 'The Prevalence of Dementia in Europe: A Collaborative Study of 1980-1990 Findings', *International Journal of Epidemiology*, 20: 736-748.

Subjective Well-being in Older Adults: Current State and Gaps of Research

Comments by Constanca Paúl

This chapter, based on Ferring and Bold (2009), reviews the relevant literature on subjective well-being (SWB), considering its cognitive (life satisfaction) and affective (happiness) components, focusing on the rather different hypothesis that looks at SWB as a result, a cause, a mediator or a moderator variable of individual or social outcomes. It explores ways to assess SWB and outlines the main limitations of existing studies. The chapter presents the major findings on the association between SWB, age, life circumstances, life-events and personality.

To extend the discussion on SWB, I would like to begin by examining life satisfaction at a country level, and investigating how life satisfaction is associated with objective living conditions. Delhey (2004) shows that life satisfaction in Europe rises with the country GDP per capita. However, when Ireland, Greece, Portugal and Spain joined the EC in 1985, it seemed that life satisfaction, after a short increase, stabilised at the respective country set point – although living conditions improved greatly during the following years. The set-point-of-happiness model (Lykkens and Tellegen, 1996) holds that external events may change the level of life satisfaction of individuals during a short period of time, but then satisfaction returns to the own set point; the same appears to happen with countries' level of happiness. Note that there is no direct effect of raising objective life conditions and SWB. The SWB range in Europe is 70-80% – or, after including non-western countries, 60-80% (Cummins, 1995, 1998).

Across Europe, the abstract idea of what constitutes life satisfaction does not differ all that much. The prevailing concerns in all countries are, according to Delhey (2004), as follows: making a living (income), family life and health – although priorities differ from country to country. These results were similar to those reported by Gundelach and Kreiner (2004), where country characteristics together with the sense of control of life are the most important predictors of life satisfaction (while happiness is mostly related with social relations, which they call social capital).

Following Ferring and Boll (2009), we can review both the hypothesis of change and the hypothesis of no change of SWB with age, with results differing across longitudinal versus cross-sectional methodologies. Blanchflower and Oswald (2007) show that there is a minimum of happiness in the fourth decade of life. Curiously, SWB fell gradually in successive cohorts in the EU area, and is rising in Europe with newer birth cohorts. Inglehart (2002), studying gender-related differences in SWB, shows that despite the fact that women have certain disadvantages in most European societies, both women and men show similar SWB within a given society. This could be explained by the aspiration-adjustment model (Campbell, Converse and Rodgers, 1976), as people adapted to the advantages and disadvantages persistent in life. Nevertheless, women suffer a decline of SWB as they age, due to the cultural tendency to devalue the social worth of older women.

Health is one of the most important variables of SWB. According to Veenhoven (2008), happiness does not predict longevity in a sick population, but it does predict longevity among those in a healthy population, functioning as a sort of protection against becoming ill. The relation between health and happiness is particularly important for old people – and if health problems disrupt daily functioning (Anger et al., 2009). Marinié and Brkljacie (2008) report that, regardless of disability, people expressed happiness and moderate life satisfaction – although satisfaction tended to be lower in the disabled group. Satisfaction with economic status seems to be an important happiness predictor for the general population, as well – but not for people with disabilities. Satisfaction with health, relationships and achievements contributes to happiness in persons with or without physical disability, while physical safety and acceptance by the community predicts happiness in people with disability. The concept of 'disability paradox' (Albrecht and Devlieger, 1999) shows the apparent contradiction between existing high levels of disability and high levels of life satisfaction in old and disabled people. Again, the findings are not consensual. It seems that there is an expected level of disability that is compatible with maintenance of SWB – but once this limit is passed, SWB decreases in step with increasing difficulties (Paúl, Ayes and Ebrahim, 2007). Perhaps there are objective/subjective limits beyond which SWB is not sustainable. To find out what the limits are on wealth, health, education and available community resources, we must turn again to country characteristics.

The models that consider SWB as a cause or a mediator, or moderator of outcomes for individuals or societies are more useful, from a political point of view. Unsatisfied people are expected to have higher morbidity and mortality, implying rising costs associated with health and social services, which will probably lead to individual and social discontentment. Satisfied and happy individuals, on the contrary, are more productive and healthy, which leads to greater social cohesion.

For the future developments in SWB studies, as a crucial area for understanding adaptation and a positive ageing process, we suggest that the following steps be taken: 1. Undertake more research to assess peoples' expectations and what standards they use when evaluating their life as a whole. 2. Focus more research on predictors and outputs of SWB. 3. Acknowledge current challenges of the oldest-old people. 4. Invest in mental health issues. Taking these steps will contribute to raising SWB, decreasing morbidity and mortality, diminishing health-related expenses, and promoting productive ageing in years to come.

References

Albrecht, G. and P. Devlieger (1999) 'The Disability Paradox: High Quality of Life Against All Odds', *Social Science and Medicine*, 48, 977-988.

Anger, E., M. Ray, K. Saag and J. Allison (2009) 'Health and Happiness among Older Adults', *Journal of Health Psychology*, 14(4), 503-512.

Blanchflower, D. and A. Oswald (2007) 'Is Well-being U-Shaped over the Life Cycle?' NBER working paper 12935.

Cummins, R.A. (1995) 'On the Trail of the Gold Standard for Subjective Well-being', *Social Indicators Research*, 35, 179-200.

Cummins, R.A. (1998) 'The Second Approximation to an International Standard for Life Satisfaction', *Social Indicators Research*, 35, 179-200.

Delhey (2004) 'Life Satisfaction in an Enlarged Europe', European Foundation for the Improvement of Living and Working Conditions. (Luxembourg: Office for Official Publications of the European Communities).

Ferring, D. and T. Boll (2009) 'Regulation of Psycho-social Well-being in Ageing', ESF Forward Look "Ageing, Health and Pensions in Europe, The Hague

Gundelach, P. and S. Kreiner (2004) 'Happiness and Life Satisfaction in Advanced European Countries', *Cross Cultural Research*, 38(4), 359-366.

Inglehart, R. (2002) 'Gender, Aging, and Subjective Well-Being', *International Journal of Comparative Sociology*, 43, 391-408.

Lykkens, D. and A. Tellegen (1996) 'Happiness is a Stochastic Phenomenon', *Psychological Science*, 7, 186-189.

Marinié, M and T. Brkljacie (2008) 'Love over Gold – The Correlation of Happiness Level with Some Life Satisfaction Factors between Persons with and without Physical Disability', *Journal of Development Physical Disability*. 20, 527-540.

Paúl, C., S. Ayis and S. Ebrahim (2007) 'Disability and Psychosocial Outcomes in Old Age', *Journal of Aging Health*, 19, 723-741.

Veenhoven, R. (2008) 'Healthy Happiness: Effects of Happiness on Physical Health and the Consequences for Preventive Health Care', *Journal of Happiness Studies*, 9, 449-469.

8

Old Age, Health and Long-term Care

Alberto Holly

1   Introduction

This chapter reviews some important issues concerning ageing, health and long-term care. It emphasises policy questions, major progress in understandings, remaining gaps in knowledge and main challenges. Also, it discusses the current state of play of European research infrastructures and networks, and what is required with regard to research infrastructures, methodological innovations, data and networks – as well as the consequences for research policy.

To provide an essential background to this chapter, it is important to highlight some demographic and epidemiological trends in industrialised countries. Demographic trends are described well in a report from the Population Division of the Department of Economic and Social Affairs of the United Nations (UN, 2002), which will be referred to extensively in the chapter. The epidemiological trends that are shown in numerous studies and briefly described below, combined with demographic changes, are of great relevance for the topic of this chapter.

At the outset, it is important to observe that two notions of 'ageing' are often confused. 'Ageing of the population' is a process by which older individuals become a proportionally larger share of the total population'. On the other hand, 'longevity' is measured by life expectancy at a given age (at birth, and at ages 65 and 80, for example). Ageing of the population depends on longevity, but also on other variables, such as birth rate and migration. Setting migration issues aside, it is important to bear in mind the fact that the demographic determinants of population ageing include fertility as well as mortality.

In the earlier stages of what is now called 'the demographic transition', fertility decline was the primary determinant of the timing and extent of population ageing. However, at the present stage of the demographic transition (where both birth- and death rates are low), mortality decline – particularly at older ages – has increasingly become the more important factor in shaping the relative size of the oldest age groups (UN, 2002).

More people will survive to older ages, and gains in life expectancy are expected to be higher at older ages in proportional terms. This implies that the older population is itself ageing. The fastest-growing age group in industrialised countries is the oldest of the old: those aged 80 years or older. By the middle of this century, one-fifth of older persons will be 80 years or older in the world (UN, 2002). In addition, although the proportion of people who live beyond the age of 100 is still very small, their number is growing rapidly in most developed countries (UN, 2002).

It is important to observe that women (because their life expectancy is greater than that of men) comprise a significant majority of the older population. Since female mortality rates are lower than male rates at older ages, the proportion of women in the older population grows substantially with advancing age. In 2000, women outnumbered men by almost 4 to 3 at ages 65 or older – and by almost 2 to 1 at ages 80 or above in the world (UN, 2002).

The decline in fertility and mortality that underlies population ageing has produced and will continue to produce unprecedented changes in the structure of all societies – notably, the historic reversal in the proportions of young and older persons.

It is also essential to note that evidence concerning recent trends in morbidity shows that people live not only longer, but also in better health, on average (Jacobzone et al., 2000; Jacobzone, 2000; Cutler 2001a and 2001b). The shift in age structure associated with population ageing, and the general increase in longevity in good health, will have a profound impact on a broad range of economic, political and social conditions. In the economic area, the shift will affect economic growth, savings, investment and consumption, labour markets, pensions, taxation and intergenerational transfers. In the social sphere, the impact will be on health and healthcare, family composition and living arrangements, housing and migration.

This chapter first addresses two main health-related issues in the economics of ageing: namely, *health and wealth*, and *housing and living arrangements*.

The chapter then examines issues related to population ageing, health and long-term healthcare expenditures. In most developed countries, as the largest population cohorts approach the age of 65, the impact of population ageing on healthcare expenditures has become a topic of growing interest in academic and policy circles.

In fact, parallel to the demographic transition, developed societies also experience a 'epidemiological transition,' characterised by the shifting burden of illness toward non-communicable diseases and injuries. For instance, the incidence of myocardial infarction (commonly known as heart attack), hip fracture and dementia increase with age.[1] In addition, because people are living longer, they are more likely to experience multiple chronic diseases.[2]

The impact of the demographic and epidemiological changes on expenditures is a function of the number of people in high-use categories, the length of time that they remain in that category, and the cost of the health services they use. Older persons use more healthcare and long-term care.

However, the notion that the ageing of the population would be one of the main drivers of healthcare- and long-term care expenditures deserves to be examined closely. It is relevant to note here that the effects of the relatively slow pace of demographic change may be overwhelmed by other factors, such as the introduction of new technologies and treatments and increased utilisation (for example, of drugs and diagnostic tests). Nevertheless, with the most rapid growth in elderly cohorts still to come, it is important to clarify how their relative spending patterns in old age are likely to compare with those of recent generations, and to examine the policy implications of these comparisons.

2   Policy questions

The chapter has as yet mentioned only a few questions of special relevance in terms of public policy, in relation to the topics that will be addressed here – in particular, housing, long-term care and informal care:

- The rapid growth of the oldest groups among the older population and its consequences.

---

1   *Incidence* in relation to disease refers to the number of new cases of a disease in a given year. *Prevalence* is the number of existing cases of this disease. Each measure can be reported as a number or as a rate – for example, per 1,000 of the population.
2   Examples of important chronic diseases include the following: Myocardial infarction, heart failure, stroke, diabetes, obstructive chronic broncopneumopathy, lung cancer, colorectal cancer, breast cancer, prostate cancer, depression, dementia of the Alzheimer type, Parkinson's disease and Parkinson's syndromes, osteoarthritis, lumbago, osteoporotic fractures, functional deficits and disability.

- The rising female share of the older population and its consequences.

We shall return to the policy issues towards the end of the chapter, after reviewing the literature concerning ageing, healthcare and long-term care.

3   Health-related issues in the economics of ageing

For the past 20 years, a growing number of papers have been published in the academic literature on health-related issues within the larger domain of the economics of ageing. These publications not only concern important and interesting topics, but also are part of a scientific research area that could be labelled as 'Health Economics of Ageing'. In fact, two demographic aspects related to ageing outlined earlier (namely, the increase in longevity [or in life expectancy], and population ageing [which is about the evolution of the dependency ratio]) have been shown to have an important impact on issues pertaining to health economics. It is often believed that the main issue in health economics centres on the evolution of the demand for healthcare and healthcare financing. But, as indicated below, this is in fact part of a much larger field.

One way to identify the growing field of 'Health Economics of Ageing' is to review the work on the Economics of Ageing conducted as part of the 'Program on the Economics of Aging' at the National Bureau of Economic Research (NBER). This work was published in a series of ten volumes (starting in 1989), edited by David Wise. The analyses published in these NBER volumes are representative of the important health-related themes in the economics of ageing, some of which are listed below.

3.1   Health and wealth

Trying to understand the explanation for the well-documented and strong relationship between health and socioeconomic status (SES) is one of the most challenging research issues in the economics of ageing – particularly, the direction of causality.

Research to date has shown that the strong relationship holds for a variety of health variables (most illnesses, mortality, self-rated health status, psychological well-being and biomarkers) and alternative measures of SES (wealth, education, occupation, income, level of social integration). In the case of wealth, this association is known as the *health-wealth gradient*.

One direction of causality is from wealth to health. This is because individuals with more wealth can afford better medical care, live in healthier environments, and so on. Another direction is from health to wealth. Healthier individuals may be able to work more than those who are ill, enabling them to accumulate more wealth. Finally, wealth and health status may be simultaneously determined by possibly unobserved common factors.

In order to design economic policy to improve welfare, health and wellbeing, it is important to distinguish these explanations of the health-wealth gradient in the context of an ageing population.

The health-wealth gradient is also important because of the relation between health, retirement, and incentives of social security benefits and health insurance. Better health is positively associated with savings, labour-force participation and earnings – and negatively with the old-age social security benefits replacement rate.

A large number of publications deal with this issue using longitudinal data in the US. However, the direction of causality remains an open issue. This is because there are several differences between these studies that do not always make them comparable.

Studies may differ in the way in which they measure health status. Some studies use a subjective measure of health status in the form of a categorical variable, whereas others use an objective measure taking the form of a continuous and possibly latent (unobserved) variable.

Another difference is whether the model looks at wealth and health of individuals within a household – the head of the household, for example – or at wealth and health of both spouses.

Finally, another important difference is whether the study uses a static- or a dynamic simultaneous equation model. The advantage of dynamic over static models is that the former may apply the procedures developed independently by Granger (1969) and by Sims (1972) to test for causality in a dynamic model. This is notably the case of Adams et al. (2003) and Michaud and Sust (2008).

Of course, to test for causality in a model for health and wealth, these variables need to be assumed to be endogenous a priori. The instruments chosen by various authors for health and wealth relate to shocks that do not have direct effects on the other outcome. For instance, Michaud and Sust (2008) use onsets of critical health conditions as instruments for health changes. Also, following the suggestion of Meer et al. (2003), Michaud and Sust (2008) use inheritances as instruments for changes in wealth. As Meer et al. (2003) note, 'receipt of an inheritance is clearly correlated with the change in an individual's wealth, but is plausibly unrelated to changes in his or her health, conditional on initial health status.'

Given that the model developed by Michaud and Sust (2008) is one of the more recently published models, and is quite elaborate, we state only some of the results they have obtained as an illustration of the conclusions that may be reached through a causality analysis. According to the authors, their 'dynamic panel data model based tests provide clear evidence of causal effects from health to wealth, but no evidence of causal effects from wealth to either the husband's or the wife's health, or from one spouse's health on the health of the other spouse'.

Another interesting set of results of Michaud and Sust (2008) is the following: 'Disaggregating health into mental and physical health shows that mental health is more important for wives, while only physical health matters for husbands. While the mental health effects are instantaneous, the physical health effects take more time and are visible only in the next wave (two years later). Insurance coverage also appears to play a role here: it is mainly if wives without employer-provided insurance experience an onset of mental conditions that household assets decline.'

Again, these results are excellent illustrations of the type of conclusions that one may reach through a causality analysis. They may, however, be sensitive to the underlying assumptions and the econometric method used to estimate the model.

Similar studies are definitely needed that use data from countries in Europe, and this should help us understand why the explanations in terms of causality may differ across countries. Comparative international work would be a very promising and productive direction to follow.

In view of future research, we would like to call attention to two issues that deserve further examination.

The first is related to the particular position of housing within household wealth. Indeed, the savings of a large proportion of the elderly are primarily in the form of housing in most of the OECD countries (Ynesta, 2008). Given the significance of housing wealth in the portfolio of the elderly, understanding how the elderly regard housing wealth and what they intend to do with that wealth as they age has been a topic of considerable interest among economists.

Most of the research to date finds that homeownership continues to be high in old age, and that home equity does not appear to fall with age (Merrill, 1984; Venti and Wise, 1989, 1990, 2004; Feinstein and McFadden, 1989).

For instance, Venti and Wise (2004) find that married and single households that stay intact continue, on average, to own a home even as they age. The authors conclude that the house is 'simply a place to live', and the elderly do not regard housing wealth as fungible wealth.

If this assessment accurately reflects how the elderly regard housing wealth, then the appropriate accounting of wealth for the elderly in models of savings and consumption should exclude housing wealth.

In addition, most of the research to date finds that the death of a spouse or entry into a nursing home increases the probability of a sale. The relationship between changes in household structure (such as widowhood, death and nursing-home entry) and housing sales has led some researchers to attribute an 'insurance' motive to housing wealth because these events are generally associated with changes to the household's economic status.

Understanding whether housing wealth is held as insurance against adverse economic outcomes in old age, or whether the elderly think of the house as simply 'a place to live', will – at the very least – provide some indication of how housing wealth can be accounted for in models of savings and consumption.

Note, however, that this result may be country- or region-specific. In the US (given its healthcare insurance system), housing wealth may be considered as insurance against high out-of-pocket medical expenses. In most European countries, the existence of more generous social health insurance should reduce the household's exposure to uninsured out-of-pocket expenses – except, possibly, for institutionalised long-term care.

To summarise, the particular position of housing in household wealth should be specifically taken into account in the studies that aim at a better understanding of the health-wealth gradient in terms of causality.

A second aspect also deserves a mention. Indeed, one must recognise that important features of the demographic and epidemiological trends discussed in the introduction to this chapter have not yet been incorporated in most of the research to date on the health-wealth gradient. The evolution of wealth, consumption and saving behaviour of the individual after retirement age is crucial in any study on the effects of demographic and epidemiological changes. How are these economic variables affected by not only the increase of life expectancy but also the increase in life expectancy in good health?

These questions are relevant – particularly with respect to bequests and inheritance. In countries where life expectancy has increased, inheritance takes place most often after retirement age, and this may be anticipated. This feature, which does not seem to be sufficiently taken into account, may have an impact, for example, on the relationship of labour-force participation with the health-wealth gradient.

Analyses delving into the many aspects of the relationship between health and wealth are definitely needed in order to advance our understanding of the reasons for the health-wealth gradient within European countries – and why it may differ across countries. Comparative international work would be a promising and productive direction for future research.

Finally, regarding data and research infrastructure, countries in Europe should try to coordinate and set up longitudinal databases that are similar – and if possible comparable – to the Panel Study of Income Dynamics (PSID) and the Assets and Health Dynamics Among the Oldest Old (AHEAD) study.

3.2   Living arrangements

NBER studies based on US data showed the striking finding of the stability of living arrangements (Börsch-Supan, 1989, 1990; Börsch-Supan, McFadden and Schnabel, 1996). Transitions to an institution or to the home of one's children are atypical. Even after the death of a spouse, or the onset of disability, or during the last five years before death, few older people change their living arrangements. While living arrangement transitions are infrequent overall, they are most common after the loss of a spouse. Almost all transitions take place in the same year as the spouse's death. Studies also showed that although living in an institution is inferior to other living arrangements, from the point of view of the oldest old people, the likelihood of institutionalisation has risen substantially during the last decades (Grundy and Glaser, 1997; Huber et al., 2009).

Similar studies would be very important to carry out in European countries – particularly in connection with formal- and informal care issues. If transitions to an institution or to the home of one's children are atypical, then the formal- and informal care issues become more and more important. Particularly deserving of further study are the impacts of these forms of care in the improvement of the well-being of the elderly and on the cost of healthcare.

4   Healthcare and long-term care

4.1   *Ageing and healthcare expenditure*

4.1.1 Study of the factors explaining the past evolution of healthcare expenditure

This section now examines the relationship between demographic and epidemiological changes and the evolution of healthcare expenditure.[3]

An initial observation is undeniable: the statistical data show that for a given year (for example, 2007), healthcare expenditure increases with age – except for the very old-age population. In particular, total health expenditure is much higher for the oldest old than the average for those aged 65 years and above. In addition, on the basis of projections of several statistical offices, one may estimate that the percentage of the population aged 65 years and above will strongly increase until 2050 (see, for example, EC, 2009; United Nations, 2001). For this reason, it is often claimed that the demographic changes will result in an acceleration of healthcare-related expenditure. But the reasoning at the base of this assertion seems to confuse the notion of correlation with that of causality. Its validity deserves to be examined more closely by studying the role of other possible factors of health expenditure growth, which can be related to the ageing of the population and the increase of longevity mentioned earlier.[4]

The first criticism of the effect of age on healthcare-related expenditure comes from research highlighting the fact that a high share of the health expenditure is concentrated around the end of life of individuals. A first category of work is based primarily on statistical analyses. Papers falling within this category compare the health expenditure of the people who die with those of the average of the population (*decedents-only studies*). Several studies using Medicare programme data in the United States observed, for example, that the 5% of recipients aged 65 or more that die each year explain 25 to 30% of the total expenditure (Lubitz and Riley, 1993). Similarly, in Japan, the amount of health expenditure during the final 12 months before death that was allocated to those who died at age 70 or over corresponds to 22% of annual health expenditure of the elderly. Also, of all health expenditure in the last year of life for the deceased

---

3   See Payne et al. (2007) for a recent review of the literature on age, mortality, morbidity and expenditures – with special emphasis on the use of time-to-death as a variable for modelling individual health expenditures.

4   Our exposition of the criticisms of the effect of age on healthcare expenditure based on end-of-life studies follows, in part, Payne et al. (2007).

elderly, 21% was spent, on average, in the last 30 days (Fukawa, 1998). One should expect that similar figures might be found in most OECD countries – with differences, however, in the proportion of healthcare-related expenditure allocated to the deceased reflecting differences in terminal care cost.

A second category of work is based on econometric models using the time interval that precedes the death as an explanatory variable for healthcare expenses (*time-to-death models*). This category of work, initiated by Roos, Montgomery and Roos (1987), allows an enriched analysis by examining the effect of age on healthcare expenditure by taking account of the time that precedes death. The researchers paid particular attention to the question of whether the positive relationship between age and healthcare expenditure mentioned above is not due to the increase of death rates with age. The results obtained in this category are not conclusive. In one of the first models, Zfeifel, Felder and Meiers (1999) used data of health insurers over the period 1981 to 1994, and found that the effect of age was not significant for the cohorts of the insured aged 65 and above – whereas the time that precedes death was significant. However, this result has not always been confirmed in other studies using other databases (Seshami and Gray, 2004a, 2004b; Stearns and Norton, 2004).

Part of the explanation for these contradictory results can be found in studies that compare the health expenditures of the people who die: not only with those of the average of the population but also with those of the same age who continue to live (*decedents-vs.-survivors studies*). This type of work (see, for example, Spillman and Lubitz, 2000) allows identification of various trends relating to the expenditure as a function of age, which, in regression models, are summarised by only one coefficient corresponding to the age variable. On the one hand, it can be observed that the costs of end of life increase with age for certain services, and decrease with age for others. In certain cases, the costs increase until a certain age and decrease afterwards. On the other hand, the healthcare costs of the people who survive generally increase with age. Since survivors constitute the majority of the population, we should expect to find that age would continue to be an explanatory variable of healthcare-related expenditure, even taking into account proximity to death.

But it is possible to carry the analysis further. Indeed, the results obtained by Lubitz and Riley (1993), for example (using data from the US), tend to show that the ratio of healthcare expenditure between survivors and decedents among people of the same age remained constant during a period of time of more than 15 years. This result is very important because it tends to show that the healthcare expenditure for these two categories results from the effect of common factors that are not related to age. One of the main factors is usually identified to be the evolution of medical technologies. Indeed, various econometric studies tend to show that medical technological advance and its diffusion is a leading explanatory factor of the increase in healthcare-related expenditure.

For instance, NBER studies analysed the growth of in-patient Medicare costs, and showed that the price that Medicare pays for admissions has been falling over time, but that the technological intensity of the treatment has been increasing. Since more-intensive technologies are reimbursed at a higher rate than less-intensive technologies, the growth of technology is at least partly responsible for the rise in Medicare costs. They also performed a detailed analysis of expenditure growth for acute myocardial infarctions (AMIs) in elderly Medicare beneficiaries. They showed that the diffusion of the technology resulted in the expansion of AMIs treatment in such a way that, as a result, spending on heart attacks rose by 4% annually – even though the price of the treatment itself was constant or even falling.

Thus, in turn, the argument according to which the costs of the last years of life replace the effect of age and constitute a determining factor of the increase of healthcare expenditure remains to be discussed. Also, a growing number of recent studies tend to completely substitute the end-of-life costs by other non-demographic factors in the explanation of the evolution of healthcare expenditure. Here again, the principal factor retained in these studies is the evolution of medical technology (Dormont et al., 2006). Thus, these studies confirm the findings of various earlier pieces of research, which were published during the last two decades.

To conclude, it is as worth noting, as it is obvious, that death as such is not an explanatory factor of the evolution of healthcare expenditure. The key factor is, rather, the medical care aiming at treating various pathologies, some of which lead to the death of the patients. From this point of view, the final years of life express a health status, and the end-of-life costs of these last years reflect the costs of healthcare, whereby some patients survive and others, unfortunately, die. It is thus natural to be interested in the first place in the medical practices and their evolution – particularly with regard to the nature of the medical technological advancement and its diffusion in the explanation of the evolution of healthcare expenditure.

Thus, this approach strongly suggests a change of paradigm, in which the ageing of the population and the increase in longevity have meagre impact on the growth of healthcare expenditure, whereas the larger part of this increase is due to the evolution of medical technology. This topic constitutes at the present time an extremely active field of research.

4.1.2 Need for cost-benefit analysis of medical care technology

The fact that technological change has accounted for the bulk of medical care expenditure over time does not imply that one should not necessarily adopt it within the healthcare sector. It is clear that in a very large number of cases, medical care technology brings benefits to society in terms of increased longevity, improved quality of life, and so on. However, the topic of the benefits that might be had from medical technology versus the costs, and what we can and want to afford, are quite relevant issues – and at the heart of the area of health economics of ageing. Indeed, the following question that underlies every medical care technology-use decision has to be addressed: *Is this technology worth its expense?*

To answer this question, and in order to be able to make welfare statements, benefits need to be compared with the costs of technology within a cost-benefit-analysis framework. The net value of medical technology change is the difference between the benefits and costs (*value for money*). A positive net value implies that the technological change is worth it in total.

This section discusses particular aspects of the cost-benefit-analysis methodology to assess the net value of medical technology for the elderly population.

Following the approach suggested by Cutler and McClellan (2001), it is preferable to measure costs and benefits at the disease level rather than at that of medical spending as a whole. This is particularly important in the context of an ageing population. Indeed, a substantial fraction of older adults with chronic diseases have multiple chronic conditions (Joyce et al., 2005), as well as some type of activity limitation.

The costs of technological change include the current and future costs of the conditions under study. The most important benefit of medical innovations, however, is the value of better health – longer life as well as improved quality of life. Following the consensus of the literature, the standard approach consists of measuring health using the quality-adjusted life year (QALY) approach.

As pointed out by Cutler and McClellan (2001), the central empirical issue in implementing this cost-benefit framework is determining the importance of medical technology changes for better health. A variety of factors – among them, medical technology – may influence health over time. One needs to isolate the medical contribution before it can be valued. Medical care is worth being valued if any of the additional increase in longevity results from improved medical care, or if medical care improves quality of life (Bunker et al., 1994).

The extent to which medical technology has contributed to better health can often be evaluated using clinical trial evidence or observational studies. The latter approach is particularly relevant in the context of an ageing population. Here we follow the arguments proposed by Avorn (2004: 102-125) in favour of observational studies. We adopt Avorn's arguments and apply them in the context of this chapter. Indeed, elderly patients may be excluded from clinical trials because of entry criteria that prohibit (for ethical reasons) enrolment by subjects older than a certain age.

In addition, the outcome measured in a clinical trial may not be the clinical problem it is designed to treat. For drugs to prevent cardiac disease, for example, regulation agencies will generally settle for evidence that the drug works better than a placebo in improving cholesterol levels on a blood test. It is assumed that improving that intermediary measure is likely to translate into real clinical benefits later. This policy is of course plausible. However, drugs to treat chronic diseases are destined to be taken for a lifetime. If improvement in the intermediary measure does not lead to the expected improvement in the real disease, in the context of long-term use of the drug, one is not able to discern it from the clinical trial. This provides the rationale for observational studies.

It is widely recognised, however, that observational studies may be seriously misleading if they are flawed, or if their results are over-interpreted. More work is needed to refine the ground rules of observational studies. Improvement of observational studies is a very active area of research in Epidemiology, Statistics and Econometrics. This refinement is possible, and needs to be further encouraged and financially supported in the future because of its importance in the context of an ageing population.

Parallel to the improvement in the methodology of observational studies, large longitudinal databases are needed in order to learn enough about risks and other outcomes from these studies. These databases should contain individual data concerning medical treatment and health status of patients followed over time.

4.2 Ageing and long-term care

Long-term care is care for chronic illness or disability instead of treatment of an acute illness. Long-term care services are needed by individuals with longstanding physical or mental disability, who have become dependent on assistance with basic activities of daily living (ADLs); many of these individuals are in the highest age groups of the population.

It is widely believed that long-term care expenditure growth will accelerate over the next 20 to 30 years, mainly as a result of larger numbers of older persons, and a steep increase in the numbers of the oldest-old. For this reason, long-term care issues are becoming increasingly important on the health and social policy agendas of developed countries.

A number of issues are related to long-term care. A detailed survey is provided by Norton (2000). This chapter is interested in two specific issues: informal care and the projection of long-term care expenditure.

Before turning to these issues, it might be useful at this stage to provide additional background on how the shift in age structure associated with population ageing and the increase in longevity in good health mentioned in the introduction impact long-term care.

As detailed earlier, the health of older persons typically deteriorates with increasing age, inducing greater demand for long-term care as the numbers of the oldest-old grow. On the other hand, more and more people in their fifties and sixties are likely to have surviving parents or other very old relatives.

An indicator of this trend is found in the parent-support ratio, which shows the number of persons aged 85 years or over in relation to those between 50 and 64 years. According to the UN (2002), at the world level, there were fewer than two persons aged 85 or older for every 100 persons aged 50-64 in 1950. By 2000, the ratio had increased to four per 100 – and it is projected to reach 11 by 2050.

The continuing increase of the parent-support ratio implies that more and more frequently the 'young-old' are expected to cope with the need to care for one or more 'oldest-old' and sometimes frail relatives. On the other hand, an important consequence of the fertility decline is a progressive reduction in the availability of kin to whom future generations of older persons may turn for support.

For this reason, concern is growing in most developed countries with regard to the long-term viability of intergenerational social support systems, which are crucial for the well-being of both the older and younger generations. This is especially true where provision of care within the family becomes more and more difficult (as family size decreases, and women, who are traditionally the main caregivers, increasingly engage in employment outside the home; UN, 2002).

4.2.1 Informal care

Informal care provided at home by family members, friends or voluntary organisations is the most important source of long-term care in all OECD countries (OECD, 2005). This is due in part to the fact that persons receiving institutional or home care are frequently required to contribute to funding long-term care – both by directly contributing to the public system, and in the form of substantial private cost-sharing. This can be a huge financial burden for the households concerned.

The level of informal care is a response to a number of factors, including the living arrangements of elderly people, the longevity of elderly husbands and wives, and trends in the labour-market participation of the groups in the labour force that are informal carers (OECD, 2005).

Presently, women over 45 years of age provide the bulk of informal care. Men are more likely to take over the role of caregiver for their spouses than in other family roles. Because more elderly people are living as couples and for a longer time, this has led to some increase in the participation of men in informal care-giving over time (Sundström et al., 2002). In fact, older persons with care needs who live together with their family or partner are more likely to receive informal help than those living alone (see Sundström, 1994). This implies that the growth in the number of older people living alone will itself increase the demand for formal care services in the future. Living alone has become a much more frequent experience for elderly people in the OECD area (OECD, 2005).

Informal care-giving is an indispensable component of care for older persons with long-term care needs. Given its importance on the health and policy agenda, more research is needed for this particular aspect of long-term care. Indeed, most of the past research done in this field focuses on the demand for nursing home care, as opposed to other forms of long-term care (see Norton, 2000, and the references contained therein). Specific aspects of informal care have not been studied sufficiently in the literature. The most important factors are health status, which determines need, and the out-of-pocket price relative to the price of close substitutes. This price depends itself on the availability of social insurance covering the different forms of long-term services. Depending on the institutional setting and the health status of individuals, informal care may be a substitute or a complement to other forms of formal long-term services (home care and institutions like nursing homes).

Other important factors are related to the topics described in the section on health-related issues in the economics of ageing (health and wealth, housing, living arrangements, financial status). In addition, a strong bequest motive may also influence the demand and supply of informal care. A parent who prefers care from his or her children to care from a nursing home may use the bequest strategically to induce the children to visit and provide help (Norton, 2000).

Another aspect of informal care that requires more attention is its impact on the use of different types of healthcare services (and hence on healthcare expenses). Development of informal care may reduce the use of some forms of healthcare services, which may reduce healthcare expenses.

For these reasons, more attention should be given to analyses of the factors that may influence the future development of informal care. From the scientific point of view, the main model of consumption and savings by the elderly (the life-cycle model) has to be extended or modified to take into account the close relationships among informal care, formal care, living arrangements, health, economic status and marital status. A number of recently published papers have devoted attention to some of these issues (see, for example, Van Houtven and Norton, 2004). However, surprisingly few economic studies focus on issues specific to non-US countries, or use non-US data. It is only recently that the situation has changed, with the availability of the data provided in the context of SHARE (Holly et al., 2008; Bonsang, 2009). These studies examine some of these issues in the context of European countries and compare the results from those obtained using US data (mainly those provided by the HRS study). Clearly, more research in the area of informal care should be encouraged.

4.2.2 Projection of long-term care expenditure

The projection of long-term care expenditure has become a topic of growing interest in European policy circles. This section surveys some of the aspects of this type of projection exercise.

As was already noted in Payne et al. (2007), the framework provided by Cutler and Sheiner (2001) may be useful for analysing aggregate expenditures in general. This section illustrates its usefulness for analysing long-term care expenditures. Cutler and Sheiner (2001) look at expenditures as the sum over all ages of the product of (1) the number of people alive in each age group, (2) the average health status at each age, and (3) the per capita medical spending conditional on health status, which also varies according to age.

Note that per capita long-term care spending grows exponentially with age, and that the bulk is concentrated on persons aged 80 years and older. This is due to the fact that the share of older persons with functional limitations increases exponentially with age and is highly concentrated in the oldest age groups. Since they have longer life expectancy, women are more likely to be in need of long-term care than men are (OECD, 2005).

Most demographically derived predictions of future health- and long-term care spending focus on the number of people that might need the care – assuming that their average health status remains constant and that per capita spending grows at a rate equal to the average rate in the past. However, models based only on demographics have serious limitations, as far as predictions are concerned.

The reliability of forecasts of future trends in life expectancy is crucial for demographic projections of ageing populations. This is particularly the case with regard to the remaining life expectancy at higher ages, as most of the additional years added to life in the past few decades of the 20th century were at higher ages (Cutler and Meara, 2001).

However, the rising number of the oldest-old is not the only factor that will drive the future demand for long-term care. It will also be influenced by the level of health and disability of future generations of elderly people.[5] Health status (defined in terms of disability or illness) may improve, together with a fall in the age-specific mortality rate. In that case, forecasts of future expenditures may be too high, as they take no account of the potential improvement in health status and the diminished need for healthcare that might be associated with a lower mortality rate. In contrast, if morbidity increases with mortality gains, the forecasts may be too low. This is an empirical issue.

The concept of *epidemiological transition,* first introduced by Omran (1971), provides a theoretical framework for conducting empirical work on issues related to trends in mortality and morbidity of the elderly. This concept characterises the ways in which social, environmental and health factors combine to change life expectancies, the most common causes of death, and the prevalence of disease among successive population cohorts.

Within this framework, three main concepts have been proposed in the literature in relation to the implications for morbidity at the end of life.

The concept of *compression of morbidity,* proposed by Fries (1980), predicts that, for the elderly, the period of morbidity preceding death would shrink over time. According to this concept, an upper bound is reached by both the process of natural ageing and human longevity. This would lead to a compression of morbidity, since poor health and disability tend to appear at later ages, on average.

---

5   Our exposition of the influence of the level of health and disability of future generations of elderly people on future demand for care follows, in part, Payne et al. (2007).

In contrast, Olshansky et al. (1991) presented a theory of *expansion of morbidity*, which associates extended longevity with extended morbidity. This is because elderly people may continue to become sick and disabled at the same ages as previously, leading to additional years of disability at the end of life.

The third possibility, known as *dynamic equilibrium*, postulates that both longevity and age of onset of poor health or disability would continue to increase, leading to deferral of disability.

A compression-of-morbidity scenario would predict a relatively short period of high expenditures in the period right before death, whereas an expansion scenario would predict a relatively long period of high expenditures associated with longer survival with a chronic disease. The evidence concerning recent trends in morbidity generally seems to favour the theory of compression. However, there is an ongoing scientific debate about which of these three concepts will prevail and can be extrapolated into the future.

The effect of trends in disability among older people on future demand for long-term care has therefore been at the centre of a number of recent studies. An overview of recent findings is reviewed in OECD (2005). Although a number of studies agree that favourable disability trends in the future could have a substantial mitigating effect on future demand for long-term care, the rapidly growing number of very old persons is nonetheless expected to increase substantially care needs – and related spending – in the future. Another possible explanation refers to the decomposition of health expenditures, formulated by Cutler and Sheiner and described above. The discussion regarding the morbidity and mortality trends we have just presented describes developments in age-specific health status, which is the second factor in this decomposition. To complete the analysis, evidence is required for the third factor: per capita medical spending conditional on health status. This would bring us to a discussion similar to the one we presented earlier about healthcare expenses, but this time it would have to do with the relationship between the evolution of the long-term care expenses and the development of medical technology intended to treat chronic diseases.

Another approach that is favoured in the medical and public health literature would involve measuring health status at each age at the disease level rather than the aggregate level. This approach is preferred because, from the point of view of public health, improvement at a disease level is more manageable than at the aggregate level. This is particularly important in the context of an ageing population. In fact, there is growing recognition that as the population ages, some diseases will make enormous demands on the healthcare system (McKee and Nolte, 2004; Rice and Fineman, 2004).

The disease-level approach requires estimates of the future age-specific incidence and prevalence of selected major diseases, and should take account of the ageing of the population. These estimates are important for evaluating the potential impact of public health interventions that prevent disease (or at least delay disease onset), and new therapies that may prolong survival.[6] The estimates are also important for predicting future needs of resources. Not all resources take the form of high-tech medical treatments. Others are related to institutional and home care needs.

Very few country studies have carried out projections of healthcare expenditures by disease. To the best of our knowledge, the studies by Goss (2008) on the Australian healthcare system, and Paccaud et al. (2006) on the Canton of Vaud in Switzerland, are among the exceptions.

Note that the recent work by Goldman et al. (2004) is concerned with the prediction of the implications for Medicare of future health and medical care spending of the elderly. They used an aggregate-level approach based on an aggregation of specific diseases. The authors explored how changes in medical technology, disease and disability would affect healthcare spending for the population aged 65 and older. Their key findings are that while medical innovations will result in better health and longer life, they will likely increase, not decrease, Medicare spending. This finding is consistent with the conclusions presented earlier.

Another very interesting aspect of the work by Goldman et al. (2004) concerns the potential effect of prevention. In fact, many Public Health researchers consider that prevention is a key instrument that should be used to reduce considerably the future demands on healthcare services that some diseases are expected to place in ageing populations. In contrast, Goldman et al. (2004) found that eliminating any one disease will not save a great deal of money, but obesity might prove to be an important exception. This result – except for the case of obesity – is controversial. This is, however, an empirical issue that needs to be examined further in other countries in Europe.

To summarise, a great deal of work must be done in this area in order to predict the future need of healthcare resources, and to evaluate the impact of preventive measures. As this stage, it is worth outlining the methodological aspects and the type of data used in Goldman et al. (2004). The authors developed a microsimulation demographic and economic model, the Future Elderly Model (FEM), to predict future costs and health status for the elderly. The model uses a representative sample of approximately 100,000 Medicare

---

6   For an interesting general methodology for projecting the incidence and prevalence of a chronic disease in ageing populations, see Brookmeyer and Gray (2000).

beneficiaries aged 65 and over, drawn from the Medicare Current Beneficiary Surveys (MCBS). MCBS are national surveys that ask Medicare beneficiaries about chronic conditions, use of healthcare services, medical care spending and health insurance coverage. Each beneficiary in the sample is linked to Medicare claim records in order to track actual medical care use and costs over time.

A similar exercise would be very useful to carry out for European countries – either at an individual level or at the level of a set of countries. The database used by Goldman et al. (2004) provides an indication of the type of data needed. It must combine, in a single database, information provided by health insurers (claims data) and medical data. This form of integrated database seems to be missing in most European countries. Given the usefulness of such databases, an effort should be made to motivate countries in Europe to construct them. A harmonisation of these databases would be necessary in order to be able to make useful comparisons between countries. This integrated form of database will provide an extremely useful infrastructure framework for carrying out a number of research studies relevant to health policies in Europe.

As far as long-term care is concerned, one of the high-priority research questions concerns future projections of care needs. These projections depend greatly upon projections of the disability trend. However, for many countries the disability trend is not clear. In addition, estimation methods for disability trends can vary substantially across countries (OECD, 2005). Potential factors that might influence disability rates among older persons are therefore not well understood at present. In particular, more work is needed to examine the impact in European countries of factors such as improvements in education, health-related behaviour, general improvement in socioeconomic status, and improvements in the treatment of chronic disease. These factors have been shown to drive trends in disability rates in the United States (Cutler, 2001a, 2001b).

5   Conclusion

This chapter outlines some of the important issues concerning ageing, health and long-term care. It identifies a number of topics of academic and policy relevance that deserve to be studied thoroughly in the context of European countries.

There is a need for more advanced research about health-related economic issues that are not directly linked to health- and long-term healthcare expenditures. Four of these topics are highlighted in this chapter: health and wealth, housing, living arrangements and financial status. Other topics, such as bequests, are also worth thorough study.

With respect to long-term care, informal care and its development is a fundamental topic that requires more research.

Projections of healthcare and long-term care are also important research topics from the perspectives of both academics and policymaking. Advances in this direction would require a convincing integration of some aspects highlighted in this chapter, which have been neglected so far in most of the published literature on these projections.

To be successful, future research on these topics requires the development of new research infrastructures – notably, the development of appropriate longitudinal databases, complementary to SHARE, allowing for cross-country comparisons at least among European countries.

These databases should contain individual data concerning medical treatment and health status, as well as socioeconomic variables of patients followed over time. Some countries have administrative data on health that could be linked to survey- or administrative data on socioeconomics – and this seems a promising development.

The results from future research on the topics mentioned above could be clarified by more effort spent in specifying policy issues. This is particularly the case for housing, long-term care and informal care. With regard to housing, one may look at solutions to increase housing-market flexibility. Also, investment in housing for the elderly may be advocated if one finds that, on the supply side, there is not enough suitable housing for the elderly.

Long-term care also poses questions with regard to housing. The very likely increase of out-of-pocket payments in this area may shift the status of housing in Europe. The elderly may not tend to think of the house as simply 'a place to live'. As in the US, housing wealth may also be considered in Europe as insurance against adverse economic outcomes in old age.

The increase of age-related chronic diseases is likely to have an impact, not only on the demand for long-term care and informal care, but also on the supply side. The labour supply of long-term care is likely to be insufficient in the near future. From the perspective of health policy, issues surrounding investments in this labour supply must be addressed.

Finally, we have seen that ageing as such is not a main driving force in health expenditure growth, but rather the diffusion of medical technology is. The continuing advance of cost-increasing technological change, particularly for treatment of conditions in the older population, raises economic and public health-related questions about the effects that these changes will entail. On average, medical technology has created value by producing benefits to society that are greater than their costs. But this does not imply that this is necessarily the case for any type of medical technology progress. In this regard, appropriate cost-benefit analysis is an important policy instrument. However, the effects on costs and the improvements to the quality of care will ultimately depend on the diffusion of new technologies to providers. Numerous studies (see, for instance, McClellan and Kessler, 2002; OECD, 2003) have shown that economic and regulatory incentives also affect the rate of adoption of new technologies. Thus, the study of patterns of diffusion, and of the incentives at work, is a major research issue with important policy implications. Indeed, if economic and regulatory incentives matter for technological change, then national health policies may have dynamic, long-term consequences on costs and improvement to quality of care – particularly for the treatment of conditions in the older population – that are far more important than their short-term, cross-sectional effects.

6   Current state of play of European research infrastructures and networks

A number of excellent researchers in Europe are working in the fields mentioned earlier in the chapter, and have carried out excellent conceptual and empirical studies. Their potential is also very high. However, they are located across several institutions in Europe. In addition, each institution is relatively small, so that the critical mass is seldom attained to carry out interdisciplinary research work of the quality and scope required by the development of the fields.

As far as data is concerned, several initiatives have recently attempted to remedy the situation. Among them is the Survey on Health, Retirement and Ageing in Europe (SHARE), which has made available a unique set of data that allows researchers to carry out interdisciplinary research of excellent quality. The prospective development of SHARE is also very encouraging, as it will provide a very rich longitudinal database for Europe.

7   Required research infrastructures, methodological innovations, data, networks and so forth, and consequences for research policy

Datasets complementary to SHARE and similar to some of those that have been developed in the US (and, in some cases, in Australia) are remarkable by their absence.

For instance, for many of the studies in the fields described above, there is a need for individual datasets containing, for the same individuals, data concerning medical treatments (hospital, ambulatory, drugs), long-term care (nursing home, home care), features of health insurance provided or bought by the individuals, and socioeconomic variables (similar to those collected in SHARE). The lack of this kind of data makes it impossible to carry out scientific studies on important issues, which may have a strong impact in formulating suitable social, economic, financial and health policies.

References

Adams P., M.D. Hurd, D. McFadden, A. Merrill and T. Ribiero (2003) 'Healthy, Wealthy and Wise? Tests for Direct Causal Paths between Health and Socioeconomic Status,' *Journal of Econometrics*, 112, 3-56.

Avorn, J. (2004) *Powerful Medicines: The Benefits, Risks, and Costs of Prescription Drugs* (New York: Alfred A. Knopf).

Bonsang, E. (2009)'Does Informal Care from Children to Their Elderly Parents Substitute for formal Care in Europe?', *Journal of Health Economics*, 28(1), 143-154.

Börsch-Supan, A. (1989) 'Household Dissolution and the Choice of Alternative Living Arrangements among Elderly Americans' in D. Wise (ed.) *The Economics of Aging* (Chicago: University of Chicago Press).

Börsch-Supan, A. (1990)'A Dynamic Analysis of Household Dissolution and Living Arrangements Transitions by Elderly Americans' in D. Wise (ed.) *Issues in the Economics of Aging* (Chicago: University of Chicago Press).

Börsch-Supan, A., D.L. McFadden and R. Schnabel (1996) 'Living Arrangements: Health and wealth Effects' in D. Wise (ed.) *Advances in the Economics of Aging* (Chicago: University of Chicago Press).

Brookmeyer, R. and S. Gray (2000) 'Methods for Projecting the Incidence and Prevalence of Chronic Diseases in Ageing Populations: Application to Alzheimer's Disease', *Statistics in Medicine*, 19, 1481-1493.

Bunker J.P., H.S. Frazier and F. Mosteller (1994) 'Improving Health: Measuring Effects of Medical Care', *Milbank Quarterly*, 72, 225-258.

Cutler, D.M. (2001a) 'The Reduction in Disability among the Elderly', *Proceedings of the National Academy of Sciences of the United States of America*, 98(12), Washington DC, 6546-6547.

Cutler, D.M. (2001b) 'Declining Disability among the Elderly', *Health Affairs*, 20(3), 11-27.

Cutler, D.M. and E. Meara (2001) *Changes in the Age Distribution of Mortality over the 20th Century*, NBER working paper 8556, Cambridge, MA.

Cutler, D.M. and L. Sheiner (2001) 'Demographics and Medical Care Spending: Standard and Non-Standard Effects' in A.J. Auerbach and R.D. Lee (eds) *Demographic Change and Fiscal Policy* (New York: Cambridge University Press).

Cutler, D.M and M. McClellan (2001) 'Is Technological Change in Medicine Worth It?', *Health Affairs*, 20(5), 11-29.

Dormont, B., M. Grignon and H. Huber (2006) 'Health Expenditure Growth: Reassessing the Threat of Ageing', *Health Economics*, 15, 947-963.

European Commission (EC) (2009) *2009 Ageing Report*, European Economy 2.

Feinstein, J. and D.L. McFadden (1989) 'Dynamics of Housing Demand by the Elderly' in D. Wise (ed.) *The Economics of Aging* (Chicago: University of Chicago Press).

Fries, J.F. (1989) 'The Compression of Morbidity: Near or Far?' *Milbank Quarterly* 67(2): 208-31.

Fukawa T. (1998) 'Health Expenditure of Deceased Elderly in Japan,' National Institute of Population and Social Security Research working paper, Tokyo, Japan.

Goldman D.P., P.G. Shekelle, J. Bhattacharya, M. Hurd, G.F. Joyce, D.N. Lakdawalla, D.H. Matsui, S.J. Newberry, C.W.A. Panis and B. Shang (2004) *Health Status and Medical Treatment of the Future Elderly*, Final Report, RAND Corporation.

Granger, C. W. J. (1969) 'Investigating Causal Relations by Econometric Models and Cross-Spectral Methods, ' Econometrica, 37, 424-438.

Grundy, E. and K. Glaser (1997) 'Trends in, and Transitions to, Institutional Residence among Older People in England and Wales, 1971-91', *Journal of Epidemiology and Community Health* 1997, 51, 531-540.

Holly, A., Th. M. Lufkin, E.C. Norton and C.H. Van Houtven (2008) 'Informal Care and Formal Home Care Use in Europe and the United States,' Institute of Health Economics and Management (IEMS) mimeo, University of Lausanne, Switzerland.

Huber M., R. Rodrigues, F. Hoffmann, K. Gasior and B. Marin (2009) *Facts and Figures on Long-term Care. Europe and North America.* European Centre for Social Welfare Policy and Research.

Jacobzone, S. (2000) 'Coping with Aging: International Challenges. What are the Implications of Greater Longevity and Declining Disability Levels?', *Health Affairs*, 213-225.

Jacobzone, S., E. Cambois and J.M. Robine (2000) 'Is the Health of Older Persons in OECD Countries Improving Fast Enough to Compensate for Population Ageing?', *OECD Economic Studies* 30 (Paris: OECD), pp. 149-90.

Joyce G.F., E.B. Keeler, B. Shang and D.P. Goldman (2005) 'The Lifetime Burden of Chronic Disease among the Elderly', *Health Affairs*, 19-29.

Lubitz, J.D. and G.F. Riley (1993) 'Trends in Medicare Payments in the Last Year of Life', *New England Journal of Medicine* 328(15), 1092-96.

McClellan, M.B. and D.P. Kessler (eds) (2002) *A Global Analysis of Technological Change in Health Care: Heart Attack* (Ann Arbor: University of Michigan Press).

McKee, M. and E. Nolte (2004) 'Responding to the Challenge of Chronic Diseases: Ideas from Europe', *Clinical Medicine* 4, 336-42.

Meer J., D.L. Miller and H.S. Rosen (2003) 'Exploring the Health-wealth Nexus,' *Journal of Health Economics*, 22, 713-730.

Merrill, S.R. (1984) 'Home Equity and the Elderly' in H. Aaron and G. Burtless (eds) *Retirement and Economic Behavior.* (Washington, DC: Brookings Institute).

Michaud, P.-C. and A. van Soest (2008) 'Health and Wealth of Elderly Couples: Causality Tests Using Dynamic Panel Data Models', *Journal of Health Economics*, 27, 1312-1325.

Mosley, W.H., J-L. Bobadilla and D.T. Jamison (1993) 'The Health Transition: Implications for Health Policy in Developing Countries' in D.T. Jamison, W.H. Mosley and J-L. Bobadilla (eds) *Disease Control Priorities in Developing Countries* (Washington, DC: World Bank; New York: Oxford University Press).

Norton, E.C. (2000) 'Long-term Care' in A.J. Culyer and J.P. Newhouse (eds) *Handbook of Health Economics, vol. 1* (Elsevier Science)

OECD (2003) *A Disease-based Comparison of Health Systems. What Is Best and at What Costs?* (Paris: OECD).

OECD (2005) *Long-term Care for Older People* (Paris: OECD).

Olshansky, S.J. and A.B. Ault (1986) 'The Fourth Stage of the Epidemiological Transition: The Age of Delayed Degenerative Diseases.' *Milbank Quarterly* 64(3), 355-91.

Omran, A.R. (1971) 'The Epidemiological Transition: A Theory of the Epidemiology of Population Change' *Milbank Quarterly* 49(4), 509-38.

Paccaud, F., I. Peytremann Bridevaux, M. Heiniger and L. Seematter-Bagnoud (2006) *Vieillissement : éléments pour une Politique de Santé Publique.* Un Rapport Préparé pour le Service de la Santé Publique du Canton de Vaud par l'Institut Universitaire de Médecine Sociale et Préventive. Lausanne: Institut Universitaire de Médecine Sociale et Préventive.

Payne, G., A. Laporte, R. Deber and P.C. Coyte (2007) 'Counting Backward to Health Care's Future: Using Time-to-Death Modeling to Identify Changes in End-of-Life Morbidity and the Impact of Aging on Health Care Expenditures,' *Milbank Quarterly* 85(2), 213-257.

Rice, D.P. and N. Fineman (2004) 'Economic Implications of Increased Longevity in the United States', *Annual Review of Public Health*, 25, 457-73.

Roos, N.P., P. Montgomery and L.L. Roos (1987) 'Health Care Utilization in the Years Prior to Death', *Milbank Quarterly* 65(2), 231-254.

Seshamani, M. and A. Gray (2004a) 'Ageing and Health-Care Expenditure: The Red Herring Argument Revisited', *Health Economics* 13(4), 303-14.

Seshamani, M. and A. Gray (2004b) 'A Longitudinal Study of the Effects of Age and Time to Death on Hospital Costs', *Journal of Health Economics* 23(2), 217-35.

Sims, C. A. (1972) 'Money, Income, and Causality, ' The American Economic Review, 62, 540-552.

Spillman, B.C. and J.D. Lubitz (2000) 'The Effect of Longevity on Spending for Acute and Long-Term Care,' *New England Journal of Medicine* 342(19), 1409-15.

Stearns, S.C. and E.C. Norton (2004) 'Time to Include Time to Death? The Future of Health Care Expenditure Predictions,' *Health Economics* 13(4), 315-27.

Sundström, G. (1994) 'Care by Families: An Overview of Trends', *Caring for Frail Elderly People: New Directions in Care; Social Policy Studies No. 14* (Paris: OECD).

Sundström, G., L. Johansson and L.B. Hassing (2002) 'The Shifting Balance of Long-Term Care in Sweden', *The Gerontologist*, 42, 350-355.

Tuljapurkar, S., N. Li and C. Boe (2000) 'A Universal Pattern of Mortality Decline in the G7 Countries', *Nature*, 405, 789-792.

United Nations (2002) *World Population Ageing: 1950-2050*, Department of Economic and Social Affairs, Population Division (New York: United Nations).

Van Houtven, C.H. and E.C. Norton, (2004) 'Informal Care and Health Care Use of Older Adults,' *Journal of Health Economics* 23(6), 1159 -1180.

Venti, S.F. and D.A. Wise (1989) 'Aging, Moving, and Housing Wealth' in D. Wise (ed.) *The Economics of Aging* (Chicago: University of Chicago Press).

Venti, S.F. and D.A. Wise (1990) 'But They Don't Want to Reduce Housing Equity' in D. Wise (ed.) *Issues in the Economics of Aging* (Chicago: University of Chicago Press).

Venti, S.F. and D.A. Wise (1991) 'Aging and the Income Value of Housing Wealth,' *Journal of Public Economics* 44, 371-97.

Venti, S.F. and D.A. Wise (2004) 'Aging and Housing Equity: Another Look' in D. Wise (ed.) *Perspectives on the Economics of Aging* (Chicago: University of Chicago Press).

Verbrugge, L.M. (1984) 'Longer Life but Worsening Health? Trends in Health and Mortality of Middle-aged and Older Persons.' *Milbank Quarterly/Health and Society* 62(3), 475-519.

Verbrugge, L.M. (1989) 'Recent, Present, and Future Health of American Adults' In L. Breslow, J.E. Fielding and L.B. Lave (eds) *Annual Review of Public Health*, 10 (Palo Alto: Annual Reviews).

Ynesta, I. (2008) 'Households' Wealth Composition across OECD Countries and Financial Risks Borne by Households', *Financial Market Trends*, OECD.

Zweifel, P., S. Felder and M. Meiers (1999) 'Ageing of Population and Health Care Expenditure: A Red Herring?' *Health Economics* 8(6), 485-96.

Old Age, Health and Long-term Care

Comments by Lou Spoor

In this chapter, the author offers a broad and useful overview on ageing and health issues and on the research related to these issues. This part of the discussion addresses some questions from the angle of policymakers. Firstly, the chapter, like most of the literature, focuses mainly on the dynamics of the demand side of healthcare services. However, as we may have serious doubts whether the supply of health services will be able to meet the demand of ageing populations over the next decades, factors driving the supply side need more attention. Secondly, improvements will have to be made at the institutional level – including the financial framework – to alleviate this potential mismatch between demand and supply. Scientific research will receive greater recognition if it is able to assist policymakers in handling these issues.

One of the relevant issues on the supply side that was not mentioned in the chapter is whether housing markets will be able to provide sufficiently adequate housing for the elderly. The literature has established the fact that elderly people live in their own homes up to very high age. However, most of these homes were never designed to meet the specific housing requirements of very elderly people. This raises several questions. The first is whether the ageing of the population will result in changing preferences of the elderly for housing. A second question is to what extent we will have to increase the supply of home care. Finally, a finance issue is involved: can we expect homeowners to use their housing equity as precautionary savings for the cost of long-term care, as we may expect higher out-of-pocket expenses in response to public budget deficits resulting from ageing? Among the institutional issues involved are the organisation and finance of long-term care (including tax incentives) and the flexibility of the housing market.

A second issue, as the author states, arises from technology being the main driver of the growth of medical care expenditure. However, as technology is not an exogenous variable only, it is relevant to look into the behaviour of medical staff. Diagnostics, surgery and therapies can be more or less aggressive, as is shown by significant differences within the medical practice. It may therefore be worthwhile to test the hypothesis that – at the same technology level and at the same quality level – healthcare systems may differ significantly in cost levels. If so, the behaviour of medical staff as well as financial incentives driving this behaviour should be addressed.

Thirdly, dementia and the supply of informal care offer serious challenges in the next decades. The chapter rightly puts emphasis on informal care as an indispensable component of elderly care. Again, the supply side deserves more attention here. The EU faces a doubling or even tripling of the current number of 6.1 million cases of Alzheimer's disease over the next four decades. Supply of long-term care services will almost certainly not be able to meet demand, as the labour force will decline. One of the challenging policy questions arising from this is whether current public subsidies and tax incentives should be shifted from childcare to informal elderly care. Further research on these issues may help policymakers in making the right choices for the long term.

My general suggestion, which arises from the topics mentioned above, is that the research programme on health and ageing will be more effective if it connects the research roadmap more closely to the policy agenda. As supply issues and institutional issues – housing for elderly, healthcare labour supply, informal care, finance, tax incentives, medical staff remuneration incentives – raise important policy questions, research may help in answering these questions by focusing research more closely on the mechanisms driving these topics.

Old Age, Health and Long-term Care

Comments by Eddy van Doorslaer

This chapter contains a very useful review – often inspired by NBER-derived publications – of some selected topical issues in the Health Economics of Ageing. In my view, it represents a serious attempt to give economic research on these topics the attention it deserves in Europe, and also provides some useful conclusions and suggestions for further research. My comments focus mainly on four issues:

(a)   Lack of a generally accepted theoretical framework for behaviour,
(b)   The still-lacking evidence base in some important domains,
(c)   Some suggestions for data collection and analysis, and
(d)   European research priorities.

Each of these will be discussed in turn.

First, in order to avoid the 'quest without a compass' problem, a growing consensus of researchers is asserting that there is an urgent need for a better theoretical model on individual behaviour with respect to economic choices affecting health, work and income over the lifetime. Neither the standard life-cycle model, nor the much-quoted Grossman model of human health capital are at present sufficiently rich to provide proper guidance for the empirical work. Fortunately, promising attempts are underway to fill this perceived gap (Galama et al., 2009) and to provide a sound theoretical basis for new empirical testing.

Second, the chapter also raises issues about (a) trends in mortality, morbidity and expenditure and (b) the important and often-studied nexus between wealth and health. On mortality, there is generally unanimous agreement that it is still declining in most of the Western world. The trends on morbidity, conditional on mortality, are less clear. There are some US claims of evidence of a decline in chronic disability among the elderly (see Crimmins, 2004; Manton,

2008), but these have not (yet) been confirmed by OECD trend analyses for many European countries (see Lafortune, Balestat et al., 2007). A similar European examination of these trends in mortality and morbidity, and their most likely implications for future expenditure on LTC and medical care is urgently required. With respect to the much-studied wealth-health linkages, evidence is mounting on both sides of the Atlantic now to suggest two important causal directions. At younger ages, much of the causality runs from wealth to health (mainly through education), while at more mature ages, there is also an important reverse causality running from health to wealth (mainly through reduced work participation; see Smith, 2007; Van Kippersluis et al., 2010). Again, if properly confirmed for other European contexts, these findings of cumulative (dis)advantage over the life course have major implications for the appropriate design of social and health policies.

Third, it is clear that the pertinent research questions raised in this chapter by Holly can only be answered if appropriate data are being collected and made accessible to the research community. The chapter mentions the Medicare Current Beneficiary Surveys (MCBS) data used for the construction of the US Future Elderly Model (FEM) to predict (or microsimulate) future costs and health status for the elderly. While a similar European dataset does not exist (at least not on a national scale; the MCBS uses a representative sample of approximately 100,000 Medicare beneficiaries aged 65 and over), the collection of longitudinal data in several European countries in SHARE offers other advantages. Notably, the variation in institutional arrangements across countries can usefully be exploited to learn from policies adopted in some countries to inform others contemplating similar moves. Moreover, some European countries, notably the Scandinavian countries (but more recently also the Netherlands), have substantially augmented the usefulness of their survey data by linking them to administrative data on mortality, income, work status, and health and LTC use and expenditures. Future combination of these three strengths (longitudinal follow-up, institutional variations, and administrative record linkage) could turn the European situation into one of advantage rather than disadvantage, compared to the US.

Finally, and derived from the above, I see three main research priorities emerging from this chapter: (i) What is fact and what is fiction concerning future expected trends in longevity, health and expenditure? (ii) What is an appropriate European model for LTC finance and organisation, and what are the public and private responsibilities in each? (iii) What policy measures and incentives (such as cost sharing or personal care budgets) work best to obtain a welfare-improving LTC system able to sustain consumption and care in old age? Although it might be said that the health economics of ageing is still in its infancy in Europe, it must also be agreed that with regard to current trends, rapid growth of interest and activity in this area are inevitable and can be expected in the very near future.

References

Crimmins, E.M. (2004) 'Trends in the Health of the Elderly', *Annual Review of Public Health*, 25, 79-98.

Fries, J.F. (1989) 'The Compression of Morbidity: Near or Far?' *Milbank Quarterly* 67(2), 208-31.

Galama, T.J., A. Kapteyn, F. Fonseca and P.C. Michaud (2009) 'Grossman's Health Threshold and Retirement', RAND working paper 658.

Lafortune, G., G. Balestat and the Disability Study Expert Group Members (2007) 'Trends in Severe Disability among Elderly people: Assessing the Evidence in 12 OECD Countries and the Future Implications', OECD Health working paper 26.

Manton, K.G. (2008) 'Recent Declines in Chronic Disability in the Elderly US Population: Risk Factors and Future Dynamics', *Annual Review of Public Health*, 29, 91-113.

Smith, J.P. (2007) 'The Impact of Socioeconomic Status on Health over the Life Course', *Journal of Human Resources*, 42(4), 739-764.

Van Kippersluis, H., O. O'Donnell, E. van Doorslaer and T. van Ourti (2010) 'Socioeconomic Differences in Health over the Life Cycle in an Egalitarian Country', *Social Science and Medicine*, 70, 428-438.

Part III:
Labour Markets and Older Workers

9

The Demand for Older Workers

Amilcar Moreira / Brendan Whelan / Asghar Zaidi

1   Policy questions

All advanced economies are experiencing rapid population ageing caused by
the combination of an increase in longevity and a sharp drop in fertility rates.
This phenomenon is reducing the share of the working-age population in many
of these countries, and is thus likely to exert negative pressure on economic
growth. The rising old-age dependency ratio also poses serious problems for
the public finances through growing pressures for expenditure on pensions,
health and long-term care. Unless participation rates rise markedly, labour force
growth will fall in the majority of European countries as well as in Japan (see
OECD, 2006). If current trends of early retirement do not improve considerably,
the exits from the labour force will begin to outnumber labour market entrants,
imposing heavy costs of adjustment on employers.

Raising labour force participation among older workers has a vital role to
play in coping with these challenges in at least three different ways:
(a)   by boosting labour force growth, thereby offsetting the negative impact
that the shrinking working-age population has on economic growth;
(b)   by reducing pressures on the public finances through later retirement and
increased tax and pension contributions; and
(c)   by slowing down the pace at which employers will have to adjust the
composition of their workforce.
Improved participation by older workers can be brought about by a variety
of actions targeted at both the supply and demand sides of the labour market.
However, perhaps reflecting the lack of evidence on the role of labour demand

– or the relatively buoyant state of European economies until recent times – many of the strategies to promote an active ageing agenda at the European level have concentrated on the supply side of the labour market.[1]

Consider, for instance, the European Commission's Communication entitled 'Towards a Europe for All Ages – Promoting Prosperity and Intergenerational Solidarity' (Commission of the European Communities, 1999), which is the first concerted effort to set the policy agenda needed to deal with demographic ageing in Europe. This document argues for the need to introduce active ageing policies as a means to secure high levels of labour supply in European labour markets (1999: 8-11).

Similarly, a later Commission Communication entitled 'Increasing the Employment of Older Workers and Delaying the Exit from the Labor Market' (Commission of the European Communities, 2004) states: *'For the economy as a whole, the increase in participation and employment rates of older workers are crucial for using the full potential of labour supply to sustain economic growth, tax revenues and social protection systems, including adequate pensions, in the face of expected reductions in the population of working age'* (2004: 3).

An overwhelming concentration on supply-side measures is also evident in policies advocated by the European Commission to reach the targets set in the Stockholm and Barcelona Declarations – having 50% of all individuals aged between 55 and 64 in employment, and increasing the effective age of labour market exit by five years by 2010. From a total of ten lines of actions proposed, only three are targeted specifically at increasing the demand for older workers in the labour market (2004: 5, 14):

- Greater awareness about ageing among employers so as to tackle age discrimination and promote the benefits of an age-diverse workforce to both individuals and firms.
- Reconsideration of the weight of seniority elements as part of pay, with a view to bringing pay more in line with productivity and performance.
- Incentives in collective agreements to recruit older unemployed persons.

Notwithstanding the references to these areas, the EU agenda does not seem to consider the full role that demand-side factors can play in promoting the participation of older people in the labour market and expanding working careers. For instance, little or no consideration is given to the effects, either positive or negative, of employment protection legislation or to reductions in wage costs as a way of encouraging firms to retain older workers.

---

1   Van Soest (chapter 11) highlights how economic, psychological and social factors as well as health play a role in determining the labour-supply decisions of older workers.

It can therefore be argued that a more effective approach for improving the participation of older workers in the labour market will require greater focus on improving the demand for older workers. Unfortunately, little is known about the factors that affect the demand for older workers in Europe. This chapter therefore aims to fill this information gap. First, it provides a survey of the current (European-based) literature on this topic;[2] second, it identifies the main areas of development of research on this topic.

The chapter is structured as follows. Section 2 reviews the existing literature on the topic. Whereas the first part of this section focuses on the key theoretical models that try to explain the demand for older workers, the remaining sub-sections focus on the empirical evidence on a number of factors that condition the demand for older workers. These factors include the impact of wages and productivity variations, macroeconomic conditions, discrimination in recruitment processes and labour market regulations. Section 3 identifies the main gaps in the literature on this topic, and specifies the key areas of work required at both the theoretical and empirical level. However, acknowledging that for this to be achieved we also need to improve the infrastructures that support the development of this research, the final sections survey the current state of the research infrastructures and networks on this topic (section 4) and recommend possible ways of improving them (section 5).

2   Major progress in understanding

This section surveys the literature on the demand for older workers in the labour market in Europe. The review of the key theoretical frameworks in the literature on this topic will be followed by a focus on the empirical literature that describes the factors that influence the demand for older workers.

2.1   *Key analytical frameworks for understanding the demand for older workers*

Surveying the relevant literature, we can identify two main approaches for explaining the demand for older workers in the labour market. The first approach focuses on how the age structure of the labour force affects labour market dynamics. Traditionally, this has been analysed using the so-called 'generation

---

2   Although the main focus has been to search for and report on the European-based literature, we will, whenever necessary, complement our survey with US-based evidence.

crowding' models (Zimmermann, 1991). These models try to explain the labour market participation of a given age-cohort in terms of its relative share in the labour force. These models rest on two basic premises. Firstly, compared with the more homogeneous capital production factor, the labour force can be differentiated by reference to workers' age. Second, a worker from a given age group cannot be perfectly replaced by an individual from a different age group (Mosca, 2009: 373-4).

Within the context of demographic change, these models predict that an increase in the availability of older workers will reduce the price that employers are willing to pay for this type of production factor. This will translate into either higher unemployment or lower wages for older workers. The degree to which this occurs depends on the substitutability between young and older workers, and on the institutional framework that regulates the labour market. If the labour market were to work without significant frictions, the adjustment would be done through the wage mechanism. If, however, rigidities exist in the labour market (such as wage-setting institutions, trade unions or employment protection laws), which limit the market to adjust through the wage mechanisms, we should expect higher unemployment in the (larger) older cohort (Mosca, 2009: 374).

Although they make no specific predictions regarding the demand for older workers, Hetze and Ochsen (2005) provide an interesting alternative approach for analysing how changes in the age structure of the labour force influence labour force dynamics. The authors try to upgrade traditional equilibrium models by incorporating the impact of age-based heterogeneity in the labour force on the formation of search equilibrium.[3] They start with the assumption that, in the decision to allocate a given job to an older or a younger worker, employers try to maximise their revenue. This revenue is determined, on the one hand, by the productivity differential between young and older workers, and on the other, by the perceived "separation risk" for both age groups. Separation risk, here, refers to the probability that younger workers will engage in job search behaviour and leave for alternative employment, or that older workers will leave to go into retirement (2005: 5).

Using these concepts, the authors show how demographic ageing affects both job creation and job destruction. If older workers are less productive than their younger counterparts, then demographic ageing will lead to a decrease

---

3   Another relevant effort here is the attempt by Christiaans (2003) to up-grade neo-classical labour demand models by introducing the assumption of age-based heterogeneity in the labour force. However, as this is not yet ready to be empirically applied (see Christiaans 2003: 2), we decided not to include it in this discussion.

in the number of vacancies, as these would reduce the company's revenues. However, if older employees are more productive than younger workers, or are willing to stay for longer with firms, then more vacancies will be created. With regards to job destruction, this is ultimately influenced by the variation in the separation risk displayed by the older generation. If older workers display a lower separation risk, then job destruction will be reduced. If, on the other hand, the separation risk for the older age group is high, then job destruction will increase (2005: 8).

This leads authors to anticipate four possible outcomes of demographic ageing (2005: 8-9):

a) When the separation risk for older workers increases, and this is not compensated by increases in productivity, then unemployment will increase.

b) When the separation risk and the productivity of the older cohort is lower than that of the younger cohort, the effect on unemployment is ambiguous.

c) When the separation risk of older workers is so low that it outweighs their productivity disadvantage with regards to younger workers, job creation will increase and job destruction will decrease, thus reducing unemployment.

d) When older workers are more productive than their younger counterparts, but also display a higher separation risk, then both job creation and job destruction increase, and the effect on unemployment is ambiguous.

The second approach for analysing the demand for older workers focuses on the hiring decisions of individual firms. Probably the best-known framework here is Lazear's 'delayed compensation contracts' model (1979). In trying to explain the existence of a mandatory retirement requirement, Lazear argues that a work contract in which the worker is paid below the value of his/her marginal product at younger ages, and above at later ages, has advantages for both workers and employers. For workers, this type of contract will increase their lifetime wealth. Employers, although they are forced to bear the higher fixed costs associated with delayed compensation, gain from improvements in performance and stronger employee commitment that are induced by the workers' fear of losing delayed compensation.[4] In this context, firms tend to avoid hiring older workers, as this will reduce the possible benefits of delayed compensation. Moreover, should they decide to hire older workers, employers

---

4   In this context, mandatory retirement is required as a way of terminating the contractual work, as the worker would not voluntarily retire in a context where the current wage is higher than his/her reservation wage (Lazear, 1979: 1265).

will offer them a lower wage than that paid to younger workers (Lazear, 1979: 1263-5, 1277; Daniel and Heywood 2007: 37-8).

Somewhat complementing Lazear's model, Hutchens (1988) highlights the role of training in reducing the incentive to hire older workers. Hutchens argues that firms that provide (specific) on-the-job training will bear a given cost, which is fixed for every new hiring. In order to reduce these costs, firms will attempt to reduce employee turnover by investing in durable relationships. The implication of this is that companies will either favour younger workers in the recruitment process or, when applicants are equally qualified, offer reduced wages to older workers (1988: 89-90).

2.2   *Empirical evidence on factors that impact on the demand for older workers*

As mentioned earlier, this section surveys the empirical literature on the factors that influence the demand for older workers. Reflecting on the existing literature, the following subsections will survey the available evidence on the following topics:

a)   The impact of age, productivity and wages on the demand for older workers.
b)   The effect of macroeconomic conditions on the demand for older workers.
c)   Discrimination against older people in recruitment processes, and
d)   The effect of labour market regulations on the demand for older workers.

2.2.1 Age, productivity and wages

An OECD study shows that there is a negative relation between (seniority) wages and the employment of older workers. For instance, as can be seen from Figure 9.1, there is a negative relationship between retention rates that occurs for men aged 55-59 relative to men aged 45-49 and the wages of older men relative to those of younger men. In the same vein, hiring rates of older workers are also negatively correlated with seniority wages (see Figure 9.2). Thus, all else equal, this cross-country evidence shows that employers are more likely to hire and retain older (male) workers in countries where wages rise less steeply with age (OECD, 2006: 69).

Figure 9.1: Seniority wages and retention rate of older workers

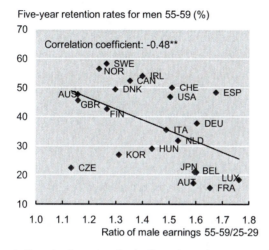

Five-year retention rates for men 55-59 (%)

Notes:
- a) The retention rate refers to the estimated proportion of all employees in 1999 that were still with the same employer in 2004.
- b) The earnings data refer to full-time workers only for various years over the period 1998-2003.

Source: OECD (2006: 69)

Figure 9.2: Seniority wages and hiring rate of older workers

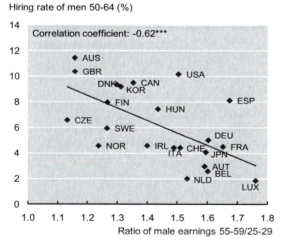

Hiring rate of men 50-64 (%)

Notes:
- a) The hiring rate refers to the ratio of employees with less than one year of tenure to all employees. The data refer to 2004.
- b) The earnings data refer to full-time workers only for various years over the period 1998-2003.

Source: OECD (2006: 69)

In addition, there are also country-specific studies that point to a negative relationship between seniority wages and employment outcomes for older workers. Hirsch et al. (2000) report that, for the US, the overall employment shares of older workers (within narrowly defined occupations) and the hiring rate of older workers tend to be negatively associated with the wage premium on experience. Similarly, Aubert (2005) also reports a negative association between the employment share of older workers and their wages relative to younger workers – especially after the age of 55. A similar negative correlation exists for hiring, except in the case of highly-qualified workers.

An important question is therefore how the patterns of wage and productivity evolve across the life-cycle. In most countries, cross-sectional data on earnings by age show a hump-shaped profile, especially for men (OECD, 2006: 66). This profile may reflect the fact that the productivity of workers initially increases with their work experience but then flattens off or declines after a given age. However, in some countries earnings rise more steeply with age or show little tendency to decline in the older age groups (see Table 9.1). Whilst, in principle, this should reflect the increasing productivity of workers as they gain more work experience, it is possible that this might also be linked to other factors.

For instance, in some European countries, these age-earning patterns reflect explicit seniority wage-setting arrangements, usually in the form of collective agreements. Such practices of seniority wages are present in the wage-setting practices of Austria, Belgium, France, the Netherlands and Spain. In Spain, for instance, about 80% of collective agreements include seniority wage clauses.

Table 9.1: Evidence of seniority wages in OECD countries

Country	What evidence is there of seniority wages?
Austria	Wages for men after the age of 40 continue to rise steeply
Belgium	Seniority wages common for non-manual workers
Finland	Non-wage costs rise with age (e.g. for disability insurance)
France	Wages rise steeply with age
Italy	Some evidence that seniority wages reduce retention of older workers
Japan	Wages rise steeply with age
Korea	Wages rise steeply with age
Luxembourg	Wages rise steeply with age
Spain	Seniority wages still important, despite some decline. High non-wage costs for part-time work
Switzerland	Both wages and non-wage costs rise steeply with age
United States	Non-wage costs rise with age (e.g. for health insurance)

Source: OECD (2006: 64, Table 3.3)

In some other countries, the share of non-wage costs also rises with age, and so the rise in total labour costs with age is steeper than in wage costs alone. For example, in Finland and Switzerland, social security contributions in the occupational schemes rise with workers' age. In the US, costs that are associated with health insurance and defined-benefit pension plans may result in non-wage costs that increase significantly with age.

In many countries, pension claims of tenured civil servants are linked to final salaries rather than the wage-base wage over the entire employment spell (for example, in Germany and the UK). Thus, there is an incentive to maximise pension claims by pushing up pay with seniority.

The earlier evidence of a steep rise in wages with age raises questions regarding the relationship between wage and productivity. Some of the studies directly contrast the pattern of productivity across a typical career with the pattern of lifetime earnings (for example, Medoff and Abraham,1981; Oliviera, Cohn and Kiker, 1989; Remery et al., 2001). Others provide evidence on the association between age and productivity (see Table 9.2 for a list of studies and their findings). The majority of these studies conclude that the age-wage gradient is different from the age-productivity gradient, which suggests a certain discrepancy in the link between wage and productivity at older ages.

Another form of evidence is drawn from a number of recent studies using matched employer-employee data. Crépon and Aubert (2003) found that the productivity of workers in France declines after the age of 55, while earnings continue to rise, although the difference in profiles is not statistically significant. Based on data for the manufacturing sector in the United States, Hellerstein and Neumark (2004) also found that workers aged 55 and over are less productive than workers in either the 35-54 age group or the younger age group (less than 35). At the same time, they found that the lower productivity of older workers is not matched by lower earnings.

A useful summary of the literature relating age to productivity is provided in Skirbekk (2003). This study reports that there is a range of evidence suggesting a decline in several aspects of physical and mental functionings from around the age of 50. This decline is often progressive, with substantial variation across individuals. The decline is particularly uneven across different types of mental ability, with stronger reductions in 'fluid' intelligence (such as reasoning power or speed of processing) than in 'crystallised' intelligence – as reflected, for instance, in verbal and communications skills (2003: 5-6).

Table 9.2:   Review of empirical studies on age and productivity

Study	Group analysed	Findings
Waldman and Aviolo (1986)	Meta-analysis of 40 studies	No clear association between age and performance. Results vary and also depend on the selected performance indicator: on average, a positive correlation in the case of indicator type 1 (peer ratings), and negative in cases of indicator type 3 (supervisors' ratings).
McEvoy and Cascio (1989)	Meta-analysis of 96 studies	No clear association between age and performance (results vary in individual studies). This conclusion holds when a separate division of studies is made according to performance and type of work (professional vs. non-professional).
Bureau of Labour Statistics (1957)	Employees in large plants in men's footwear and household furniture industries	Decline in output per hour in older people, starting noticeably after about 45 years of age.
Kutscher and Walker (1960)	Office workers, USA	Very little difference in output per hour between age groups.
Walker (1964)	Mail sorters, USA	Very little difference in output per hour between age groups.
Stephan and Levin (1988)	Researchers within the physics, geology, physiology and bio-chemistry sectors	Negative association between productivity and age.
Oster and Hamermesh (1998)	Researchers in Economics	Negative association between productivity and age.
Miller (1999)	Artists, painters, musicians and writers	The peak ages for productivity are in the 30s and 40s.
Oliviera, Cohn and Kiker (1989)	Self-employed in the US (compared with employees)	Productivity increases strongly for younger people but declines in older people (parabolic). Earnings tend to stabilize in older people for employees in similar functions.
Medoff and Abraham (1980, 1981)	White-collar employees in a number of large American corporations	Seniority is either unrelated or negatively associated with performance evaluations. So wage growth with experience cannot be explained by growing productivity.
Flabbi and Ichino (2001)	Employees in a large Italian firm	Seniority is unrelated to performance evaluations.
Remery et al. (2001)	Dutch private and public companies/ organizations	Companies expect increases in wage costs because of ageing of their workforce, but no increases in productivity.

Study	Group analysed	Findings
Gelderblom, de Koning and Kroes (2004)	Dutch private and public companies/ organizations	A large group of managers and HR managers do not see differences in the average productivity of those 55 and older, compared to younger workers.
Gelderblom and de Koning (1996)	Civil servants in the Netherlands	Older workers have more problems with work pressure, but perform well on social skills.
Haegeland and Klette (1999)	Manufacturing companies in Norway	Productivity decline for those with more than 15 years experience (late 30s and over).
Hellerstein et al. (1999)	Manufacturing companies in the US	Increase/decrease over life cycle according to model specification.
Ilmakunnas et al. (1999)	Manufacturing companies in Finland	Peak at age 40, declining thereafter.
Crépon et al. (2002)	Manufacturing and non-manufacturing companies in France	Peak during the age 25-34, and lowest for those over the age of 50.
Gelderblom and de Koning (2002a)	Manufacturing and non-manufacturing companies in the Netherlands	Productivity is rising until somewhere between 40-50, and declining afterwards. Productivity is relatively high compared to wages at middle ages. The younger and older have less favourable productivity-wage ratios.
Gelderblom, de Koning and Kroes (2004)	Manufacturing and non-manufacturing companies in the Netherlands	Productivity is rising until the age of around 50, after which a strong decline occurs.

Sources: European Commission (2006: 77-79), which is based on Gelderblom and Vos (1999) and Skirbekk (2003).

Overall, the evidence suggests that individual productivity does decline in some dimensions with age. However, this decline can be partly compensated for by experience, personal aids and suitable workplace adjustments. Volkoff et al. (2000) argue that older workers can often rely on their professional experience to adapt to, and compensate for, the decline in physical and mental ability – especially when assisted by suitable workplace adjustments. Based on analysis of a wide range of studies, Warr (1993) takes the view that the age-related decline in job performance is small, and states that, *'there is no significant difference between the job performance of older and younger workers'* (1993: 238).

2.2.2 The effect of business-cycle and technological innovation

Besides the payroll costs that this might imply, a firm's decision to hire older workers is influenced by the stage of the business cycle. A usual approach to capture the state of the business cycle is to look at the level of unemployment. For instance, using data from the Spanish Current Population Surveys, Blanco and De La Rica (2002) found that the higher the unemployment rate the lower the probability of re-employment for older workers. Even more interestingly, their study also suggests that older unemployed persons encounter greater difficulties than their younger counterparts in re-entering jobs – even when the economy is growing (2002: 19-20).

A more accurate way of analysing this relationship, in the sense that it captures the technological innovation component associated with periods of growth, is to look at variations in productivity across the business cycle. Aghion and Howit (1994) suggest that the impact of productivity growth on labour demand is the product of two competing effects. If a company is able to integrate a new technology into an existing job, then productivity growth can increase the returns from creating a new job, thereby creating an incentive for further hiring – the capitalisation effect. On the other side, if a firm needs to shed old jobs in order to introduce a new technology, this can have an overall negative effect on the demand for labour – the creative-destruction effect[5] (see Aghion and Howit, 1994: 478, and Pissarides and Valanti, 2007: 608).

Langot and Moreno-Galvis (2008) argue that, because the horizon to recover the investment in new technologies is shorter for older workers, this reduces the firms' incentives to upgrade the technology employed in the jobs they hold – which makes these older works more vulnerable to creative-destruction effects. However, this effect is moderated by the level of productivity of older workers. Thus, even if the horizon is shorter, it is more profitable for a firm to train a highly productive worker than to replace him/her (2008: 4-6).[6]

---

5   This must be understood in the context of Aghion and Howit's (1994) criticism of the view – espoused by Pissarides (2000) and others – that a higher productivity rate raises the return on the creation of new vacancies and so induces a faster exit rate from unemployment. According to Aghion and Howit (1994: 478), this model neglects the reallocation processes that are associated with the growth in productivity, which sometimes involves the creation of new jobs at the expense of the old, less productive, jobs.

6   However, we should be cautious in accepting Langot and Moreno-Galvis's (2008) argument. While attempting to test their model, they found that the predicted elasticities of the relation between the growth of total factor productivity and the level of employment for young and older workers fall short of the elasticities produced using labour-market data from OECD countries (2008: 22-5).

2.2.3 Age discrimination in recruitment

An obvious explanation for the lack of demand for older people in the labour market is the existence of discrimination in recruitment processes. At the theoretical level, this has been explained by two distinct analytical frameworks: Becker's hypothesis of a 'taste for discrimination' and the theories of 'statistical discrimination' proposed by Phelps (1972) and Arrow (1973). In his seminal work on discrimination in the labour market, Becker (1957) assumes that employers have a preference for a certain category of worker – that is, they have a taste for discrimination. He further assumes that the employer's utility is an increasing function of both profits and the number of workers from the favoured group. If the wages of workers from the favoured (F) and disfavoured (D) groups are the same, the employer will tend to hire more individuals from the favoured group[7] (see Johnes and Sapsford, 1996: 12).

Phelps (1972) and Arrow (1973) start from the assumption that discrimination in the labour market does not necessarily arise from an employer's taste for discrimination. Instead, discriminatory practices derive from the fact that it is difficult and costly to obtain accurate information about the likely differences in productivity between two prospective workers. In order to avoid these costs, employers make recruitment decisions by using a rule of thumb, based on assumptions about the productivity of certain categories of worker, such as those of a certain age, gender or racial group. In this sense, discriminatory practices emerge as rational ways of maximising profits in hiring (Arrow, 1973: 23-32; Baumle and Fosset, 2005: 1259-60).

There seems to be relatively little evidence in EU countries that would allow one to assess the validity of these two standpoints. To a certain degree, this reflects the significant shortcomings of the evidence on the extent and impact of the discrimination against older workers in recruitment processes in Europe. The existing literature has focused on two key themes. The first concerns the prevalence of discriminatory attitudes toward older workers in the labour market. For instance, a special Eurobarometer survey (European Commission, 2007) shows that almost half of the surveyed population feels that a candidate's age, together with his/her appearance and the existence of a disability, will put a candidate at a significant disadvantage when competing for a job against someone with the same qualifications (2007: 18). Moreover, the

---

7   However, if the salaries of D are lower, then the higher the percentage of F the higher the costs – and therefore the profits will drop. In the longer run, this means that in a competitive market, non-prejudiced employers (as they have lower costs) will drive prejudiced employers out of the market.

study shows that 78% of respondents feel that a person aged 50 and over is less likely to get a job, to be accepted for training or be promoted, than someone aged under that age (2007: 19).

More specific information is given by surveys that map employers' perceptions of the productivity and skills of older workers. In a survey of 500 large employers (employing 500 people or more) in the UK, Taylor and Walker (1994) showed that a sizable group of respondents had negative stereotypes of older workers, especially with regard to their openness to training and their ability to adapt to new technologies (1994: 581). A more recent comparative study, covering Greece, Spain, the Netherlands and the UK, confirms this finding[8] (van Dalen, 2006: 29). This despite significant cross-country variation, with UK employers being the most positive about the productivity of older workers, and Dutch employers the least.

A second (but rather small) stream in the literature has focused on measuring the impact of discriminatory attitudes in hiring processes. Busch and Konigstein (2001) conducted a laboratory experiment[9] on age discrimination in job interviews. The authors asked 174 German students to evaluate three hypothetical applicants with regard to their working skills and adequate wage levels, and to decide whom to hire (2001: 4-5). Their study suggests that, overall, there were not many differences in how they evaluated job applicants. However, negative age stereotypes did emerge in the decision on who to hire (2001: 12-13).

Whereas laboratory experiments occur in a tightly controlled environment, field experiments are able to bring some features of the standardised laboratory experiments to an environment that is closer to reality.[10] Riach and Rich conducted a set of field experiments in England (2007a), Spain (2007b) and France (2007c). Overall, the authors found high levels of discrimination against older applicants. Curiously, whereas in Spain there is little variation in the levels

---

8   Nonetheless, in all countries, a significant majority of surveyed employers felt that workers above the normal retirement age could still make a valuable contribution to the firm (Van Dalen, 2006: 29).

9   In laboratory experiments, participants are asked to play the role of interviewers/job selectors and to evaluate randomly assigned job applications where qualifications are artificially minimised and the impact of the age variable is artificially maximised.

10   The existing literature features two alternative approaches. The first consists of 'audit' (or 'situation') studies (Riach and Rich, 2002: 481), where matched pairs of individuals (which differ only with regards to one particular characteristic – such as age, race, gender, etc.) are asked to attend job interviews or to apply over the phone. The second consists of correspondence studies (2002: 484), which involve sending pairs of written job applications that are closely matched in terms of individuals' qualifications and experience (but, again, differ only with regards to one particular characteristic).

of discrimination found (Riach and Rich, 2007a: 7), in France and (especially) in England discrimination against older individuals is much higher in large metropolitan areas (Riach and Rich, 2007b: 7; Riach and Rich 2007c: 13).

Despite its relevance, this type of field experiment is not without problems. Focusing on audit studies (see footnote 9), Heckman criticizes the assumption that all the relevant characteristics that influence productivity are captured in job applications, and that unobservable factors are the same for all applicants. Depending on the distribution of unobserved characteristics, he argues, audit studies would identify discrimination or reverse discrimination, even if this was not actually present (1998: 107-11).

The potential limitations of experimental methods are further highlighted by the meta-analyses conducted by Morgeson et al. (2008) and Gordon and Arvey (2004), which seem to suggest that the effective level of discrimination against older people in the labour market is much lower than is commonly assumed (see Morgeson et al., 2008: 230; Gordon and Arvey, 2004: 485). Morgeson et al. (2008) suggest that job-related information and job/applicant fit are more important factors in predicting hiring decisions than age. Moreover, they found that laboratory experiments, which tend to involve less experienced recruiters, seem to show higher levels of discrimination than field studies, which tend to use more experienced evaluators (2008: 230).

2.2.4 Labour market regulation, recruitment and retention

Since it was published, Morten and Pissarides' (1999) theory of equilibrium unemployment has set the standard for the analysis of the impact of labour market policies. Looking at the variety of interventions available to policy-makers, the authors argue the following:

- that unemployment compensation and employment/payroll taxes, as they increase the price of labour supply, will decrease labour market tightness and increase the reservation wage, and consequently augment unemployment;
- that employment protection legislation, to the extent that reductions in job destruction outweigh any possible increases in unemployment duration, will reduce unemployment (1999: 41-2);
- that the effect of hiring subsidies is ambiguous. On the one hand, hiring subsidies reduce the cost of job creation, thus promoting job creation. On the other hand, they reduce the duration of jobs which, in turn, increases labour market tightness and job destruction.

Cheron, Hairault and Langot (2007) argue that this model, despite its significance, neglects the structuring influence of age in the formation of labour market equilibrium. Contrary to Morten and Pissarides' model (1999), which assumes that workers remain indefinitely in the labour market, the authors argue that hiring and firing are determined by reference to an (exogenous) age at which all workers are expected to leave the labour market (2007: 4).

This would explain the lower employment rate amongst the older population. First, because they have a shortened horizon, older workers are more exposed to firings than their younger counterparts. This, in turn, decreases the willingness of firms to recruit older workers, as the expected returns to this type of vacancy will be lower. Second, older workers tend to disinvest in job-search activities because the expected return decreases the closer a worker is to his/her expected exit date from the labour market (2007: 5-17).

This has implications for the analysis of the impact of different forms of labour market regulation. For instance, Cheron, Hairault and Langot (2007) argue that because firms can avoid firing an older worker by just waiting for him/her to retire, employment protection legislation would produce a stronger reduction in firings than predicted by Morten and Pissarides (1999). Also, because the period for which a firm must bear higher costs is smaller for older workers, unemployment benefits will also have less of a negative impact for older workers (2007: 17).

Thirdly, Cheron, Hairault and Langot (2007) argue that the potential employment gains arising from employment protection are high for older workers, but that higher firing taxes for these workers increase job destruction rates for the younger generations. Age-decreasing firing taxes, however, can lead to lower job destruction rates at all ages. These effects are strengthened by distortions related to unemployment benefits and bargaining power (2007: 20-21).

Besides this theoretical debate, there is a (rather limited) body of research on the effect of labour market interventions in the demand for older people. For instance, using matched employer-employee data from Germany, Bookman et al. (2007) found that the expansion of eligibility to the Integration Supplement (a hiring subsidy) to unemployed workers aged 50 plus during the period between 2002 and 2004, led to a decrease in the duration of unemployment for East-German women – with this effect disappearing once entitlement was terminated in 2004. However, this had no significant effects for other groups of older unemployed individuals. Benefiting from the experimental approach adopted in this study, the authors were also able to identify the presence of deadweight effects for older jobless males (2007: 12-14).

Krugler, Jimeno and Hernanz (2002) adopt an experimental approach to evaluate the impact of hiring subsidies. Using data from the Spanish Labour Force Survey, the authors found that the reduction of payroll taxes and dismissal costs prompted by the reform of labour legislation in 1997, had a (large) positive effect on the probability of transition from unemployment to permanent employment for older men – much stronger than that for their younger counterparts. The results for older women, however, were insignificant (2002: 18-20).

Whereas there is little evidence on hiring subsidies, several relevant studies have explored how employment protection legislation affects the demand for older workers. For instance, Bassanini and Duval (2006), using cross-country / time-series data from 21 OECD countries over the period 1982-2003, found that there is a positive relationship between the level of employment protection legislation and the employment rate of workers aged 55 to 64 (2006: 47-8).[11] Unfortunately, this study does not differentiate between the effect of employment protection legislation on job creation and the impact on job destruction.

A number of studies examined the impact of employment protection legislation on labour market flows. Using cross-sectional data from labour force surveys in OECD countries, a recent OECD study suggests that there is a statistically significant negative relation between employment protection (here measured through an Index of Employment Protection Legislation [EPL]) and the hiring rate of men aged 50 to 64 (OECD 2006: 71-72). Despite benefiting from significant cross-national variation, this study is limited by the fact that it controls neither for individual characteristics that might influence participation in the labour market, nor for the effect of the economic cycle.

Another example is the study conducted by Behagel, Crepon and Sedillot (2008), which looked at the effects of the reform of the Delalande Tax, a measure that imposes a penalty on firms that fire workers aged 50 plus. Using data from the French Labour Force Survey, the authors found that the introduction of an exemption for firms that lay-off older workers that were hired after age 50, increased the chances of finding a job for unemployed male workers aged 50 plus, compared with those below that age. For women, the results were statistically insignificant (2008: 704-10).

These studies are somewhat limited by the fact that they do not control for firm-level factors that condition the impact of labour market interventions.

---

11   In contrast, a recent OECD study suggests that there is a statistically significant negative relation between employment protection (here measured through an Index of Employment Protection Legislation [EPL]) and the employment rate of people aged 50 and older (OECD 2006: 71-72). However, as there is no sufficient methodological information on how this study was conducted, it is not possible to evaluate the reasons behind this dissonance.

In this context, the study conducted by Daniel and Heywood (2007) provides an interesting alternative. Using matched individual/establishment data from the 1998 Workplace Employment Relations Survey (WERS), the authors found strong evidence that the existence of delayed compensation mechanisms, such as mandatory pensions, reduces the likelihood of companies hiring older workers (2007: 49).

In addition to the effect of employment protection laws on hirings, there is also some evidence on its impact on firms' decision to shed older workers. For instance, the aforementioned OECD (2006) study found that more dynamic – and presumably, less regulated – labour markets tend to be more disadvantageous for older workers, as they tend to be more affected by job destruction (2006: 71-72).[12]

Using a more sophisticated approach, Behagel, Crepon and Sedillot (2008) found that a higher tax penalty on the laying-off of older workers reduces the probability of male older workers being fired. The estimates, however, lack robustness and precision (2008: 717-18). Moreover, Krugler, Jimeno and Hernanz (2002) found that the reduction of dismissal costs approved in 1997 increased the probability of making the transition from permanent employment to unemployment – which practically cancels out any increases in hirings[13] (2002: 19-21).

Unfortunately, there is no European evidence on the labour market effectiveness of anti-discrimination legislation, for which we will have to rely on US-based evidence. Using data from the Census of Population, and adopting an experimental (difference-in-difference) approach, Neumark and Stock (1997) found that the introduction of anti-discrimination laws increased employment in the group of individuals age 60 and over that were protected by anti-discrimination – the effect being stronger when federal legislation was introduced (1997: 21). A weakness of this study is that it ignores the differential impact of anti-discrimination laws on job creation and job destruction dynamics.

Lahey (2006) adopts a more differentiated approach. Using data from the Current Population Surveys, she found that older workers in states with anti-discrimination laws are less likely to be hired by 0.2% (2006: 24).[14] Her use of a large time interval for estimation purposes (1978-1991) leaves her results open to the charge that confounding factors might still be at work. Her results

---

12   The study showed a negative relation between EPL and the retention rate for men aged 55-59 – although this was not statistically significant.

13   Effects for women are, once again, insignificant (2002: 19-21).

14   In addition to this, Lahey also found that workers covered by anti-discrimination laws were less likely to be separated from their jobs – but this was not statistically significant (2006: 22).

conflict with earlier work by Adams (2004), who had used a tighter time interval, and concluded that there was no evidence that the introduction of anti-discrimination laws had any significant effect on the hiring of older workers – especially for those who were in active job search (2004: 236-7).

3   Remaining gaps in knowledge and main challenges

Reflecting on the evidence presented in the previous sections, it can be concluded that there are notable gaps in the literature on the factors that drive the demand of older workers in Europe. At the theoretical level, we cannot but highlight the inadequacy of existing theoretical frameworks to deal with the issues raised by the ageing of the labour force. At the same time, we have a set of recent theoretical developments (see Hetze and Ochsen, 2005; Aghion and Howit, 1994; or Hairault, Cheron and Langot, 2007) that, despite prompting some interesting theoretical advancements, need to be taken forward. In this context, the following theoretical issues require further enquiry, both at the fundamental and applied level:

- **The existence of heterogeneity in the labour force:** As mentioned above, the presence of heterogeneity in the labour force has only recently begun to be integrated in the models that try to understand labour demand (see Hetze and Ochsen, 2005; or Hairault, Cheron and Langot, 2007). In particular, it is important to understand the degree to which young and old cohorts can be seen as competing or complementary in the labour market (see Ermish, 1988: 77), and how the perceived differences in the productivity of old and younger workers affect firms' hiring decisions and wage-setting procedures.

- **The relationship between age, productivity and wages:** Looking at the literature, it can be said that Lazear's seminal work on the 'delayed compensation contracts' model (1979) has set the benchmark for analysing the relationship between age, productivity and wages – and the way that this affects the demand for older workers. However, as our review has shown, some evidence suggests that age-related declines in productivity can be compensated by experience, personal aids and suitable workplace adjustments. This suggests that more research needs to be done on how productivity varies across the life cycle, and on how both firms and workers anticipate, and adapt to, such changes. This, in turn, should inform new theoretical developments in the analysis of the factors that drive the demand for older workers.

- **The influence of the timing of retirement on employers' hiring decisions:** As in the previous case, Lazear's model of delayed compensation contracts provided an interesting insight into how the expectation of a short horizon for older workers can influence employers' hiring decisions. Nonetheless, the model is limited by the fact that it is based on the assumption that older workers would not want to leave in a context where their wages are higher than their productivity (which is obviously contradicted by the prevalence of early retirement, especially in Europe). Not only that, the model seems to assume that younger workers would not move to a different firm when offered a higher salary (Skirbekk, 2003: 18). In this context, the focus by Hetze and Ochsen (2005) on the differentiated separation risk for younger and older workers provides a better option for analysing the impact of the timing of retirement on employers' hiring decisions.

At the methodological level – despite the increased focus on a life-cycle approach – a great deal of effort needs to be made to improve the quality and accuracy of the studies on the demand for older workers:

- More effort needs to be put into the analysis of processes over time, especially on the relation between age, productivity and wages, the impact of business cycles and technological innovation, and the employment effectiveness of labour market policies aimed at improving the demand for older workers.
- Also, although some work has been done to study the interaction between individual- and macro-level factors (see Blanco and De La Rica, 2002; and Krugler, Jimeno and Hermanz, 2002), between individual- and meso- (firm) level factors (see Crepon and Aubert, 2003; Hellerstein and Neumark, 2004; Bookman et al., 2007; and Daniel and Heywood, 2007), little has been done to capture variations at these three levels simultaneously. To some extent, this reflects the limitations in the existing research infrastructures (see sections 4 and 5).
- Finally, significant improvements are required in the analysis of the extent and impact of age-related discrimination on the hiring of older workers. As this review has shown, in order to improve the potential contribution to the literature, significant improvements need to be made to the design and implementation of field studies (see Heckman, 1998). Moreover, alternative approaches need to be explored. For instance, vignettes have been proven a good tool to uncover hidden attitudes and stereotypes (see Erber and Long, 2006), and could provide a useful tool in the analysis of hiring practices.

A more detailed look at the various domains in the literature on the demand for older people allows us to identify a number of areas where gaps are evident:

- The evidence on the relationship between age and productivity, although abundant, is marred by methodological limitations that limit its accuracy. This would suggest that more research on the relationship between age, productivity and wages is needed – especially for European countries – and preferably using more reliable evidence such as employer-employee matched data.
- The evidence on the impact of business cycles and technological innovations on the demand for older workers is very scarce. Such research needs to take careful account of cross-national variations so as to avoid over- or under-estimating the impact of economic cycles. In particular, it would be useful to see how older workers may have been at a greater or lower risk of losing their job due to the economic crisis, and how they face difficulties in becoming re-employed during the recovery period.
- The evidence on the extent and impact of age discrimination in the labour market is scarce, and its identification is marred by methodological issues. Besides exploring the methodological innovations mentioned above, more research on this topic is needed.
- The evidence on how labour-market regulations affect the demand for older workers is rather scarce. This would suggest that more research on this topic is also needed (especially of a comparative nature), making use where possible of 'natural experiments' such as policy changes across European countries.

4   Current state of play of European research infrastructures and networks

Reflecting on the research needs identified in the previous section, it seems that the problem relates to deficiencies in research capacity rather than to a lack of suitable data for research. For instance, there is now a growing set of longitudinal databases, such as the Survey of Health, Ageing and Retirement in Europe (SHARE), the English Longitudinal Study of Ageing (ELSA) or the (forthcoming) Irish Longitudinal Study on Ageing (TILDA). Currently, these studies have two great strengths:

a) They are multi-disciplinary, allowing better insights into the various domains of older people's lives.

b)   They are designed to be internationally comparable both within Europe and with studies elsewhere in the world.

In addition, there is a set of databases with matched employee-employer data, such as the IDA (Integrated Database for Labour Market Research; Denmark), the LIAB database (Germany), the Workplace Employment Relations Survey (WERS; UK), and other matched databases in France, Sweden, and Slovenia that can be used for various type of research in this domain. Alongside these national studies is the European Union Structure of Earnings Survey 2000, carried out by the Eurostat, which could provide some opportunities for research that covers variations at the macro-, meso- and micro-level. However, this survey does not provide information on worker flows in firms. In this context, the Micro-Dyn project[15], which aims to develop a comparable micro-level dataset for most European countries on job- and worker flows, might provide a more interesting alternative for research on this topic.

Finally, a number of international organisations have organised cross-national databases, including MISSOC, the Labour Market Reforms Database, the OMC for pensions (European Commission), and the OECD Labour Market Policy database. These data sources will be essential for the development of cross-national comparisons – of both cross-sectional and longitudinal nature. EUROSTAT surveys, such as the EU-SILC and the European Labour Force Survey (EU-LFS), also contain data that can be used to conduct research in this important domain.

In contrast with the relative wealth of data available, there is a deficit in the research infrastructures and networks that support research on this topic. To a certain extent, this reflects the disparity of factors that influence the demand for older workers, and the sheer infancy of ageing research in Europe. In light of this, besides the set of individual researchers that have been identified in this review, we were able to identify a set of networks that could potentially hold the necessary resources for conducting research on these topics. For instance, the consortium responsible for the SHARE survey combines resources from leading research institutes on ageing in Europe, and it could serve as a key resource from where a research network on the demand for older workers could be developed. The institutions that have been carrying out longitudinal studies of ageing (such as ELSA or TILDA) also incorporate researchers that could be attracted to conduct research on some relevant issues such as the relation between age and productivity, or the discrimination of older workers in the labour market.

---

15   See www.micro-dyn.eu

In addition to this, there are established research networks that, although not focusing specifically on ageing-related issues, conduct research activities in areas that overlap some of the topics discussed in this chapter. For instance, the ASPEN Network (Active Social Policies European Network) integrates a set of established social researchers that conduct high-level research on the role of labour market policies in helping the unemployed in getting back to work. Another relevant resource is the RECOWE network (Reconciling Work and Welfare in Europe), which brings together a group of high-level economists and other social scientists with an interest in issues related to employment and labour market regulation in Europe.

5   Required research infrastructures, methodological innovations, data and networks – and consequences for research policy

Reflecting on the state of the literature and of the existing research infrastructures, we believe that future development efforts should be prioritised. Firstly, bearing in mind the wealth of existing datasets available, the most immediate priority should be to direct funding towards sponsoring the development of research programmes that have the following aims:

- to explore the introduction of relevant theoretical developments, such as the assumption of age-based heterogeneity in the labour market, the possibility that productivity does not necessarily decrease with age, and the way that the decisions of economic actors are influenced by the effective age of withdrawal from the labour market;
- to adopt appropriate methodological approaches (namely, longitudinal studies, or studies that adopt a life-cycle approach, or that capture the interaction between macro-, meso- and micro-level labour market dynamics). Increased funding for the analysis and evaluation of policy interventions in the various domains of labour demand should preferably go to studies that adopt an experimental approach;
- to focus on areas where our review has highlighted significant gaps (such as the dynamics of job creation and destruction, and the way this affects the demand for older people; the relationship between age, productivity and wages; the impact of business cycles and technological innovation; or the extent and impact of ageing discrimination). In particular, as this review shows, funding should be directed to studies that focus on the evaluation of

public-policy interventions (such as anti-discrimination legislation, hiring subsidies, employment protection legislation, unemployment benefits), as this would have significant returns in terms of the cost-effectiveness of these instruments.

Secondly, accepting that there is a pool of data immediately available for research, further effort should nonetheless be put into reinforcing the existing databases:

- Support towards the emerging longitudinal studies should be continued, if not enhanced. Moreover, support should be given to existing efforts to promote the harmonisation of European and other international studies.
- Support should be given to existing efforts to improve the comparability of existing matched worker/employer datasets.
- Funding should go into the development of primary data in the areas that our review has highlighted as particularly limited – notably, on the extent and impact of age discrimination.

Finally, further effort should be made in the development of the research infrastructures that can support the development of research on the demand factors for the employment of older workers. However, for cost-effectiveness purposes, these efforts should be integrated within a broader effort to promote research infrastructures on the labour market implications of demographic ageing. It would be most useful to promote meetings and workshops where people working on this topic could discuss the most recent developments in this domain and promote future collaborations.

References

Adams, S. (2004) 'Age Discrimination Legislation and the Employment of Older Workers', *Labour Economics*, 11, 219-241.

Aghion, P. and P. Howit (1994) 'Growth and Unemployment', *Review of Economic Studies*, 61(3), 477-494.

Arrow, K. (1973) 'The Theory of Discrimination' in O. Ashenfelter and A. Rees (eds) *Discrimination in Labor Markets* (Princeton: Princeton University Press), 3-33.

Arrow, K. (1998) 'What Has Economics to Say about Racial Discrimination?', *Journal of Economic Perspectives*, 12(2), 91-100.

Atkinson, J., C. Evans, R. Willison, D. Lain and M. van Gent (2003) 'New Deal 50plus: Sustainability of Employment,' Report WAE142 (Department for Work and Pensions).

Aubert, P. (2005) 'Les salaires des seniors sont-ils un obstacle à leur emploi?', in INSEE, *Les salaires en France* (Paris: INSEE), 41-52.

Baumle, A. and M. Fosset (2005) 'Statistical Discrimination in Employment: Its Practice, Conceptualization, and Implications for Public Policy', *American Behavioral Scientist*, 48, 1250-74.

Becker, G. (1957) *The Economics of Discrimination* (Chicago: University of Chicago Press).

Behagel, L., B. Crepon and B. Sedillot (2008) 'The Perverse Effects of Partial Employment Protection Reform: The Case of French Older Workers', *Journal of Public Economics*, 92, 969-721.

Bergmann, B. (1971) 'The Effect on White Incomes of Discrimination in Employment', *Journal of Political Economy*, 79(2), 294-313.

Bergmann, B. (1974) 'Occupational Segregation, Wages and Profits when Employers Discriminate by Race or Sex', *Eastern Economic Journal*, 1(2), 103-110.

Blanco, A. and S. De la Rica (2002) 'Unemployed Older Workers versus Prime Age Workers: Differences in their Re-employment Determinants in Spain,' Working Paper N°. 8 (Universidad del País Vasco – DFAE-II).

Blank, R., M. Dabady and C. Citro (2004) 'Measuring Racial Discrimination: Panels on Methods for Assessing Discrimination (Washington DC: National Academies Press).

Bookmann, B., T. Zwick, A. Ammermuller and M. Maier (2001) 'Do Hiring Subsidies Reduce Unemployment among the Elderly? Evidence from Two Natural Experiments,' ZEW discussion paper 07-001.

Büsch, V. and M. Königstein (2001) 'Age Discrimination in Hiring Decisions: A Questionnaire Study,' Mimeo. Retrieved from http://www.cepii.fr/anglaisgraph/communications/pdf/2001/enepri07080901/busch-konig.pdf

Cheron, A., J.-O. Hairault and F. Langot (2008) 'Age-Dependent Employment Protection,' IZA discussion paper 3851.

Christiaans, T. (2003) 'Aging in a Neoclassical Theory of Labor Demand,' Universitaet Siegen – Fachbereich Wirtschaftswissenschaften discussion paper 112-03.

Crépon B. and P. Aubert (2003) 'Productivité et Salaire des Travailleurs âgés,' *Economie et Statistique*, 368, 157-185.

Daniel K. and J. Heywood (2007) 'The Determinants of Hiring Older Workers: UK Evidence', *Labour Economics*, 14, 35–51.

Duncan, C. and W. Loretto (2004) 'Never the Right Age? Gender and Age-based Discrimination in Employment', *Gender, Work & Organization*, 11, 95-115.

Erber, J. and B. Long (2006) 'Perceptions of Forgetful and Slow Employees: Does Age Matter?', *The Journals of Gerontology Series B: Psychological Sciences and Social Sciences*, 61, 333-39.

Ermisch, J. (1988) 'British Labour Market Responses to Age distribution Changes' in R.D. Lee, W.B. Arthur and G. Rodgers (eds) *Economics of Changing Age Distributions in Developed Countries* (Oxford: Clarendon Press), 76-86.

European Commission (2006) *European Handbook on Equality Data*, (Brussels: European Commission – Directorate-General for Employment, Social Affairs and Equal Opportunities).

European Commission (2007) *Discrimination in the European Union, Special Eurobarometer 263*.

Gelderblom, A. and G.J. Vos (1999) 'Age and Productivity', in R.M. Lindley (ed.) *The Impact of Ageing in the Size, Structure and Behaviour of Active Age Population and Policy Implications for the Labour Market* (Coventry: Institute for Employment Research).

Gordon, R. and R. Arvey (2004) 'Age Bias in Laboratory and Field Settings: A Meta-Analytic Investigation', *Journal of Applied Social Psychology*, 34(3), 468-492.

Hairault, J.-O., A. Cheron and F. Langot (2007) 'Job Creation and Job Destruction over the Life Cycle: The Older Workers in the Spotlight,' IZA discussion paper 2597.

Heckman, J. and P. Siegelman (1993) 'The Urban Institute Audit Studies: Their Methods and Findings,' in M. Fix and R. Struyk (eds) *Clear and Convincing Evidence: Measurement of Discrimination in America* (Washington DC: Urban Institute Press).

Hellerstein, J.K. and D. Neumark (2004) 'Production Function and Wage Equation Estimation with Heterogeneous Labor: Evidence from a New Matched Employer-Employee Data Set,' NBER working paper 10325.

Hetze, P. and C. Ochsen (2005) 'How Aging of the Labor Force Affects Equilibrium Unemployment,' Thünen-Series of Applied Economic Theory working paper 57.

Hirsch, B., D. Macpherson and M. Hardy (2000) 'Occupational Age Structure and Access for Older Workers', *Industrial and Labour Relations Review*, 53(3), 401-418.

Hutchens R. (1988) 'Do Job Opportunities Decline with Age?', *Industrial and Labor Relations Review*, 42(1), 89-99.

Johnes G. and D. Sapsford (1996) 'Some Recent Advances in the Economic Analysis of Discrimination', *International Journal of Manpower*, 17 (1), 10-25.

Krugler, A. J. Jimeno and V. Hernanz (2002) 'Employment Consequences of Restrictive Permanent Contracts: Evidence from Spanish Labour Market Reforms, Universitat Pompeu Fabra – Economics working papers 651.

Lahey, J. (2006) 'State Age Protection Laws and the Age Discrimination in Employment Act,' NBER working paper 12048.

Langot, F. and E. Moreno-Galbis (2008) 'Does the Growth Process Discriminate against Older Workers,' IZA discussion paper 3841.

Lazear, E. (1979) 'Why Is There Mandatory Retirement?,' *Journal of Political Economy*, 87, 1261-1284.

Medoff, J.L. and K.G. Abraham (1981) 'Are those Paid More Really More Productive? The Case of Experience', *Journal of Human Resources*, 16(2), 186-216.

Morgeson, F., M. Reider, M. Campion and R. Bull (2008) 'Review of Research on Age Discrimination in the Employment Interview', *Journal of Business Psychology*, 22, 223-232.

Myrdal, G. (1944) *An American Dilemma* (New York: Harper).

Neumark, D. and W. Stock (1999) 'Age Discrimination Laws and Labor Market Efficiency', *Journal of Political Economy*, 107(5), 1081-1125.

OECD (2006) *Ageing and Employment Policies: Live Longer, Work Longer* (Paris: OECD).

Oliveira, M.M. de, E. Cohn and B.F. Kiker (1989) 'Tenure, Earnings and Productivity', *Oxford Bulletin of Economics and Statistics*, 51(1), 1-14.

Olson, P. (2007) 'On the Contributions of Barbara Bergmann to Economics', *Review of Political Economy*, 19(4), 475-496.

Phelps, E. (1972) 'The Statistical Theory of Racism and Sexism', *American Economic Review*, 62(4), 659-661.

Pissarides, C. (2000) *Equilibrium Unemployment Theory* (Massachusetts: MIT Press).

Pissarides, C. and D. Morten (1999) 'New Developments in Models of Search in the Labour Market', CEPR discussion paper 2053.

Pissarides, C. and G. Valanti (2007) 'The Impact of tfp Growth on Steady-state Unemployment, *International Economic Review*, 48(2), 607-640.

Remery, C., K. Henkens, J. Schippers, J. van Doorne Huiskes and P. Ekamper (2001) 'Organisaties, Veroudering en Management: Een Onderzoek onder Werkgevers' (Organisations, Ageing and Management: Employers' Views), NIDI Report 61.

Riach, P. and J. Rich (2002) 'Field Experiments of Discrimination in the Marketplace', *Economic Journal*, 112, 480–518.

Riach, P. and J. Rich (2007a) 'An Experimental Investigation of Age Discrimination in the Spanish Labour Market', IZA discussion paper 2654.

Riach, P. and J. Rich (2007b) 'An Experimental Investigation of Age Discrimination in the English Labour Market', IZA discussion paper 3029.

Riach, P. and J. Rich (2007c) 'An Experimental Investigation of Age Discrimination in the French Labour Market,' IZA discussion paper 2522.

Skirbekk, V. (2003) 'Age and Individual Productivity: A Literature Survey', Max Planck Institute for Demographic Research – WP 2003-028.

Taylor, P. and A. Walker (1994) 'The Ageing Workforce: Employers' Attitudes Towards Older People', *Work, Employment and Society*, 8(4), 569-591.

Van Dalen, H., K. Henkens, W. Henderikse and J. Schippers (2006) 'Dealing with an Ageing Labour Force: What do European Employers Expect and Do?', NIDI Report 73.

Volkoff, S., A. Molinié and A. Jolivet (2000) 'Efficaces à Tout âge: Vieillissement Démographique et Activités de Travail', Dossier du Centre d'Études de l'Emploi 16.

Warr, P. (1993) 'In What Circumstances does Job Performance Vary with Age?', *European Work and Organizational Psychologist*, 3(3), 237-249.

The Demand for Older Workers

Comments by Rob Euwals

Only three out of ten lines of action of those set by the Stockholm and Barcelona declarations to reach the EU targets are aimed at the demand side of the labour market. The study by Moreira, Whelan and Zaidi claims that policy pays too little attention to the demand side and too much to the supply side. The statement clearly holds at the Dutch national level – and observing the focus on retirement age in many EU countries, the statement seems to hold at the national level for many other countries as well. The following discussion first explores some reasons for the focus on the supply side. Next, the discussion explores the issues raised in the Moreira, Whelan and Zaidi study, and then ranks them according to their relevance.

Supply and demand

Labour supply and labour demand play equally important roles in economic theory of the labour market. So why has policy been focusing on the supply side? An obvious reason may have been that the supply side is relatively easy to reach by policy. Through the tax-, social security- and early retirement system, policy can directly influence individual decisions on participation and retirement. Many countries have been reforming the early retirement- and pension system, and the Dutch government even introduced a tax credit for older workers, in order to encourage their participation.

The demand side of the labour market is more difficult to influence by policy. Of course, the government may decide to subsidize employers who hire older workers. Demand is influenced not only by taxes and subsidies, however, but also by labour-market institutions, such as the wage negotiation system and employment protection legislation. The impact of such institutions on employment and welfare is still heavily debated among economists. This should not, however, provide an argument for policy to disregard the demand side – and at the end of my discussion I return to the importance of institutions.

Five issues on the demand of labour

The Moreira, Whelan and Zaidi study discusses five issues involving the demand for older workers. First, the study discusses theory – of which, in particular, the theories put forward by Lazear are well-known. Although the list of relevant studies may indeed be too long to mention, I believe that Shimer (2001) should also be mentioned. Shimer discusses the impact of the presence of youngsters on job creation and vacancies. The lack of mobility of older workers decreases the profitability of vacancies, thereby reducing production and increasing unemployment. Through this channel, the ageing of the population may have a considerable impact on the labour market.

Second, the chapter discusses wages and productivity. In many countries, the labour market for older workers doesn't seem to be functioning very well, as job durations are long and hiring rates of older workers are low. A central question in this respect is whether this is the result of restrictive institutions or optimal institutions. Studies like Teulings (2009) and Euwals, de Mooij and van Vuuren (2009) explore international comparisons of the wage profiles and discuss policy conclusions. Again, institutions play an important role, and I will return to the issue.

Third, innovation plays a central role in job creation and job destruction for older workers. In this regard, there is a strong link with human capital formation, which is discussed in more detail in chapter 10 of this book. Nevertheless, I am puzzled by a question on the speed of technological progress and the obsolescence of skills. In public discussions it is often claimed that the speed of technological progress is high nowadays – that it is even accelerating. My claim, however, is that at (almost) any point in history people have claimed that technological progress was accelerating. I wonder, therefore, whether the current time period is really all that noteworthy in this respect, compared to previous periods? The acceleration of technological progress can be said to make a difference, in the sense that some skills nowadays are really outdated within a life course – while until recently this was not the case. But is this really so? And for how many occupations is this true?

Fourth, research shows that age discrimination is a serious issue in many countries. While I believe that laws should be clear on the issue, I doubt, as an economist, whether laws alone will be sufficient to ban age discrimination. A gap between wages and productivity of older workers is always going to have some adverse effects on the labour-market position of older workers.

Fifth, the study discusses the role of institutions. Policy has a substantial impact on the demand for labour, as it determines institutions such as unemployment insurance, employment protection legislation and hiring subsidies. This is a difficult issue for policy. Institutions do not come out of the blue. In practice, they always serve one or more policy goals. These goals should be traded off against possible negative effects of the institutions on the functioning of the labour market.

Ranking the issues

Two of the discussed issues are crucial elements of the other three issues. First, the relationship between knowledge, productivity and wages is important for the functioning of the labour market for the elderly, the incentives of older workers to acquire new skills and improve productivity, and the discrimination of older workers. Second, institutions have an important impact on all of these issues, including employment and welfare. Discussions on these issues make little sense if we do not understand better the relationship between knowledge, productivity and wages, and the impact of institutions. Policy should be keenly interested in scientific progress on these issues, as the time of effortless cherry picking at the supply side of the labour market is over. Many countries have introduced measures to increase the labour supply of older individuals. Relying on the popular version of Say's law ('every supply creates its own demand') is too easy. Policy should start to be concerned about the demand for older workers and about designing institutions in such a way that older workers end up in productive jobs with the appropriate knowledge.

References

Euwals, R., R. de Mooij and D. van Vuuren (2009) 'Rethinking Retirement', CPB special publication 80.

Shimer, R. (2001) 'The Impact of Young Workers on the Aggregate Labour Market', *Quarterly Journal of Economics*, 116, 969-1007.

Teulings, C. (2009) 'Earnings Profiles, Employment Protection and the Labour Market for Older Workers', Conference paper presented at Netspar Annual Conference, April 24, 2009.

The Demand for Older Workers

Comments by Jørgen Mortensen

Main findings and arguments

A main argument of the chapter by Moreira, Whelan and Zaidi is that a more effective approach for improving the participation of older workers in the labour market will require greater focus on improving the demand for older workers, but, unfortunately, little is known about the factors that condition the demand for this category of workers in Europe.

Reviewing various studies, they find a negative relationship between seniority wages and employment outcomes for older workers. This finding could be explained by a hump-shaped curve for productivity through the life cycle associated with seniority wage; although it is possible that for some tasks and in some branches the productivity of workers will not necessarily decline from a certain age. Furthermore, the relation between productivity and wage cost is also determined by the regulatory environment and the incentives to early retirement, which may either promote employment of the elderly or, alternatively, actually lead to shedding of older workers.

They argue that further research is needed to examine the degree to which young and old cohorts are seen to be competing or complementary, the variations of productivity over the life cycle and, not least, the interaction between the pension systems and the individual worker's decisions to change job and to retire.

The thesis of Moreira, Whelan and Zaidi is also confirmed in the two other chapters in this part, by Jacobs and van Soest, respectively. However, the former puts additional emphasis on the need to undertake deeper enquiry into the interfacing between human capital investment, notably training and learning-by-doing, in general and the pattern of life-cycle employment while the latter argues more in favour of additional exploration of the role of financial incentives, as regards early retirement and part-time employment.

Educational attainment and employment rates

In general, in the EU, the rate of employment (employment as % of the population in the age group) is higher with the level of education. This is in particular the case for the age group 55-64, where the rate of employment for those with only elementary and primary education in 2007 was about 30 percentage points lower than for those with tertiary education. The overall rate of employment was higher and the level-of-education gradient was slightly lower for males than for females but for both genders the level of education clearly was a main determinant of the rate of employment in this age group (Table 1).

Table 1:   Employment rates by level of education, EU25

	Total	ISCED 0-2	ISCED 3-4	ISCED 5-6
55-64				
Total	44.7	35.5	47.4	65.4
Males	53.9	44.9	54.1	69.5
Females	36.0	28.4	39.9	59.9

Source: Eurostat.

However, of importance is also that, for the EU, the level of educational attainment for the age group 55-64 on average for the EU is considerably lower than for the younger age classes. In fact, in 2007 for the 19 EU countries being members of the OECD, the level of educational attainment, measured by years of schooling, for the age group 55-64 was, for males at 11.5 years, 1.4 year lower than for the age group 25-34, and for females, at 10.4 years, 2.5 years lower than for those aged 22-34.

The strong age gradient of educational attainment in the EU contrasts sharply with that observed in the United States, where, for males the level of educational attainment for the age group 55-64 in fact was slightly higher than for those aged 25-34. For females the educational attainment of females in the age group 55-64 was 0.5 years lower than for males but nevertheless as much as 3.1 years higher than for the EU.

Table 2:   Educational attainment (years of schooling) by age

	Total	25-34	35-44	45-54	55-64	Em-ployed	Unem-ployed	Not in the labour force
Males								
United States	13.8	13.7	13.8	14.0	13.8	14.0	13.0	12.9
EU19 (x)	11.9	12.5	12.2	11.7	11.1	12.3	11.4	10.7
Females								
United States	13.9	14.0	14.0	13.9	13.5	14.2	13.2	13.1
EU19 (x)	11.8	12.9	12.2	11.4	10.4	12.5	11.6	10.6

Note: (x) EU15 (old member states) plus Czech Republic, Hungary, Poland and Slovak Republic.
Source: OECD: *Education at a Glance.*

So combining the two preceding tables, the low level of educational attainment of the age group 55-64 (Table 2) is the main determinant of the low level of employment of this group in certain Member States. In this respect, however, disparities are very large within the EU, with Denmark recording a level of educational attainment close to that of the United States and with the same level for the age group 55-64 as for those in the 25-34 age group while in Portugal the level of educational attainment for those aged 55-64 was, at 7.2 years, below that of Mexico (7.7). (Space constraints do not allow a more detailed analysis of country differences.)

The prospective improvement in human capital endowment

However, with respect to the endowment with human capital, in many or all of the EU Member States the overall level of educational attainment is increasing steadily due essentially to the fact that the elderly with a low level of education are gradually replaced by cohorts with a higher level. As, indeed, is already apparent from Table 2, the age group 25-34 has a higher level of educational attainment than the age group 35-44 and so forth. In addition, during the next decade those who are now still in education or just entering the labour market already have a higher rate of enrolment in tertiary education and will boost the overall average level of educational attainment further towards the best performers in this respect.

Thus, according to projections prepared by the European Centre for the Development of Vocational Training (CEDEFOP), the share of the working-age population with only low level of education (ISCED 0-1) over the 20 years from 2000 to 2020 will have declined from 36.3% to 22.1% of the total. Over the same period, those with high education will have increased from 17.2% in 2000 to 28.5% in year 2020. Strikingly, as a result of the steep rise in female enrolment in higher education, the share of females with high level of education will in 2020 be some 3 percentage points higher than for males (Table 3).

Table 3:   Population by level of education

Males	2000	2007	2013	2020
Low	34.2	29.9	26.3	23.4
Medium	44.3	44.8	46.2	46.8
High	16.1	19.5	22.4	26.0
All	100.0	100.0	100.0	100.0
Females				
Low	37.2	31.8	25.2	20.1
Medium	41.9	44.4	45.2	44.8
High	15.5	20.5	24.5	29.2
All	100.0	100.0	100.0	100.0
Total				
Low	36.3	31.6	26.8	22.1
Medium	43.6	45.5	46.5	46.5
High	17.2	21.0	24.8	28.5
All	100.0	100.0	100.0	100.0

Source: CEDEFOP (2008, 2009).

Some conclusions for the research agenda

Whereas the comments presented above in no way invalidate the research agenda on employment of older workers presented in the chapter by Moreira, Whelan and Zaidi and the other two chapters in this section, there is a strong case for extending the agenda in two directions:

1. Taking much more account of the role of educational attainment as a determinant of employment performance in general and employment of older workers in particular.

2. Taking as much account as possible of the dynamic aspects of the "equation", considering notably that the endowment of human capital of the working population has changed importantly during past decades and will do so over the next couple of decades before, possibly, reaching a plateau, at least as far as formal educational attainment is concerned.

A satisfactory level of insight into the process of labour market transition can hardly be achieved without longitudinal data, allowing research to follow individual workers and firms through a certain time span. Fortunately the availability of data, notably the Community's survey of income, social inclusion and living conditions (SILC) opens new doors for research of the temporal dimension of employment performance. The EU SILC thus is the starting point for the analysis of labour market transition referred to above but can also be exploited more extensively for longitudinal studies of other aspects of employment performance and retirement behaviour of the elderly workers.

References

CEDEFOP (2008) *Future Skill Needs in Europe. Medium-term forecast*, Luxembourg: Office for Official Publications of the European Communities.

CEDEFOP (2009) *Future Skill Supply in Europe: Medium-Term Forecast up to 2020. Synthesis Report* Luxembourg: OOPEC

OECD (2009), *Education at a Glance*, Paris.

10

Human Capital, Retirement and Saving

Bas Jacobs[1]

1   Introduction

Understanding the life-cycle interactions between investments in human capital, retirement choices and pension savings is highly policy-relevant. Most Western governments will be confronted with the consequences of demographic ageing in the upcoming decennia. Tax bases will shrink, due to the retirement of older generations of workers. Outlays on state pensions and healthcare will rise substantially. Pension systems with strong intergenerational risk sharing face difficulties as well, since it will become more difficult and costly to smooth pension risks over different generations by means of contribution adjustments. At the same time, individuals do not invest in skills, because they expect to retire early. And, individuals retire early because they have not invested in skills. As a result, many European countries are confronted with a vicious circle of low investments in on-the-job training of older workers and strong incentives to retire early.

Given these developments, policymakers are considering a range of policies to increase investment in skills, promote later retirement and pension

1   My thanks to Lans Bovenberg, Elsa Fornero, Daniel Hallberg, Pierre Pestieau and seminar participants at the ESF Forward Looks Workshop in Dublin, November 8, 2008, and in The Hague, April 22, 2009, for their helpful comments and suggestions. This chapter covers a broad research area that crosses numerous subdisciplines. I apologise for not being able to do justice to the scientific work of the many scholars that I have not mentioned. All errors and omissions are mine. I thank Netspar and the Netherlands Organisation for Scientific Research for financial support (Vidi Grant No. 452-07-013, 'Skill Formation in Distorted Labour Markets').

savings. For example, all European countries have subscribed to the Lisbon agenda. One of its main targets is that the EU average level of participation in life-long learning should be at least 12.5% of the adult working-age population (25-64 age group) by 2010. Most countries have started to implement 'life-long learning' policies to promote investments in on-the-job training (OJT), so as to raise labour productivity and to improve the employability of especially older workers. In addition, governments aim to promote labour-market participation of older workers and to improve their employability so as to broaden tax and premium bases. In particular, (early) retirement schemes and labour markets are reformed in order to stimulate later retirement. Until recently, implicit taxes on continued work have often been so high that individuals were 'thieves of their own wallet' if they did not retire early. Furthermore, many governments stimulate private pension saving, for example, through tax-favoured saving schemes, so as to reduce the dependency of pensioners on state pensions and collective occupational pension schemes.

Unfortunately, too little is known about life-cycle interactions between learning, retirement and saving, both theoretically and empirically. Generally, training, (pension) saving and wage determination are separately analysed, and no generally accepted theories are available to address these issues simultaneously. The consequence is that human capital policies are considered in isolation from retirement and pension policies. This chapter closely follows the theoretical structure of Jacobs (2009a) and Heckman and Jacobs (2010) to provide an analysis of the interactions between human capital investments in OJT, retirement choices and pension saving. In particular, retirement and pension saving affect the incentives to invest in human capital over the life cycle. By extending the time horizon over which investments in skills materialise, a higher retirement age promotes investments in OJT. Later retirement and OJT investment are therefore complementary. Generous support for early retirement therefore indirectly discourages investment in OJT. Individuals also make a life-cycle portfolio choice by investing in both financial and human capital. Stimulating retirement savings implies that savings in human form are discouraged. The intuition is that the opportunity return at which future labour earnings are discounted increases. Equivalently, arbitrage between financial and human investments ensures that both assets must earn equal returns. Hence, human capital and financial capital are substitutes over the life cycle. Simulations of various tax and retirement policies illustrate the importance of life-cycle interactions between human capital, retirement and pension saving.

Labour-market institutions and welfare-state arrangements appear to be crucial to understand economic incentives for OJT investments. Labour-market institutions – such as employment protection, wage setting of unions, efficiency wages, deferred payment schemes and minimum wages – could rotate the wage profile over the life cycle and may result in wage compression. This may be the consequence of various welfare-state arrangements that affect wage setting such as benefits for sickness, disability and unemployment as well as pensions and early retirement schemes. In addition, the value of outside options for workers typically increases as they become older. Both wage compression and rotation of wage profiles can harm the incentives to invest in skills. In particular, by reducing the wage differences between skilled- and unskilled workers, the incentives to become skilled diminish. Further, by rotating the wage profile over the life cycle, younger workers will invest too much, and older workers too little, in human capital.

The main message of this chapter is that any policy reform should take into account the dynamic interactions of OJT investment, retirement and pension saving. The following policy-relevant implications appear from the analysis:

- Promoting life-long learning or later retirement will not be effective if strong disincentives caused by labour-market institutions, early retirement schemes and incentives for pension savings remain in place.
- Promoting private savings for old age may inadvertently create implicit taxes on skill formation and indirectly stimulate early retirement, thereby worsening the ageing problems.

After having elaborated on the theoretical structure from which these arguments are derived, the chapter will discuss in more detail i) its underlying assumptions, ii) its empirical content, and iii) various competing theories. This exercise will reveal a number of important gaps in our knowledge. Attention will be paid to the assumptions regarding the functioning of labour and financial markets. Labour-market distortions due to, for example, unions or minimum wages, are expected to affect the incentives for OJT training. Similarly, borrowing constraints or non-insurable risks affect the incentives to invest in human capital.

The chapter will elaborate on various theoretical, empirical and methodological issues when bringing the theory to the data and will demonstrate that the empirical evidence is very much in line with the theoretical framework. However, certain important data limitations prevent us from directly proving empirically that the standard, neoclassical human capital model causally explains life-cycle earnings. The most pressing problem is that investments in OJT are hard to measure by the analyst, and that most proxies used in empirical analyses have substantial shortcomings. Other, competing theories of life-cycle earnings determination and investments in human capital could explain salient features of the data as well. The chapter will therefore discuss theories of specific investments in human capital (Becker, 1964), and general training in distorted labour markets (Acemoglu and Pischke, 1998, 1999), incentive theories and deferred payment schemes (Lazear, 1976, 1979, 1981), and learning-by-doing theories (Killingsworth, 1982; Heckman et al., 2002). The argument will be made that theories of specific investments and training in non-competitive labour markets have some empirical implications that are counter-factual. Moreover, learning-by-doing and on-the-job training models are hard to distinguish from each other once general equilibrium feedbacks have been taken into account. Theories on deferred payments are not concerned with human capital investments, but do probably explain part of the patterns in earnings over the life cycle.

The chapter contends that the remaining gaps in knowledge are large. In order to close the gaps, future research should be directed towards using structural models that aim to identify non-observable human capital investments by imposing theoretical structure on the data. However, identification of these non-observables is as good as the theoretical structure used. Hence, better theories are needed to understand investments in human capital as well as retirement and pension choices over the life cycle. In particular, development of models with labour-market distortions appears to be key in understanding life-cycle choices in European-style labour markets. As regards the data, micro-panel data are needed to properly estimate life-cycle models. Little can probably be learned from cross-country panel studies, since the time-series variation is often too limited, and identification of effects on the cross-sectional dimension of the data is often rather problematic, from an econometric point of view. Researchers from multiple subdisciplines should join forces to make scientific progress. In particular, structural micro-econometricians and micro-, macro-, and labour theorists should cooperate.

2   A stylised theory of training, retirement and saving

Jacobs (2009a) develops a standard life-cycle model with a representative household to analyse training, retirement and saving decisions. His analysis closely follows Heckman and Jacobs (2010) and Jacobs (2009b) by adding an endogenous retirement decision to the otherwise standard Ben-Porath (1967) model of OJT investments (see also Heckman, 1976; and Weiss, 1986). This is the canonical model to analyse OJT. Although savings are made to ensure consumption smoothing over the life cycle, most savings will be made for the retirement period, in which individuals have no labour earnings. The individual starts his or her life without financial assets and can borrow and lend on a perfect capital market. There is no endogenous (initial) education choice, and there are no labour-supply decisions on the intensive margin (that is, hours of work). The model is deterministic and there is no risk in earnings, longevity, and so forth. A partial equilibrium set-up is chosen in which the paths of the rental rates for human capital and the interest rate are exogenously given.[2] Labour markets are perfectly competitive and frictionless. Upon entering the labour market, the individual may devote some time to training on-the-job in order to augment his or her stock of human capital, which raises his or her future earnings potential. The individual optimally chooses i) consumption at each moment of the life cycle, ii) human capital investment at each moment while active in the labour market, and iii) the date of retirement so as to maximise lifetime utility.

The saving decision is governed by the standard Euler-equation for consumption. If the rate of time preference is lower than the real after-tax return on financial saving, consumption features an upward-sloping profile over the life cycle. A larger intertemporal elasticity of substitution results in a stronger upward-sloping consumption profile and a stronger sensitivity of saving with respect to net after-tax returns. A larger tax on saving reduces the slope of the consumption profile if the substitution effect in saving dominates the income effect (the empirically relevant case).

Optimal retirement choices ensure that the marginal willingness to pay for an additional year in retirement should be equal to the marginal costs of an extra year in retirement. The marginal costs are the net forgone labour earnings in the last year on the labour market. The retirement choice is distorted not only by the implicit tax on retirement, but also by the explicit labour tax levied on all labour earnings. This direct tax is often overlooked in retirement studies. Due to wealth effects, richer individuals retire earlier. In addition, a larger tax on (pension) saving gives stronger incentives to retire later. Intuitively, as individuals accumulate fewer assets, a higher tax on (pension) savings provokes a wealth effect, which delays retirement. The individual has stronger incentives to retire later if he or she has acquired more human capital on-the-job, since this raises forgone labour earnings while being retired. Thus, better-skilled workers retire later, when the income effect of higher skills is outweighed by

---

2   This would be the case in small, open economies with perfect capital mobility and perfect substitution between labour types in labour demand.

the substitution effects of higher skills. Similarly, if individuals do not train, and end their career with low levels of human capital, the incentive to retire will be stronger, since the opportunity costs of doing so diminish.

Investment in on-the-job training is such that the marginal costs of an hour devoted to OJT human capital investment (that is, net forgone labour earnings) should be equal to the discounted value of the marginal benefits in terms of higher future wages. Investment in OJT increases if the individual has a higher level of initial education before entering the labour market. Intuitively, initial education raises the productivity of OJT investments, since initial education and OJT are complementary. Investment in human capital falls continuously over time, until it becomes zero at the date of retirement. The reason is that the time horizon over which the returns to the investments can be reaped diminishes, as individuals grow older. At the date of retirement, investments have no remaining value, since the returns on OJT are zero if individuals do not work anymore. The net return on the investment in human capital (that is, after depreciation) must be equal to the net return on financial saving. The labour tax does not affect the net return to human capital, since all opportunity costs and benefits from investments in human capital receive a completely symmetric tax treatment (Heckman, 1976). A higher tax on financial saving makes human capital investment more attractive by lowering the effective rate at which future wage increases are discounted, and by delaying retirement.

Jacobs (2009a) simulated the model for a reasonable set of parameters. Extensive details on the simulations and sensitivity analyses can be found there. All simulations below use the same benchmark values of the parameters. In particular, individuals start working at age 20 and die at age 80. The pure rate of time preference is assumed to be 2.5%, and the real interest rate equals 5%. The elasticity of the human capital production function is 0.6 (see Trostel, 1993). The depreciation rate of human capital is 2%.[3] The uncompensated elasticity of labour-force participation of older workers with respect to the implicit tax on retirement (thus, including wealth- and income effects) takes a value of 0.2.[4] The tax rate on labour income amounts to 50%.[5] The tax rate on interest income

---

3   Depreciation of human capital appears to be modest, since most earnings profiles do not
     tend to level off much at the end of the life cycle (Heckman et al., 1998).
4   The estimates in Gruber and Wise (1999, 2002), OECD (2004) and Duval (2004) imply that
     the uncompensated elasticity of labour-force participation of older workers with respect to
     the implicit tax on retirement (thus, including wealth- and income effects) is approximately
     one-third.
5   Total marginal tax wedges on labour income (including employer contributions and local
     taxes) are on average 51% for 16 advanced OECD countries (Jacobs, 2009b).

is set at 30%, and the implicit tax on retirement is 30%.[6] The model parameters are calibrated such that the individual retires at age 60, he or she invests 71% of his or her time endowment at the start of the career in human capital, and the individual's gross labour earnings per year are 30.6 (thousand euro), on average, during working life.

The baseline time paths of consumption ($C(t)$), the value of total investment in human capital ($WI(t)H(t)$), total labour earnings ($W(1 - I(t))H(t)$), and total human capital ($WH(t)$) are plotted in Figure 10.1, where human capital at date t is denoted by $H(t)$, investment in human capital (as a fraction of total time) is denoted by $I(t)$, and the constant rental rate per efficiency unit of human capital is given by W. Investment in human capital is high at the beginning of the working career, and declines monotonically until the retirement age is reached. The reason is that the payback time of these investments continuously decreases. Hence, returns on investments fall over time. Indeed, labour earnings drop to zero at the retirement age of 60. The life-cycle profile of labour earnings steadily increases until it peaks at age 53, and then decreases slightly afterwards. This reflects both the investment in OJT before the peak and the depreciation of human capital after the peak. There would be no decline in labour earnings at the end of the life cycle in the absence of depreciation of human capital. Also, the total value of the time endowment is plotted ($WH(t)$). This is a measure for total labour productivity, since rental rates are constant over time. It peaks at age 46, before the peak in earnings (Ben-Porath, 1967; Heckman, 1976). The intuition is that at age 46, individuals are still investing about 10% of their time endowment in OJT. Consequently, total labour productivity peaks earlier in the life cycle than total earnings do. The individual also has a valuable time endowment after retirement, although it is steadily depreciating. Investment in human capital drops to zero at retirement, since the marginal value of investment in human capital has become zero at that date. Finally, the life-cycle path of consumption is increasing. The reason is that the net interest rate is larger than the pure rate of time preference. Note that the consumption path is substantially lower than the earnings path, since the latter are denoted in gross terms (that is, before 50% income taxes).

6　Gruber and Wise (1999), OECD (2004) and Duval (2004) show that the implicit tax on retirement amounts to around 30% for an older worker aged 55-65 years, although there are substantial cross-country differences.

Figure 10.1:   Labour earnings, consumption, OJT investment and human capital over the life cycle

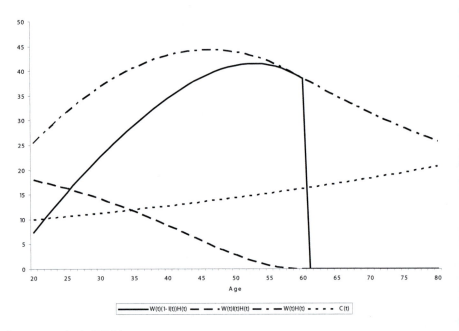

Source: Jacobs (2009a).

Figure 10.2 plots the simulated patterns of OJT investment and life-cycle earnings for different values of the labour tax rate and the capital tax rate. Life-cycle investments in OJT are affected by the labour tax rate through its impact on retirement only (recall that all OJT costs are deductible). Since retirement is distorted by the presence of the implicit tax, a higher explicit tax on retirement reduces OJT investments to a considerable extent, since the payback period of investment in human capital falls substantially. As a result, life-cycle earnings profiles shift towards the origin. As OJT investments fall, the peak of earnings will be earlier. Moreover, since less time will be invested in OJT, earnings when young increase slightly. However, at later ages this is more than offset by lower stocks of human capital, so that earnings decline. This, in turn, makes earlier retirement more attractive, as the opportunity costs of retirement are lower when wages in the final year of work are lower. This graph indirectly shows that policies stimulating earlier retirement can have important consequences for OJT investments (more on this below).

Figure 10.2: Labour earnings and OJT investment over the life cycle for varying labour- and capital income taxes

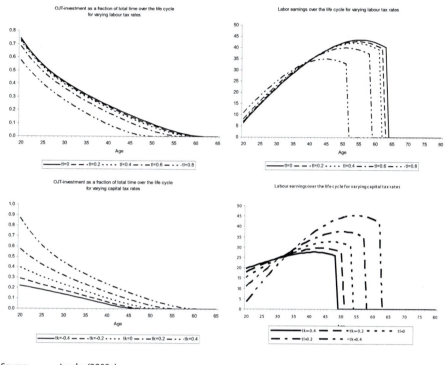

Source: Jacobs (2009a).

A higher capital tax boosts investments in human capital, since saving becomes less attractive compared to investment in OJT. Again, as was seen under the labour tax, earnings profiles rotate, but now in the reverse direction. Especially at the beginning of working careers, OJT investment increases; hence, total gross labour earnings fall. Over time, however, this fall in earnings will be compensated by rising levels of human capital, which result in increasing labour earnings at later ages. The peak in the earnings profile shifts to later ages, and individuals end their working careers with substantially higher earnings. This graph demonstrates the fundamental interactions between saving policies and OJT investments. Indeed, human capital investments can be seriously affected if governments want to boost saving by lowering the capital tax (or even offering tax incentives for saving). Consequently, OJT policies cannot be seen in isolation from pension- and saving policies.

Figure 10.3:   Labour earnings and OJT investment over the life cycle for varying implicit retirement taxes and depreciation rates of human capital

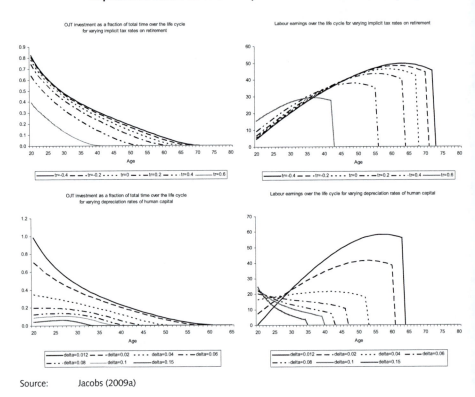

Source: Jacobs (2009a)

Figure 10.3 plots the investment and earnings profiles for various implicit tax rates on retirement and depreciation rates of human capital. A higher implicit tax on retirement – much like the labour tax – provides stronger incentives to retire early from the labour market. Indeed, investment in human capital falls during all ages. Although this increases earnings temporarily (as workers have higher labour earnings at the beginning of the life cycle), their wage growth over the life cycle will be substantially lower. Since less human capital will be accumulated, workers end up with lower wages at the end of their careers. This makes retirement also more attractive, as the opportunity costs of retirement have fallen. Thus, when retirement schemes are actuarially very unfair, thereby causing large distortions on retirement, this seriously impairs investments in OJT, as well. As a result, our theoretical model confirms the notion that individuals do not invest in skills because they retire early, and they retire early because they do not invest in skills.

A larger rate of depreciation of human capital has similar effects as a higher implicit tax on retirement, except that the consequences of higher depreciation rates are more severe. Indeed, the higher depreciation rate makes saving in financial capital relatively more attractive at all times; hence, investments in human capital decrease throughout the life cycle. Indeed, at relatively modest depreciation rates (5% and higher), earnings profiles even become downward sloping over the life cycle. The reason is that the depreciation rate has become larger than the real interest rate, so that human capital decumulation has become optimal.

The messages from the retirement-augmented Ben-Porath model are clear and simple. Investment in OJT shifts the wage profiles upwards, which implies that there are positive returns to OJT investments. Investment in OJT increases if the retirement date increases (lower explicit and implicit taxes on retirement), if the opportunity return on saving decreases (higher capital taxes), and if the depreciation rate is lower. The life-cycle earnings profile is typically 'hump-shaped'. Moreover, policies that boost investment in human capital depress earnings at the beginning of the life cycle and boost earnings at later ages. This is because the cost of investment is forgone working time. Finally, the policy environment is critical to understand life-cycle patterns in OJT investment, labour earnings, retirement ages and savings behaviour. Indeed, financial saving and human capital investments are substitutes, whereas retirement and human capital investments are complements. However, caution should be exercised in drawing strong quantitative conclusions. All simulations are driven by the particular assumptions on the parameters of the model. Jacobs (2009a) provides extensive sensitivity analyses for widely differing parameters.

Jacobs (2009a) thus provides a parsimonious theory of investment in human capital, saving and retirement that contains a number of empirically testable implications:

- Earnings tend to be 'humped-shaped', and labour productivity peaks before earnings.
- Investment in human capital decreases with age.
- Investment in training increases if productivity of training is larger (due, for instance, to larger investment in initial education.
- Retirement ages decrease with the implicit or explicit tax rate on continued work, which in turn reduces OJT.
- (Retirement) savings decrease with a larger tax on savings, which in turn boosts OJT.

3   Empirical content

This section argues that the empirical evidence is in line with the stylised features of the theory described above.

3.1   Earnings profiles and OJT

Age-earnings profiles are indeed hump-shaped, which follows from the commonly estimated Mincer wage equation, with experience (age) and experience squared (age squared) (see Card, 1999).

Direct measurements of productivity over the life cycle are indeed quite suggestive of a hump-shaped pattern of productivity of the life cycle as well. Note, again, that productivity does not equal labour earnings, because of investments in training. Skirbekk (2005) surveys the literature and finds the following stylised facts. Cognitive abilities decline after some stage in adulthood. Older workers compensate withering cognitive skills with sufficient working experience (for example, by OJT or learning-by-doing). Based on subjective evaluations of managers, age-productivity profiles do not seem to display systematic patterns. Evaluations by workers suggest that worker productivity indeed falls at older ages. Objective evaluations (based on measured outputs) suggest that quantity and quality of output show a hump-shaped pattern with age. Importantly, Skirbekk also presents empirical evidence that labour productivity measures peak before labour earnings, which is also theoretically predicted by Jacobs (2009a).

However, from the hump-shaped pattern of earnings one cannot conclude that they are *caused* by investments in OJT. Indeed, other theories of wage determination over the life cycle could also be relevant (deferred payments, learning-by-doing, wage-setting institutions, and so forth). Skirbekk resorts to Lazear's (1976) theory of deferred payments to explain the earnings profiles. This theory will be discussed later in more detail as well.

Direct estimates of the effect of training activities on wages generally give positive wage returns (Leuven, 2005; Bassanini et al., 2006). Allocating time to training activities is correlated with rising wages over the life cycle. However, the empirical evidence also seems fragile, due to selectivity problems in the estimations. Moreover, some serious measurement issues prevent drawing strong conclusions (see below).

3.2 Time horizon and complementarity with initial education and OJT

Given a finite horizon, younger workers are expected to participate more in training, since the payback period of their investments is larger. Furthermore, better-educated workers are also expected to invest more in training, since training increases with the productivity of training activities. Both are indeed found to be stylised facts in the data (Leuven, 2005; Bassanini et al., 2006).

3.3 Participation and OJT

Another stylised fact is that male workers have higher participation rates in training than female workers do. One obvious explanation is that men work more hours and have higher labour participation. Consequently, their 'utilisation rates' of OJT human capital are higher. Jacobs (2009a) does not allow for an endogenous work/participation decision (hence, this feature has been missing thus far in the discussion). However, Heckman and Jacobs (2010) extend a similar model with endogenous labour supply and find that workers with less labour supply utilise their human capital less and therefore invest less in OJT. Women could be outside the labour market because they invest more in the human capital of children, which is something that we abstracted from.

3.4 Retirement and OJT

Gruber and Wise (1999) show that the labour-force attachment of the average worker is rapidly declining with age. Many workers retire long before statutory retirement ages via all kinds of early-retirement schemes. Pension benefits can be generous as well. Pension-replacement incomes in Continental Europe are quite high, and about 60-80% of pre-retirement earnings for an average worker (OECD, 2005). In most countries, pensioners receive the main part of their pension incomes from PAYG schemes. Exceptions are the Anglo-Saxon countries, the Netherlands, Sweden and Denmark, which also heavily rely on substantial private funding – either through DB/DC occupational pensions or individual saving schemes (OECD, 2005). It is not easy to make international comparisons because the institutional details vary from country to country. Gruber and Wise (1999) summarise the impact of early retirement schemes on the labour market by the implicit marginal tax rates imposed on an additional year of work. Duval (2004) and OECD (2004) demonstrate that early retirement schemes do indeed cause very high marginal tax rates on pre-retirement incomes. Moreover, retirement ages and benefit generosity are very negatively related. Gruber and Wise (1999, 2002) present strong evidence that this is a causal relation. In recent years, some countries have attempted to reform their pension schemes. The Netherlands, Germany, France and Italy are examples.

Bassanini et al. (2006), using a simple cross-country panel analysis, suggest that OJT investments and later retirement are indeed positively correlated. This is consistent with the findings here. Moreover, skilled workers typically retire much later than unskilled workers (OECD, 2006). Since education and training are complementary activities, this should come as no surprise, either.

3.5   Pensions

Little is known about the impact of saving or pension policies on the incentives for OJT investments. There doesn't seem to be any empirical evidence that directly estimates the impact of saving and pension policies on OJT investment. At least theoretically, saving and investing in human capital are substitutes for a given level of overall (that is, human and financial) saving. Hence, a higher tax rate on financial saving tends to boost human capital investments. However, also the level of saving can be affected by taxes on savings, depending on offsetting income and substitution effects. Clearly, tax incentives are important for financial saving decisions (see Bernheim, 2002). The earlier empirical literature found only small effects of tax incentives on saving. On balance, however, most of the recent empirical evidence clearly points to a dominant substitution effect in saving (Bernheim, 2002).

4   Theoretical, empirical and methodological issues

Does Jacobs (2009a) provide the correct framework to analyse the interactions between human capital, retirement and pensions? This cannot be answered, for various reasons. First of all, his model made a number of important assumptions, which may not be warranted, from an empirical point of view. Second, the standard human capital model may not be the right model to capture life-cycle earnings. Various competing theories are available. Third, both measurement and methodological problems prevent the direct testing of the model. These issues will now be discussed in more detail.

4.1 Modelling assumptions

The most important modelling assumptions include the following:

- Perfectly competitive labour markets: the wage rate per unit of skill is constant and equal to productivity per unit of skill.
- Life-cycle earnings profiles are driven by investments in human capital.
- Perfect capital markets: all assets are liquid, borrowing and lending at common risk-free rate is possible.
- Perfect insurance markets: there is no risk/uncertainty, and life expectancy is certain.
- No heterogeneity: there are no differences in abilities and life spans.
- Rational expectations, individuals are blessed with perfect foresight: there is no myopia, no hyperbolic discounting, and so forth.

Subsection 4.2 is entirely devoted to a discussion of the assumption of perfectly clearing labour markets. This seems to be the most important assumption made so far, and will be discussed extensively below. In addition, subsection 4.2 discusses various competing theories that could also explain observed patterns in the data. The remainder of this subsection briefly discusses the other assumptions.

Capital-market failures

Naturally, capital markets can fail, and not all assets are perfectly liquid. Illiquid housing wealth, for example, represents a large fraction of total financial wealth in household portfolios in most Western countries. Moreover, individuals can be borrowing-constrained. In some countries, such as the Netherlands, individuals are obliged to save for retirement in collective labour agreements. This also generates borrowing constraints if individuals cannot collateralise their pension wealth. Both liquidity and borrowing constraints can, in theory, impede socially desirable investments in human capital.

The micro-econometric literature contains ample empirical evidence that borrowing constraints could be important for consumption choices (see Hall and Mishkin, 1982; Hayashi, 1985; Mariger, 1987; Zeldes, 1989; Attanasio, 1995; Browning and Lusardi, 1996; Blundell, 1988). Empirically, direct evidence is missing of borrowing constraints on investments in on-the-job training (see Bassanini et al., 2006). The prime reason is that both costs and returns are hard to verify for the analyst – an important issue that will receive attention below. A large literature identifying the role of liquidity and borrowing constraints for initial education only finds small effects for the lower end of the income distribution (see Carneiro and Heckman, 2003; Cunha et al., 2006, and the studies they cite). There is thus some temptation to conclude that borrowing constraints should probably not be the primary focus of future research. Indeed, in the simulations, savings are made mainly for the purpose of saving for retirement (Jacobs, 2009a). Hence, if this theory is only roughly plausible, then binding borrowing constraints are expected to affect the results for training, although probably not to a very large quantitative extent.

Risk and uncertainty

We also abstracted from non-insurable risk and uncertainty, thereby ruling out any precautionary savings or any effects of risk on human capital investments. Browning and Lusardi (1996) argue that precautionary saving, in particular, is an empirically important component of the financial savings of households.

How risk affects human capital investment is critically determined by the ways in which human capital affects the risk to which individuals are exposed (Jacobs et al., 2008). If human capital investment increases risk in labour earnings, then risk-averse individuals will underinvest so as to reduce their exposure to risk. However, if human capital investment reduces the exposure to risk, the opposite holds true: that is, risk-averse individuals will overinvest (see also Levhari and Weiss, 1974). Empirically, little is known about the risk properties of human capital (see Jacobs, 2007; and Jacobs et al., 2008, and the references cited there). Indeed, better-skilled individuals seem to have both a larger variance in earnings and a lower incidence of unemployment, sickness and disability. Moreover, better-skilled individuals participate more and retire much later. Consequently, investment in human capital can both increase and decrease earnings risk. The impact of longevity risk on human capital investment is shown to be ambiguous in theory, as Sheshinksi (2009) recently showed.

Neither theory nor empirical research seems to be available that simultaneously addresses uncertainty in human capital returns and risky returns on savings. Note that labour earnings do not directly measure the marginal return to investing one unit of resources in human capital. Therefore, it is not clear how returns to human capital co-vary with the returns on risky assets, and how the covariance structure changes over the life cycle. Allowing for uncertainty seems to be a potentially important avenue for future research – especially when it comes to understanding financial planning, skill maintenance, and retirement and pension choices. However, a thorough analysis of risk in life-cycle settings with endogenous human capital formation requires much more advanced theory and empirical research.

Heterogeneity and distributional issues

Thus far, the model has completely abstracted from distributional issues. Instead, the focus has been on a positive analysis of understanding the behavioural interactions between OJT, pension saving, and retirement. It has been argued that promoting pension saving or introducing early retirement schemes may have (unintended) adverse consequences for investments in human capital. However, this does not imply that these policies are socially undesirable. Naturally, many public policies could be justified by distributional concerns. For example, it may well be worthwhile to introduce distortions in retirement choices if this helps to redistribute resources to the lifetime poor – those who have been born with low ability or have been hit by adverse skill shocks during the life cycle (Bovenberg and Jacobs, 2009). This could be applied more generally. Heterogeneity and finite lives render the taxation of labour income optimal (Mirrlees, 1971) and could also make taxation of capital income optimal (Jacobs and Bovenberg, 2008). However, the implications of life-cycle considerations for the optimal setting of tax rates on labour or capital income over the life cycle are not yet fully crystallised (Diamond and Banks, 2009).

Two other potentially important sources of heterogeneity could originate from differences in depreciation rates for human capital or differences in life expectancy. Typically, less-skilled individuals seem to have higher depreciation rates of human capital (due to physically more-demanding jobs) and have a lower life expectancy (due to more unhealthy lifestyles). A higher life expectancy could also be viewed as the result of investment in human capital (that is, investment in health). Future research should dwell upon these issues in more detail.

Rational and forward-looking behaviour

Like any life-cycle model, also Jacobs (2009a) relies heavily on the forward-looking- and rational behaviour of households. Recent developments in behavioural economics have pointed to the weaknesses of this traditional framework for analysing saving- and investment decisions over long time horizons – especially when it comes to retirement and pension decisions. Many of the arguments raised in the behavioural economics literature could also be applied to investment in human capital over the life cycle. Certainly, concepts from behavioural economics could be fruitfully applied in this area.

Health, fertility and bequests

The theory presented above also abstracted from various important aspects such as health, fertility and bequests. This is not to say that these issues are not important – on the contrary. However, one must start somewhere in considering life-cycle interactions between human capital, retirement and pensions. Naturally, health conditions are an important ingredient of both labour-market outcomes and retirement choices. Again, we expect that interactions between human capital investments and health are important. Health could even be seen as a specific form of human capital. Also, longevity risk can be endogenous and (partly) determined by human capital investments. Similarly, the model paid no attention to fertility choices – and it is well-known that these are importantly associated with skill levels of individuals. How fertility choices, human capital investments and pension policies interact is less well-known, however – let alone, what optimal policies should look like. The model, moreover, abstracted from bequests and issues on intergenerational redistribution (see also Cremer and Pestieau, 2006). In real life, these are importantly intertwined with the design of pension and retirement policies. Moreover, financial bequests could be substitutes for bequests in the form of investment in human capital (Grossman and Poutvaara, 2009).

4.2 *Imperfectly competitive labour markets and alternative theories for life-cycle earnings*

The most important assumption in the theory outlined above is that the labour market is perfectly competitive and frictionless. The rental rate of human capital equals the productivity per efficiency unit of human capital.[7] However, a perfectly competitive labour market is unlikely to serve as a valid approximation for many European labour markets, which are characterised by all sorts of frictions, institutions and government interventions. In non-competitive labour markets, the theoretical connection between the productivity per efficiency unit

---

7 In addition, workers with different vintages of OJT human capital are perfect substitutes in production, so that rental rates per unit of human capital are equalised across all individuals with different levels of OJT human capital. At first sight, it would seem implausible that this would indeed be the case. Empirical evidence on these matters, however, is currently lacking.

of human capital and the price per efficiency unit of human capital is generally lost. Wages (or more precisely, rental prices per unit of human capital) then do not purely reflect productivity, but also market frictions. Moreover, the rental rates do not need to be constant over the life cycle.

In addition, our model had assumed that the Ben-Porath (1967) human capital model is the correct explanation for the life-cycle earnings patterns observed in the data. Nevertheless, other theories can also explain why earnings profiles are hump-shaped. Some of these alternative theories are directly tied to frictions in the labour market. These are discussed in this section as well. The main message of this section is that understanding the working of the labour market is key to understand the interactions between training, saving and retirement.

Minimum wages

A wage floor in an otherwise competitive labour market destroys employment for all workers that have labour productivity below the wage floor. This results in involuntary unemployment among these workers. Wage floors increase the wages of unskilled workers relative to skilled workers. Consequently, incentives to invest in OJT diminish. Minimum wages may also generate general equilibrium effects on the wage structure by changing relative supplies of workers (Teulings, 2003). In addition, the employment probabilities of low-skilled workers diminish, and incentives to become skilled improve. Hence, the effect of wage floors on skill formation is ambiguous. If the adverse employment effects on the low-skilled generate sufficiently strong incentives to counter the decline in the skill premium, then investment in human capital might even increase (Gerritsen and Jacobs, 2010).

Unions, efficiency wages, frictions and insider-outsider problems

In a wide class of models with unions, efficiency wages, search frictions or insider-outsider problems, equilibrium wages are typically characterised by a mark-up equation relating the equilibrium wage to the outside options of workers (see Booth, 1995; Mortensen and Pissarides, 1999; Akerlof and Yellen, 1986; Lindbeck and Snower, 1998, 2002). Equilibrium unemployment results because wages are pushed above the market clearing level. The wage mark-up generally increases with greater bargaining power of workers, a lower elasticity of labour demand, a higher replacement rate, lower marginal- and higher average income tax rates, higher firing costs and better employment protection (see Layard et al., 1991; Pissarides, 1998; Sørensen, 1999; Lindbeck and Snower, 2002; Bovenberg, 2006; Van der Ploeg, 2006).

Labour-market frictions will not only have static effects, but also affect the wage structure over the life cycle. Employment protection legislation typically protects older workers better than it does younger workers. Labour turnover costs increase with workers' experience, due to higher firing costs, stricter employment protection legislation, seniority rules ('last in, first out'), and other terms of employment. Older workers may have more bargaining power than younger workers, which is relevant for labour markets with unions, search frictions, and insider-outsider problems. Typically, entitlements to social benefits increase with work experience and with income. Hence, outside options become more valuable as workers get older. All theories on non-competitive labour markets (unions, search frictions, efficiency wages, insiders-outsiders) then imply that wages are pushed more above market clearing levels as workers age. The actual design of labour-market policies, tax systems and social benefits is therefore critical in understanding how the outside options of workers are affected over the life cycle (see Bovenberg and Van der Ploeg, 1994). Most analyses in the training literature pay insufficient attention to the tax treatment of both earnings and outside options, the way in which entitlements to benefits are built up over time, whether benefits are related to final earnings, and so forth.

Worker incentives

The wage profile rotates also in Lazear's (1976, 1979, 1981) incentive theories of deferred payments, mandatory retirement and hour restrictions. By changing the earnings over the life cycle, the firm can provide incentives to workers if the firm cannot observe their productivity levels. Typically, an optimal contract features lower wages than labour productivity at the beginning of the life cycle and higher wages than labour productivity at the end of the life cycle. As such, also incentive issues can explain a hump-shaped pattern of earnings. Given the above market-equilibrium wage at the end of the life cycle, it is optimal to have mandatory retirement (Lazear, 1979). And, given that wages are not constant across years, it is optimal to have hour restrictions in order to avoid welfare losses of distortions in labour supply (Lazear, 1981).

Effect of non-competitive wage setting on OJT

One might be tempted to conclude that in non-competitive labour markets, investments in OJT will be reduced, as wages (the main cost of the investment) will be driven above market-clearing levels. Investing in human capital thus becomes less attractive. However, also here some individuals will be priced out of the labour market and become unemployed / inactive. Employment rates are indeed much higher among the better-skilled workers, and better-skilled individuals retire much later (OECD, 2006). Therefore, investment in OJT might also be boosted in non-competitive labour markets if workers want to lower the probability of becoming unemployed or inactive. As a result, the impact of labour-market institutions on OJT appears to be ambiguous from a theoretical perspective.

If the wage profile indeed tilts in favour of older workers – due to labour-market frictions, institutions, or deferred payment schemes – then the incentives to invest in schooling and training can be considerably affected. Older workers face weaker incentives to maintain skills and will invest less in second careers because the opportunity costs of doing so increase. Younger workers, on the other hand, have stronger incentives to invest in their careers early. However, if the tilting wage profile also affects unemployment rates, then older (younger) workers might get also stronger (weaker) incentives to invest in human capital so as to avoid unemployment. The tilting of the wage profile can promote steeper depreciation of human capital over the life cycle. Incentives to retire early increase, and employment rates of older workers decrease (see also the model simulations in the previous section). This is not necessarily efficient, and may be costly in terms of labour supply. As a corollary to Lazear (1979, 1981), binding limits on training for younger workers and compulsory OJT programmes for older workers could be optimal – for a given retirement age – to avoid distortions in human capital accumulation over the life cycle if an increasing wage profile is used to provide work incentives. This is conjecture, however.

The direct evidence on the effect of labour-market imperfections on training is rather inconclusive (Bassanini et al., 2006). There indeed appears to be some evidence that increased opportunity costs (due to minimum wages, for example) reduce investments in OJT. However, most empirical testing typically suffers from sample attrition biases. The reason is that more productive workers have positively selected into jobs, whereas unproductive workers would have become unemployed and vanish from the data samples being analysed. Empirical testing of different labour-market settings on cross-country data is also highly problematic. Institutions are slowly varying over time, and the econometrician has to rely on cross-country differences to identify the effects. However, allowing for country-specific effects generally destroys any cross-sectional correlations found in cross-country panel analyses (see, for instance, Heckman and Pages, 2003). Moreover, estimates relying on the cross-sectional dimension could be biased, due to cohort effects. Ideally, micro-panel data are needed to identify life-cycle impacts of various labour-market settings, but this is not often done.

Monopsony

Both non-competitive labour markets and deferred payments could result in wage distributions that will not be 'compressed', but 'decompressed' over the life cycle, since earnings at the end of the life cycle increase, and those at the beginning decrease. This contrasts sharply with many modern training theories that emphasise the monopsonistic nature of labour markets (see Acemoglu and Pischke, 1998, 1999). Similar to the literature on minimum wages in monopsonistic labour markets (see Manning, 2003), this line of research essentially argues that wages are driven *below* productivity levels by firms that exert monopsonistic or oligopsonistic wage-setting powers. Consequently, firms may even pay for general training, a finding that contrasts with Becker (1964). The intuition is that productivity of workers increases faster than the wages that the firm will pay: hence, firms benefit from investing in general skills that increase the productivity of workers.

Since the labour market is typically inefficient (due to wages that are set below labour productivity), minimum wages, unions and other wage-increasing mechanisms may in fact be second-best optimal. Monopoly-like behaviour on the labour-supply side is a countervailing power to monopsonistic behaviour of firms, so that wages can be better aligned with labour productivity (see, for example, Booth and Chatterji, 1998; Acemoglu and Pischke, 1999, 2003).

An important empirical issue is whether wages (or, more precisely, rental rates of human capital) would indeed be driven *below* market-clearing levels – and the more so for better-trained workers. All unemployment or underutilisation of human capital would then be voluntary. Moreover, a 'compressed' wage distribution would not only increase employment, but also boost investment in human capital. A priori, this seems hard to believe, given the apparent lack of skills of many workers who (involuntary) end up as being unemployed.

Welfare-state interventions are indeed associated with compressed wage structures (Freeman and Katz, 1995; Blau and Kahn, 1996; Gottschalk and

Smeeding, 1997). Non-employment is generally higher in countries with 'compressed wage structures', in comparison with those featuring more competitive labour markets. Wages are raised above market-clearing levels in corporatist labour markets, especially at the low-end of the wage distribution and for older workers (given the much larger prevalence of non-employment among these groups). At the same time, corporatist countries with stronger labour-market regulations and more extensive welfare states have more steeply increasing age-earnings profiles compared to the countries with more competitive labour markets (Brunello, 2000; CPB, 2009). Hence, life-cycle earnings profiles 'decompress', rather than 'compress', due to various labour-market interventions. It is therefore important to distinguish between age-earnings profiles and cross-sectional wage distributions. Cross-sectional wage distributions can indeed be compressed, but age-earnings profiles need not.

Monopsony-based theories struggle to explain unemployment, especially among the older workers. Indeed, if monopsony were the true characterisation of labour-market imperfections, then employment rates of elderly workers would be much *higher* than employment rates of younger workers (since firms extract more monopsony rents from older than younger workers because they accumulated more human capital through OJT). Monopsony-based labour-training theories could therefore be a red herring, empirically.

Specific investments

Not all OJT investment is general, as was stressed by Becker (1964). Some investments in human capital are specific to the employer-worker relationship. If the labour market is perfectly competitive, the firm pays for all costs and benefits of the investment. This provides an explanation why firms seem to pay for most OJT investments of workers (Bassanini et al. 2006). Since the firm is the residual claimant of the specific investment, one could say that the firm 'owns' all specific human capital. The worker just receives the spot wage rate in the labour market that would be obtained without any specific investments (see also Leuven, 2005). As a result, firm-specific investments in human capital cannot explain the hump-shaped age-earnings profiles. If the spot wage rate would be flat – as we assumed in the model above – then the labour-earnings profile would be flat, too. More generally, specific investments would typically flatten age-earnings profiles, which go in the opposite direction of explaining the hump-shape in earnings over the life cycle.

With specific investments in human capital, labour earnings must be higher than labour productivity at the beginning of the life cycle, and lower than labour productivity at later stages of the life cycle, if specific human capital is to be accumulated. The intuition is as follows. Perfect competition between firms ensures that profits are driven down to zero in equilibrium. Moreover, assuming perfect mobility across jobs, the present value of earnings in a job with specific investments in human capital must be equal to the present value of a job without specific investments in order to attain equilibrium in the labour market. Thus, as long as labour productivity increases over time, the job with more investment in specific human capital pays higher wages than productivity at the beginning of the life cycle and lower wages than productivity at the end.

Empirically, it is therefore not clear whether specific investments in human capital can go a long way in explaining age-earnings profiles and relatively low employment levels of older workers. Indeed, firms would find their older workers who acquired a lot of specific human capital attractive, as they pay them less than their productivity.[8] It is also practically difficult for the analyst to distinguish specific from general training. Moreover, it is not so clear whether firms really pay for most of the costs of OJT, once the general equilibrium feedbacks in the labour market have been taken into account. Indeed, the workers may pay for the investments by accepting a lower earnings profile in a job with a lot of specific OJT investment. Most empirical analyses abstract from these general equilibrium feedbacks.

Only if labour turnover is introduced into models of specific investments will both workers and firms typically share the costs and returns to the investment in human capital. The intuition is that the firm does not find it attractive to invest in specific investments if there is a probability that the worker will quit the firm. Then, wages will be increasing over the life cycle. However, the presence of exogenous labour turnover must be due to some form of contract incompleteness or some form of market friction. For example, it is generally impossible for firms to claim part of the wages of workers once they quit the firm. Alternatively, there can be various sources of asymmetric information or differences in bargaining power between the employee and the firm. As a result, various types of hold-up problems emerge, which may result in inefficient levels of OJT investment and inefficient quits (Hashimoto, 1980; Leuven and Oosterbeek, 2001; Leuven, 2005).

The empirical implications of specific OJT are similar to those of the monopsony models. Indeed, monopsony power is also driven by specificity in worker-employer relationships (Acemoglu and Pischke, 1998, 1999). Consequently, also

---

8   Note that firms are generally not interested hiring in older workers with a high level of specific human capital acquired in other firms.

theories on specific investments cannot explain why especially older workers would be more unemployed than younger workers.

Learning-by-doing

Wage profiles might not be generated by OJT, but by learning-by-doing (LBD). The basic idea is simple. As long as workers are employed, they accumulate work experience. Since older workers have accumulated more work experience, their productivity levels will be higher, and – in competitive labour markets – their wages will rise over the life cycle. The distinguishing feature of learning-by-doing models is that there is no trade-off between current and future earnings, as in the standard human capital models. In the latter, working time and investment in OJT are rivalrous. In LBD models they are not; current earnings raise future earnings, as higher current earnings reflect more labour effort, which implies that there is more learning-by-doing (see also Killingsworth, 1982; Heckman et al., 2002).

However, learning-by-doing theories resemble standard OJT theories once a general equilibrium perspective is adopted (Heckman et al., 2002). In a partial equilibrium setting, the acquisition of human capital appears as manna from heaven in LBD theories. However, this is a problematic feature in general equilibrium. Jobs that feature a lot of LBD would have a larger present value of earnings than jobs without human capital accumulation through LBD. Equilibrium in the labour market would then require that jobs without LBD must have the same present value in earnings as jobs with LBD, as long as competition drives the firms' profits to zero. Suppose that a job without LBD pays a flat spot wage rate, then the job with LBD must pay lower wages at the beginning of the life cycle and higher wages at the end of the life cycle for the present value of wages in the LBD job to be equal to the job without LBD. Hence, the LBD model is observationally equivalent to the standard Ben-Porath model, and under some conditions the models might even become identical (Killingsworth, 1982; Heckman et al., 2002). Learning-by-doing models are therefore empirically hard to distinguish from standard human capital models. Indeed, both the time invested in OJT investments and the time spent accumulating work experience are hard to measure. As such, there appears to be no clear-cut way to empirically discriminate between the two theories.

4.3   Measurement of investment and returns

A major empirical problem in the training literature is that investments in OJT (flow) or human capital (stock) are extremely difficult to measure precisely. Neither is easily directly verifiable to the econometrician. Indeed, Heckman (2000) and Carneiro and Heckman (2003) argue that most training is informal, rather than formal. This fundamental non-verifiability of OJT investments severely limits the applicability of commonly employed training measures, which are often based on subjective data (firms or employees) on formal investment in OJT. Generally, regression analyses employ dummy variables that indicate whether workers have participated in (some) training. Moreover, the intensity of training is not always known with much precision. Further, firms and employees seem to have different views on the participation/intensity of training. See also Leuven (2005) for an elaborate review.

Not only the costs (that is, the investment in OJT), but also the returns (future wages) are difficult to measure empirically. The reason is that earnings are not equal to labour productivity – even if labour markets are perfectly competitive –, since time investment in OJT drives a wedge between gross labour productivity and gross labour earnings. This is something that is often overlooked (see, for example, Skirbekk, 2005: 16-18; Bassanini et al., 2006: 9). Clearly, time costs are the most important ingredient of investment in human capital (Mincer, 1958, 1962; Schultz, 1963; Becker, 1964; Trostel, 1993). Thus, worker productivity cannot directly be inferred from labour earnings. As a result, the returns to OJT are quite difficult to measure. Heckman et al. (1998) do obtain estimates, however, by identifying skill prices per unit of human capital from the earnings of the older workers who are in the final years of their careers. Indeed, human capital investments would approximately be zero for these workers, so that labour earnings indeed reflect productivity.

5   Remaining gaps in knowledge: main challenges

The main question can thus be formulated as follows: how can we understand, both qualitatively and quantitatively, the life-cycle interactions between investing in human capital, retirement and pension saving? This chapter started by arguing that answers to these questions are highly policy-relevant, but that no framework is available to understand these interactions, with the exception of Jacobs (2009a) and Heckman and Jacobs (2010). These frameworks help to shed light on a number of potentially important life-cycle interactions. Although these assertions answered some questions, they also raised numerous new ones. The previous section pointed out an important number of gaps in our knowledge. To address these gaps, this final section attempts to sketch a research agenda for the future. This research agenda can be summarised as follows:

- Theory: developing life-cycle models of human capital investment in distorted labour markets.
- Empirics 1: employing structural econometrics to identify non-observable investment in human capital.
- Empirics 2: exploiting quasi-experimental evidence to identify institutional impacts.
- Data: using micro-panel data.

The remainder of this section explains the research agenda in more detail.

5.1 Theory

It is yet unknown what the most appropriate theory is for describing human capital formation and earnings over the life cycle. This chapter started from the Ben-Porath (1967) model of general OJT investments, which is firmly grounded in neoclassical human capital theory. This is a useful benchmark, given that the empirical evidence is completely in line with the predictions of the theory.

However, competing theories could provide alternative explanations for the patterns that can be seen in the data. The learning-by-doing theories are observationally equivalent from a general equilibrium perspective. Hence, it does not seem to matter much for practical purposes whether human capital is accumulated through training on-the-job or learning-by-doing. The theories on specific training and training in monopsonistic labour markets are clearly not compatible with standard human capital models. However, these theories have some predictions that are more difficult to reconcile with the data. While incentive theories (as developed by Lazear) do describe some real-world features of earnings profiles, they say little about human capital accumulation. Hence, for the time being, it seems most practical to start with standard human capital models as developed in this chapter.

Market failures and institutions are likely to be very important, but little is known on their impacts. Although some work on this has been done in static or one-shot models of investment in OJT, the literature in the field shows a completely scattered picture of the impacts of different labour market settings and institutions on OJT or life-cycle earnings.

5.2   *Structural estimation*

Employing even the simplest human capital framework to analyse human capital investments over the life cycle involves a host of methodological issues and data problems. Indeed, the data are likely to remain a substantial bottleneck, because training in firms is hard to verify/measure by the analyst. Also, the returns to OJT are difficult to quantify, given the non-verifiability of investments (flows) and human capital levels (stocks).

Developing structural models appears to be the most promising – and possibly the only – route for future research. Time investment in human capital is mostly informal, and cannot, by definition, be precisely measured by researchers (Carneiro and Heckman, 2003). The estimation of structural models allows the identification of non-observables such as time invested in OJT (see, for example, Heckman et al., 1998; Heckman et al., 2002). It seems unwise to continue on the path of using very soft, noisy, and often subjective data on training efforts by workers and firms. Bassanini et al. (2006) and Leuven (2005) identify major problems with this line of research.

However, structural empirical models need to be firmly grounded in theory. The identification of non-observables is as good as the theoretical structure that is imposed on the data. In particular, the modeling of the market structure is key. Before any serious structural estimation can be done, it is therefore urgent to theoretically analyse labour-market imperfections, capital markets and various institutional details in dynamic human capital models.

5.3   *Quasi-experiments*

The empirical literature has produced disappointingly little evidence on the impacts of labour-market institutions on investment in human capital. The difficulty involved in measuring costs and returns of investment in human capital is, again, one of the culprits. However, also identification problems in estimating the impact of various market structures on OJT investments are pervasive, since many of the impacts of labour market and institutional details may not be individual-specific, and may change only slowly over time. Consequently, structural methods (to identify OJT investment) should be combined with quasi-experimental evidence (due to policy changes, discontinuities in policies, and so forth) or instrumental variables to estimate the impact of institutions, labour and capital markets for the life-cycle patterns of earnings, OJT investments, saving and retirement.

5.4   Micro-panel data

Panel data should ideally be used to identify life-cycle interactions. Estimates based on cross-sectional data could be biased, since life-cycle patterns for individuals generally do not coincide with cross-sectional patterns. Moreover, panel data allow the econometrician to eliminate some of non-observed individual heterogeneity. Finally, panel data are suitable to estimate the impact of quasi-experiments.

A fundamental empirical problem is that most empirical analyses are confined to working individuals only. Hence, most data samples suffer from potentially severe attrition problems, since they do not include non-working individuals that could have been priced out of the labour market. Consequently, the identification of the impact of various labour-market imperfections and institutions could be seriously flawed. Moreover, the role of capital markets, saving and pension policies for human capital investment is a seriously under-researched area. Data collection should therefore take into account the fact that labour-market frictions may result in censored samples. However, non-employed workers need to be included for any meaningful empirical assessment of the impact of labour-market distortions and labour-market institutions.

Gathering more aggregate cross-country evidence would probably be ineffectual in gaining a better understanding of labour markets and life-cycle behaviour of individuals. Indeed, empirical cross-country analyses have produced little, if any, empirical evidence, due to limited time-series variation within countries, and large sensitivity of estimation results to country fixed effects.

6   Current state of play of European research infrastructures and networks

The main problem is that there is no 'current state of play of European research infrastructures and networks'. Various research groups operate within their own disciplines. A large group of mainly microeconomists has done extensive theoretical and empirical work on training (see the authors of Bassanini et al., 2005, and the papers they cite). However, the theoretical focus of this line of research is mainly on stylised static or one-shot human capital investment models. The empirical work is microeconomic in nature, and emphasises instrumental variables and quasi-experimental evidence. Only James Heckman and his co-authors have so far developed structural models of training in life-cycle settings (see for example, Heckman et al. 1998; Heckman et al., 2002).

Similarly, numerous researchers have also been working on retirement, with prominent examples among those participating in the project of Gruber and Wise (1999, 2002). The latter group of researchers adopts mainly a micro-econometric approach. There is hardly any theory on retirement behaviour. Retirement is often seen as a corner solution in labour-supply choices. Alternatively, retirement is modelled according to the Stock and Wise (1990) retirement-option model (for an overview, see De Hek and Van Erp, 2007). Instrumental variables, quasi-experimental evidence and structural methods are all commonly used in this literature. Some authors develop structural dynamic models of retirement and estimate them (Rust, 1989; Van der Klaauw and Wolpin, 2005; Gustman and Steinmeier, 2005; French, 2005; Blau, 2007). Human capital formation plays no discernable role in this literature.

A number of researchers have extensively analysed saving behaviour (for example, Hall and Mishkin, 1982; Hayashi, 1985; Mariger, 1987; Zeldes, 1989; Attanasio, 1995; Browning and Lusardi, 1996; Blundell, 1988). Particularly in the research group of Richard Blundell at UCL/IFS in London, a great deal of research is carried out on life-cycle behaviour in consumption and labour. Human capital formation is generally ignored in these life-cycle models of consumption behaviour.

Europe lacks a unified single research group analysing the joint impacts of labour and capital markets and institutions on the incentives for on-the-job training, pension saving and retirement.

7   Required research infrastructures, methodological innovations, data, networks and consequences for research policy

The requirements to fully understand interactions between human capital, retirement and pensions are demanding. The policy questions raised in the introduction can only be answered by an innovative combination of theory, structural econometrics, quasi-experimental evidence and micro-panel data. Despite the high policy relevance, the complexity of all this may easily become too large, thereby hindering important results being obtained anytime soon. Theorists should develop better life-cycle theories of human capital investment that address the role of labour markets (and their imperfections), capital markets and various institutional details. Empirical economists should start to use more structural models to identify non-observable investment in human capital. They should try to develop empirical strategies to test the relevance of competing theories under different labour-market conditions. Identifying the role of institutions requires quasi-experiments. Only micro-panel data appear to be useful in order to fully identify life-cycle interactions, to obtain unbiased life-cycle profiles, and to make quasi-experimental evaluations. Cross-fertilisation between different subdisciplines in labour theory and econometrics appears to be critical, and achieving this cooperation among different research groups will be vital.

References

Acemoglu, D. and J.S. Pischke (1998) 'Why do Firms Train? Theory and Evidence', *Quarterly Journal of Economics*, 113, 79-119.

Acemoglu, D. and J.S. Pischke (1999) 'The Structure of Wages and Investment in General Training', *Journal of Political Economy*, 107, 539-572.

Akerlof, G.A. and J. Yellen (1986) *Efficiency Wage Models of the Labor Market* (Cambridge: Cambridge University Press).

Attanasio, O.P. (1995) 'The Intertemporal Allocation of Consumption: Theory and Evidence', *Carnegie-Rochester conference series on public policy*, 42, 39-89.

Bassanini, A., A.L. Booth, G. Brunello, E. Leuven and M. De Paola (2006) 'Workplace Training in Europe', in G. Brunello, P. Garibaldi and E. Wasmer (eds) *Education and Training in Europe* (Oxford: Oxford University Press), ch. 8-13.

Becker, G.S. (1964) *Human Capital: A Theoretical and Empirical Analysis with Special Reference to Education*, 3rd edn 1993 (Chicago: University of Chicago Press).

Ben-Porath, Y. (1967) 'The Production of Human Capital and the Life Cycle of Earnings', *Journal of Political Economy*, 75(4), 352-365.

Bernheim, B.D. (2002) 'Taxation and Saving', in A.J. Auerbach and M. Feldstein (eds) *Handbook of Public Economics* (North-Holland: Elsevier), vol. 3, ch. 18.

Blau, D.M. (2007) 'Retirement and Consumption in a Life Cycle Model', IZA discussion paper 2986.

Blau, F.D. and L.M. Kahn (1996) 'International Differences in Male Wage Inequality: Institutions versus Market Forces', *Journal of Political Economy*, 104(4), 791-837.

Blundell, R. (1988) 'Consumer Behaviour: Theory And Empirical Evidence - A Survey', *Economic Journal*, 98(389), 16-65.

Booth, A.L. (1995) *The Economics of the Trade Union* (Cambridge: Cambridge University Press).

Booth, A.L. and M. Chatterji (1998) 'Unions and Efficient Training', *Economic Journal*, 108, 328-343.

Bovenberg, A.L. (2006) 'Tax Policy and Labor Market Performance', in: J. Agell and P.B. Sørensen (eds), *Tax Policy and Labor Market Performance* (Cambridge: CESifo and MIT Press).

Bovenberg, A.L. and F. van der Ploeg (1994) 'Effects of the Tax and Benefit System on Wage Formation and Unemployment', University of Amsterdam/Tilburg University unpublished manuscript.

Bovenberg, A.L. and B. Jacobs (2009) 'Optimal Taxation of Retirement', mimeo.

Browning, M. and A. Lusardi (1996) 'Household Saving: Micro Theories and Micro Facts', *Journal of Economic Literature*, 34(4), 1797-1855.

Brunello, G. and S. Comi (2000) 'Education and Earnings Growth: Evidence from 11 European Countries', IZA discussion paper 140.

Card, D. (1999) 'The Causal Effect of Education on Earnings' in O. Ashenfelter and D. Card (eds) *Handbook of Labor Economics* (Amsterdam: Elsevier-North Holland), vol 3A, 1801-1863.

Carneiro, P. and J.J. Heckman (2003) 'Human Capital Policy' in J.J. Heckman and A.B. Krueger (eds) *Inequality in America: What Role for Human Capital Policy?* (Cambridge: MIT Press).

CPB (2009) *Rethinking Retirement* (The Hague: CPB Netherlands Bureau for Economic Policy Analysis).

Cremer, H. and P. Pestieau (2006) 'Wealth Transfer Taxation: A Survey of the Theoretical Literature' in S.C. Kolm and J.M. Ythier (eds) *Handbook of the Economics of Giving, Altruism and Reciprocity* (Amsterdam: Elsevier Science), vol. 2, 1107-1134.

Cunha, F., J.J. Heckman, L.J. Lochner and D.V. Masterov (2006) 'Interpreting the Evidence on Life Cycle Skill Formation' in E. Hanushek and F. Welch (eds), *Handbook of Economics of Education* (Amsterdam: Elsevier-North Holland).

Diamond, P.A. and J. Banks (2009) 'The Base for Direct Taxation' in *Reforming the Tax System for the 21st Century: The Mirrlees Review* (Oxford: Oxford University Press).

Duval, R. (2004) 'Retirement Behavior in OECD Countries: Impact of Old-Age Pension Schemes and Other Social Transfer Programmes', *OECD Economic Studies*, 37, 2003/2, 7-50.

Freeman, L.B. and L.F. Katz (1995) 'Introduction and Summary' in L.B. Freeman and L.F. Katz (eds), *Differences and Changes in Wage Structures* (Chicago: University of Chicago Press).

French, E. (2005) 'The Effects of Health, Wealth, and Wages on Labour Supply and Retirement Behaviour', *Review of Economic Studies*, 72(2), 395-427.

Gerritsen, A. and B. Jacobs (2010) 'Optimal Minimum Wages and Optimal Redistribution in Competitive Labor Markets with Endogenous Skill Formation', Erasmus University Rotterdam mimeo.

Gottschalk, P. and T.M. Smeeding (1997) 'Cross-National Comparisons of Earnings and Income Inequality', *Journal of Economic Literature*, 35(2), 633-687.

Grossman, V. and P. Poutvaara (2009) 'Pareto-Improving Bequest Taxation', *International Tax and Public Finance*, forthcoming.

Gruber, J. and D. Wise (1999) *Social Security and Retirement around the World* (Chicago: University of Chicago Press).

Gruber, J. and D. Wise (2002) 'Social Security and Retirement around the World: Microestimation', NBER working paper 9407.

Gustman, A.L. and T.L. Steinmeier (2005) 'The Social Security Early Entitlement Age in a Structural Model of Retirement and Wealth', *Journal of Public Economics*, 89, 441-463.

Hall, R.E. and F.S. Mishkin (1982) 'The Sensitivity of Consumption to Transitory Income: Estimates from Panel Data on Households', *Econometrica*, 50(2), 461-482.

Hashimoto, M. (1981) 'Firm-Specific Human Capital as a Shared Investment', *American Economic Review*, 71(3), 475-482.

Hayashi, F. (1985) 'The Effects of Credit Constraints on Consumption: A Cross Section Analysis', *Quarterly Journal of Economics*, 100, 183-206.

Heckman, J.J. (1976) 'A Life-Cycle Model of Earnings, Learning and Consumption', *Journal of Political Economy*, 84, S11-S44.

Heckman, J.J. (2000) 'Policies to Foster Human Capital', *Research in Economics* 54(1), 3-56.

Heckman, J.J., L.J. Lochner and C. Taber (1998), Explaining Rising Wage Inequality: Explorations with a Dynamic General Equilibrium Model of Labor Earnings with Heterogeneous Agents', *Review of Economic Dynamics*, 1, 1-58.

Heckman, J.J., L. Lochner and R. Cossa (2002) 'Learning by Doing vs. On-the-Job Training: Using Variation Induced by the EITC to Distinguish between Models of Skill Formation', NBER working paper W9083.

Heckman, J.J. and C. Pagés (2003) 'Law and Unemployment: Lessons from Latin America and the Carribbean', NBER working paper 10129.

Heckman, J.J., and B. Jacobs (2010) 'Policies to Create and Destroy Human Capital in Europe' in H.W. Sinn and E. Phelps (eds), *Perspectives on the Performance of the Continent's Economies* (CESifo and MIT Press), forthcoming.

Hek, P. de and F. van Erp (2007) 'Analyzing Labor Supply of Elderly People: A Life-Cycle Approach', The Hague: CPB Netherlands Bureau for Economic Policy Analysis mimeo.

Jacobs, B. (2007) 'Real Options and Human Capital Investment, *Labour Economics*, 17(6), 913-925.

Jacobs, B. (2009a) 'A Theory of Human Capital, Saving and Retirement', Erasmus University Rotterdam mimeo.

Jacobs, B. (2009b) 'Is Prescott Right? Welfare State Policies and the Incentives to Work, Learn and Retire', *International Tax and Public Finance*, 16, 253-280.

Jacobs, B. and A.L. Bovenberg (2008) 'Human Capital and Optimal Positive Taxation of Capital Income (revised version)', mimeo.

Jacobs, B., D. Schindler and H. Yang (2009) 'Optimal Taxation of Risky Human Capital', CESifo working paper 2529.

Killingsworth, M.R. (1982) 'Learning by Doing and Investment in Training: A Synthesis of Two Models of the Life Cycle', *Review of Economic Studies*, 49(2), 263-271.

Klaauw, W. van der and K. Wolpin (2005) 'Social Security and the Retirement and Savings Behavior of Low Income Households', PIER working paper 05-020.

Layard, R., S. Nickell and R. Jackman (1991) *Unemployment* (Oxford: Oxford University Press).

Lazear, E.P. (1976) 'Age, Experience and Wage Growth', *American Economic Review*, 66(4), 548-558.

Lazear, E.P. (1979) 'Why Is There Mandatory Retirement?', *Journal of Political Economy*, 87(6), 1261-1284.

Lazear, E.P. (1981) 'Agency, Earnings Profiles, Productivity, and Hours Restrictions', *American Economic Review*, 71(4), 606-620.

Leuven, E. (2005) 'The Economics of Private-Sector Training: A Review of the Literature', *Journal of Economic Surveys*, 19(1), 91-111.

Leuven, E. and H. Oosterbeek (2001) 'Firm-Specific Human Capital as a Shared Investment: Comment', *American Economic Review*, 91(1), 342-347.

Levhari, D. and Y. Weiss (1974) 'The Effect of Risk on the Investment in Human Capital', *American Economic Review*, 64, 950-963.

Lindbeck, A. and D.J. Snower (1988) *The Insider-Outsider Theory of Employment and Unemployment* (Cambridge: MIT Press).

Lindbeck, A. and D.J. Snower (2002) 'The Insider-Outsider Theory: A Survey', IZA discussion paper 534.

Manning, A. (2003) *Monopsony in Motion: Imperfect Competition in Labor Markets* (Princeton: Princeton University Press).

Mariger, R.P. (1987) 'A Life-Cycle Consumption Model with Liquidity Constraints: Theory and Empirical Results', *Econometrica*, 55(3), 533-557.

Mincer, J. (1958) 'Investment in Human Capital and Personal Income Distribution', *Journal of Political Economy*, 66, 281-302.

Mincer, J. (1962) 'On-the-Job Training: Costs, Returns and Some Implications', *Journal of Political Economy*, 70, 50-79.

Mincer, J. (1974) *Schooling, Experience and Earnings* (New York: Columbia University Press).

Mirrlees, J.A. (1971) 'An Exploration in the Theory of Optimum Income Taxation, *Review of Economic Studies*, 38, 175-208.

Mortensen, D.T. and C.A. Pissarides (1999) 'New Developments in Models of Search in the Labor Market' in O. Ashenfelter and D. Card (eds), *Handbook of Labor Economics* (Amsterdam: Elsevier-North Holland), *vol 3B*, 2567-2627.

Nickell, S. (1997) 'Unemployment and Labor Market Rigidities: Europe versus North America', *Journal of Economic Perspectives*, 11(3), 55-74.

OECD (2004) 'The Labor Force Participation of Older Workers: The Effects of Pension and Early Retirement Schemes', OECD Economics Department working paper.

OECD (2005) *Pensions at a Glance* (Paris: OECD).

OECD (2006) *OECD Labor Force Statistics Database* (Paris: OECD).

Ploeg, F. van der (2006) 'Do Social Policies Harm Employment? Second-Best Effects of Taxes and Benefits on Labor Markets' in J. Agell and P.B. Sørensen (eds), *Tax Policy and Labor Market Performance* (Cambridge: CESifo and MIT Press).

Rust, J. (1989) 'A Dynamic Programming Model of Retirement Behavior' in D. Wise (ed.) *The Economics of Aging* (Chicago: University of Chicago Press), 205-224.

Schultz, T.W. (1963) *The Economic Value of Education* (New York: Columbia University Press).

Shefrin, H.M. and R.H. Thaler (1988) 'The Behavioral Life-Cycle Hypothesis', *Economic Inquiry*, 26, 609-643.

Shesinski, E. (2009) 'Uncertain Longevity and Investment in Education ', CESifo working paper 2784.

Skirbekk, V. (2005) 'Age and Productivity: A Literature Survey', International Institute for Applied Systems Analysis mimeo.

Sørensen, P.B. (1999) 'Optimal Tax Progressivity in Imperfect Labour Markets', *Labour Economics*, 6(3), 435-452.

Stock, J.H. and D.A. Wise (1990) 'Pensions, the Option Value of Work, and Retirement', *Econometrica*, 58(5), 1151-1180.

Teulings, C.N. (2003) 'The Contribution of Minimum Wages to Increasing Wage Inequality', *Economic Journal*, 113(490), 801-833.

Trostel, P.A. (1993) 'The Effect of Taxation on Human Capital', *Journal of Political Economy*, 101(2), 327-350.

Weiss, Y. (1986) 'The Theory of Life-Cycle Earnings' in O. Ashenfelter and R. Layard (eds), *Handbook of Labor Economics* (Amsterdam: Elsevier-North Holland), *vol 1*.

Zeldes, S.P. (1989) 'Consumption and Liquidity Constraints: An Empirical Analysis, *Journal of Political Economy*, 97(2), 305-346.

Human Capital, Retirement and Saving

Comments by Daniel Hallberg

Population projections suggest that increased longevity and decreased fertility will result in a shrinking tax base and an increased number of dependants in old age. Increasing costs for not only pensions but also (and perhaps to a greater extent) public old-age care are therefore predicted. In particular, government reforms that stimulate later retirement, labour productivity and private pension saving are therefore suggested to deal with this problem.

Jacobs' chapter focuses on life-cycle interactions in investment in skills via on-the-job-training (OJT), retirement, and private savings for pensions. The chapter argues that policies stimulating private pension saving can be harmful by reducing human-capital investment. The reason is that such policies may ignore important life-cycle interactions between learning, retirement and saving. However, human capital investments, retirement choice and assets accumulation should not be considered in isolation – the reason being that earlier retirement reduces the time over which human-capital investments pay off. Moreover, the need to invest in human capital competes with the need to invest in assets. Indeed, Jacobs argues that human capital and financial capital are substitutes.

Jacobs offers two main implications of the analysis. First, disincentives for labour caused by labour-market institutions, early retirement schemes and tax-favoured private pension savings will render life-long learning policies and later retirement policies ineffective. Second, policies stimulating private saving for retirement may unintentionally exacerbate the ageing problem, since they discourage workers from investing in skills. In short, individuals retire too early because they did not invest very much in developing skills, and individuals under-invest in skills because they expect to retire early.

Jacobs' chapter is well structured, written and documented. At both a theoretical and empirical level, this chapter represents a valuable contribution to the analysis of the determinants of elderly workers' activity rates, savings behaviour, and on-the-job training choices. The policy questions raised are highly relevant and the implications intuitive. The chapter discusses underlying assumptions in detail and with good reference to the existing literature. It backs the theoretical implications by empirical findings in a convincing way.

However, there are some drawbacks to the analysis. To those who are interested in the micro retirement literature, the model adopted from Jacobs (2009) may seem overly simplified in several aspects – in addition to those assumptions already discussed by Jacobs. Indeed, the remainder of my comments focuses on Jacobs (2009), which is the paper on which the chapter bases its analysis. I recognise that some simplification is necessary in complex dynamic models that optimise over the whole life cycle. To illustrate, the model assumes that retirement is well defined, that work and retirement are not combined, and that a return from retirement to work does not occur. More importantly, all retirement is assumed to be voluntary. There are no labour-demand or health shocks that may cause (an early) retirement. Whereas the labour demand side of the retirement decision has been quite under-investigated, there is a growing body of literature pointing to its importance (see chapter 9 in this volume). If investments in on-the-job training by the employer on behalf of the employee were costly and uncertain in terms of future pay-off, the employer would make an assessment about such investments. This would affect the alternatives that would face the older worker. These effects should also be explored.

In the model, financial and human capital are treated as competing forms of assets – but the two forms of capital are in fact quite different. First, human capital cannot be saved and stored or transferred to another agent or time; second, investments in human capital presumably transfer to investments in health (see below); and, third, human capital is something rather indeterminate that cannot be measured with precision. The uncertain measurement of human capital is indeed pointed out by Jacobs as problematic when assessing the theoretical implications of the adopted model. It may be too restrictive to treat human capital and financial capital as substitutes.

We know that defined-benefit schemes represent an important part of retirement income, at least in Europe. In addition, public or occupational pension systems feature (often minor) funded parts – but agents often do not have discretion over saving levels. Therefore, a model that focuses on voluntary private saving towards retirement has less to do with existing policies, as private-pension saving (for example, through tax-favoured saving schemes) is only small. One may wonder how policy recommendations would change if the model would assume a more realistic pensions system as a benchmark. In addition, the financing of these defined-benefit systems through contributions levied on employers may affect the demand for older workers.

Higher human capital may be interpreted as better health. But in the discussion of the model implications, human capital or health is never affecting the consumption bundle of the retiree, which seems quite unrealistic. With more human capital accumulated, and thus more health built up, we would expect ageing to be less problematic in terms of old-age support. What are the greatest costs in terms of ageing? The impression one gets from reading Jacobs's chapter is that pensions comprise the major costs of ageing. However, an equally important cost may well be that of old-age care services.

Last but not least, there are no bequests or intergenerational transfers in the model. At the time of death, all assets must be run down. The possibility of bequests or resource transfers at the end of life seems important to include in a model that has the aim of explaining lifetime choices of saving, work and education. One can point to several reasons. The policy conclusions may well change if (as is the case in many countries) the elderly receiving public transfers make large private transfers to other generations. Motives might involve such things as familial transfers to support the young within the extended family, with an effect on the initial human capital via investment in private education. This might be important in countries where assets and bequests form a large portion of inter-generational transfers, and, in the end, are important for social mobility.

References

Jacobs, B. (2009) 'A Theory of Human Capital, Saving and Retirement', Erasmus University Rotterdam, mimeo.

11

Labour Supply and Employment of Older Workers[1]

Arthur van Soest

1   Introduction

A key concern of European policymakers is how to raise the labour-force participation of older workers. This is necessary, for example, to guarantee the sustainability of pension systems in many countries – one of the challenges posed by population ageing. Retirement and labour-supply decisions of older workers are core topics in microeconomic research. While there is consensus that economic factors (such as the generosity of early retirement benefits and pensions) play an important role, a crucial role is also played by psychological and social factors (such as quality of work and work satisfaction, social networks and retirement decisions of family members and peer groups). Health also plays a major role, through work disability, chronic diseases (morbidity) or the expected remaining lifetime (longevity). Public policy affects the economic environment under which labour-supply decisions are made, through eligibility rules and levels of state pensions, and through taxation of occupational pensions and other savings.

Obviously, labour-force participation and labour-supply decisions are not the only factors driving employment; employers and firms also play a role. Demand-side adjustments will be necessary to accommodate increased

1   This chapter benefited from the comments of Lans Bovenberg, Tim Callan, Maarten Lindeboom, Ruud de Mooij, Asghar Zaidi, and other participants in the ESF workshops in Dublin and The Hague.

supply of older workers, and research progress can be made on such issues as hiring and firing policies, attitudes towards older workers, training of older workers, use of alternative exit routes by employers and employees, accommodating workers who are experiencing health problems, and creating more opportunities for gradual retirement. These adjustments, which may require changes in labour-market institutions, will have substantial macroeconomic implications. Still, this chapter focuses on the supply side – as the demand for older workers and (investing in) their productivity are covered in chapters 9 and 10 of this volume.

This chapter first sketches the background and policy questions relevant for this topic (section 2), and presents an overview of the existing literature (section 3). Section 4 discusses the remaining knowledge gaps and research challenges. Section 5 discusses what is needed to face these challenges, and section 6 concludes.

2   Background and policy questions

Keeping labour supply and retirement patterns constant, population ageing will increase the dependency ratio (the ratio between economically inactive and economically active people). This will likely have negative macroeconomic consequences, hindering economic growth and leading ultimately to a lower average standard of living – through either lower (state or occupational) pensions or higher taxes or (pension) premiums. There are also other reasons why increasing labour-force participation at older ages is desirable. For example, increasing the number of labour-force participants leads to more social cohesion (Burniaux, Duval and Jaumotte, 2004), reduces old-age poverty, and increases the economic welfare of the elderly. Also, some evidence suggests that keeping at work increases well-being (Hartlapp and Schmid, 2008), and helps in maintaining cognitive skills (Bonsang et al., 2007). Many countries have consequently introduced policies aimed at increasing labour-force participation and hours worked by older people (see, for example, Sigg, 2007).

Labour-force participation rates of older men and women vary greatly across European countries. Zaidi, Makovec and Fuchs (2007) present participation rates in 2005 for the age group 50-64 in EU countries ranging from below 50% to above 75% for men, and from below 20% to around 70% for women. These differences have a number of potential explanations. First, countries differ substantially with regard to the institutional arrangements they have concerning (early) retirement: the generosity of state and occupational pensions, and, in particular, how the levels of these benefits vary as a function of the age at which people stop working or start claiming pension benefits. Kapteyn and Andreyeva (2008) emphasize the importance of financial incentives for retirement decisions and summarise the implications of this for policy aimed at increasing labour-force participation:

- Pension benefits should be adjusted so that the additional annual pension benefit given for each extra year of work is a fair compensation for postponing pension benefit receipt and for the extra contribution made (in technical terms: make pensions actuarially fair).
- The eligibility ages for pension benefits (for example, index the eligibility age by life expectancy) should be increased.
- Defined-benefit pensions should be replaced by defined-contribution pensions, so that benefits are a function of one's contribution (and of the return on the investment). This automatically leads to actuarially fair benefit adjustments when retirement is postponed.

There also appears to be common agreement that financial incentives cannot explain all features of retirement patterns, leaving a substantial role for other factors (see, for example, Lumsdaine and Mitchell, 1999). One of the non-economic factors recently emphasised is health. According to Kapteyn and Andreyeva (2008), the importance of health implies a role not only for health policy but also for educational policy – since both theoretical and empirical work suggest that the health capital of individuals with low education depreciates faster than that of individuals with high education. This implies the following policy options:

- Improve education to raise human capital.
- Make jobs healthier.
- Promote healthy behaviour (diet, physical activity, no smoking).

The role of education is also apparent from the data: Zaidi et al. (2007, Table A3) find a strong positive association between education and labour-force participation in EU countries.

While the literature has focused mainly on labour-force participation and retirement, the number of hours worked by participants is also relevant. The central concept here is *gradual retirement* (Reday-Mulvay, 2000): a stepwise transition from full-time, full-effort employment (usually, the 'career job') to complete retirement – typically involving a period of part-time work, but sometimes also full-time work of a different nature, requiring less effort (a 'bridge job'; see, Ruhm, 1990). Gradual retirement may take two forms: without change of employer (phased retirement) or with a change of employer (partial retirement). While the latter seems more popular in the US, the former seems to be more common in Europe, where job mobility among older workers is limited (Kantarci and van Soest, 2008).

Table 11.1, taken from Kantarci and van Soest (2008), is based upon several waves of the European Community Household Panel. In almost all cases, part-time employment is more prevalent in the older age category (51-64 years) than in the younger category (35-50), suggesting that workers reduce their work effort later in life. Substantial variation exists in the prevalence of part-time work across countries – for both sexes and in both age groups. In the Netherlands, part-time work among older men and among women in both age groups is much higher than in most other countries. Again, institutional differences may play a significant role here (for example, concerning possibilities of combining part-time work and receiving a pension). Many European countries have taken policy measures to stimulate gradual retirement – first with the purpose to create more jobs for younger workers, and later with the aim to increase participation of older workers (see Belloni et al., 2006). This corresponds with the notion that facilitating gradual retirement has two opposite effects on labour supply: people who otherwise would keep working full-time may reduce their hours of work, while others will keep working part-time instead of retiring completely. For macroeconomic policy, the most important question is which of these effects dominates, and what will be the net effect on total labour supply.

To illustrate the fact that labour-supply considerations are not the only relevant factors affecting actual hours worked, Figures 11.1 and 11.2 (see Kantarci and van Soest, 2008) compare the fractions of male and female workers in the Netherlands who *actually* work part-time with the fraction of workers who would *prefer* to work part-time rather than full-time (keeping the hourly wage rate constant). The fact that the latter fractions are larger than the former suggests that many people face constraints with regard to the number of hours they must work – and cannot simply reduce hours of work on their current job. This applies to younger workers as well as to older workers for whom part-time work would be a form of gradual retirement. The figures therefore suggest that access to gradual retirement is limited – even in the Netherlands, where part-time jobs are much more common than in most other European countries, implying that public policies should consider not only labour supply- but also demand-side or institutional restrictions.

Table 11.1: Part-time work of total employment (%)

	1994		1998		2001	
	35-50	51-65	35-50	51-65	35-50	51-65
			Men			
Germany	2.1	7.3	2.3	8.6	2.8	2.8
Denmark	2.5	6.8	2.4	8	2.1	4.8
Netherlands	4.6	13.4	5.3	11.6	5.7	10.4
Belgium	2.3	6.9	1.7	5.2	0.8	(5.1)
France	5.8	9.4	1.3	(3.9)	1.3	3.6
UK	5.1	8.9	2.5	6.4	2.5	4.8
Ireland	5.3	8.8	5.7	12	5.3	11.9
Italy	5.9	10.4	2.1	5.7	1.6	4.1
Greece	6.6	9	2.1	2.6	1.2	2.8
Spain	3.7	6.5	1.9	3.4	2.2	2.9
Portugal	3	9	1.3	6.8	1.1	5.3
Austria	*0.7*	*3.5*	1.8	(4.2)	1.9	5.8
Finland	*3.3*	*8.5*	3.2	9	3.1	7.5
Sweden	*2.3*	*6.4*	1.7	6.2	1.5	4.6
			Women			
Germany	37.9	39.7	32.5	37.1	30	33.4
Denmark	19.4	37.3	16.9	28.6	14.7	28.6
Netherlands	63.8	68.3	63.2	64	61.6	60.4
Belgium	28.6	32.3	30.8	29.4	32.6	35.1
France	23.3	27.8	17	19.7	14.3	20.3
UK	43.7	45.3	15.8	23.7	14	26
Ireland	46.4	42.6	44.6	55.5	40.3	46.3
Italy	26.2	29.9	12.4	11.1	13.3	10.6
Greece	17	21.6	9.6	18.9	7.6	17.4
Spain	21.5	23.4	16.2	24.1	19.7	22.4
Portugal	12.5	25.6	11.7	24.8	10.8	25.2
Austria	*27.6*	*(25.9)*	29.3	31.5	30.6	27.5
Finland	*8.6*	*12.8*	9.1	15.7	8	17.3
Sweden	*15.9*	*22*	16.8	20.2	13	19.4

Notes:
1. Based upon self-assessed labour-market status.
2. For Germany, Sweden and the UK, the presented numbers are from national surveys converted into the ECHP format.
3. The numbers in italics refer to the closest survey year for which data is available: for Austria, it is 1995, for Finland, it is 1996, and for Sweden, it is 1997.
4. The numbers in parentheses indicate that data are missing for various ages within the age category.
5. The sample is weighted with cross-sectional weights for interviewed persons.

Source: Kantarci and van Soest, 2008.

Figure 11.1:   Employees in the Netherlands working 1-35 hours per week

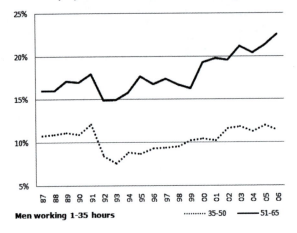

Men working 1-35 hours   ·········· 35-50   ——— 51-65

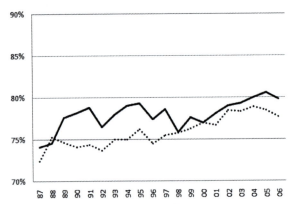

Women working 1-35 hours

Note: Employees working 1-35 hours.

Source: *Enquete Beroepsbevolking*, Statistics Netherlands; about 60,000 observations for each cross-section.
               Observations are weighted with cross-sectional weights. The percentages represent the share of
               those working 1-35 hours in those working any number of hours.

Figure 11.2:   Employees in the Netherlands desiring to work 1-35 hours per week

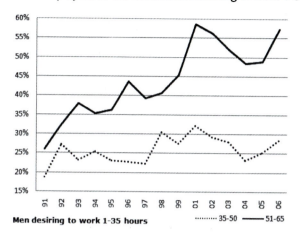

Men desiring to work 1-35 hours   ·········· 35-50 ——— 51-65

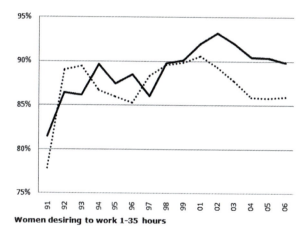

Women desiring to work 1-35 hours

Note: Employees desiring to work 1-35 hours.

Source: *Enquete Beroepsbevolking*, Statistics Netherlands; 5,000 to 9,000 observations for each cross-section. Observations are weighted with cross-sectional weights. The variable is not available for 2000 and for before 1991. The percentages represent the share of those who desire to work 1-35 hours in those who desire to work any number of hours.

All of this suggests a number of research questions relevant for public policy, including the following:

- What is the role of financial incentives in explaining labour-force participation of older age groups and differences that are seen in participation rates across European countries?
- Which other factors explain labour-force participation of older age groups? What is the role of health problems that limit ability to work, job characteristics and satisfaction with work, family considerations (such as labour-force participation of the spouse or responsibilities for grandchildren or elderly parents), cultural factors or peer-group effects?
- How effective are current national pension policies to maximise labour-force participation of older age groups? And what about additional policies, such as improving work conditions for people with work disabilities?
- How desirable is gradual retirement for workers and employers? And how does it affect the macroeconomic supply of labour?
- What is the interplay between supply- and demand factors driving participation, employment, desired hours and actual hours? Are supply policies sufficient, or should we also consider policies focused on employer attitudes towards older workers and enhancing the productivity of older workers?
- What are the consequences of increasing the labour supply of older workers for the economy as a whole, and for the economic and non-economic well-being of the older population?

3   Existing literature[2]

Following the overview of Kapteyn and Andreyeva (2008), at least three explanations for the large variation in retirement patterns across countries have been given. The first focuses on financial incentives. A second explanation points at differences in preferences or culture. A third explanation considers the role of institutions – particularly the power of unions. Finally, health seems an important explanation, since retired individuals often give poor health as the primary reason for why they exited the labour force. This section reviews the main studies with these four explanations.

---

2   This section strongly hinges on an excellent recent literature review in Kapteyn and Andreyeva (2008).

3.1   *Financial incentives*

Numerous studies have documented that, until recently, workers were leaving the labour force at increasingly younger ages – while at the same time real earnings have risen and the average length of the working year has declined. Increased private income and higher public spending make it possible to maintain or even raise consumption levels at reduced hours of work. In a static neoclassical model of labour supply, this happens if leisure is a normal good and the income effect of higher real wages dominates the substitution effect. In reality, there will be considerable heterogeneity across individuals – with some choosing more leisure and others favouring more consumption – and financial incentives are more complicated than the stylised model suggests.

Gruber and Wise (1999, 2004) investigate the links between state pensions (social security) and labour-force participation patterns in 12 OECD countries. They describe the relationship between social security incentives to retire and the proportion of older people out of the labour force, emphasising that social security programmes often create strong incentives to leave the labour force. They find a close correspondence between the eligibility age for early- and normal retirement benefits and departure from the labour force. Moreover, disability and unemployment programmes often provide ('quasi early retirement') benefits before the early (or standard) retirement age. The large cross-country variation in the tax burden on work affects the incentives to withdraw from the labour force, and this explains a large part of the variation in retirement patterns.

Using micro-simulation models of retirement incentives, the Gruber and Wise studies show that the retirement incentives inherent in most social security programmes are strongly linked to withdrawal from the labour force – although the size of the effect varies across countries. To illustrate this, Figure 11.3 reproduces Figure 21 from the summary of Gruber and Wise (2004). It summarises the consequences of two hypothetical pension reforms in 12 countries, based upon one of the models estimated for all countries (the Option Value [OV] model). The first reform ('Three-Year Delay') shifts all entitlements by three years (if the actual entitlement at age X is Y, then the simulation assigns entitlement Y to age X+3). The second ('Common Reform') replaces the actual system by an approximately actuarially fair system with normal retirement age 65 and corresponding replacement rate 60%, and the possibility to retire as of age 60, with a 6% compensation for each year of postponing retirement. The figure shows, for each country, the change in the percentage out-of-labour-force ('OLF') for a country-specific age group: the first age at which at least 25% of the workforce is out of the labour force, plus the next four years.

Figure 11.3 shows that retirement behaviour is indeed sensitive to the financial incentives created by the pension system in most countries. In all countries except Canada, non-participation in the age group considered falls by more than 40% if entitlements are shifted by three years. The effect of the common reform depends on how different the actual system is from the (hypothetical) actuarially fair system of the reform. This reform would have a particularly large effect in the Netherlands, where actual early retirement arrangements were extremely generous.

Figure 11.3:   **Simulated change in % OLF in the year OLF attains 25% plus the next four years**

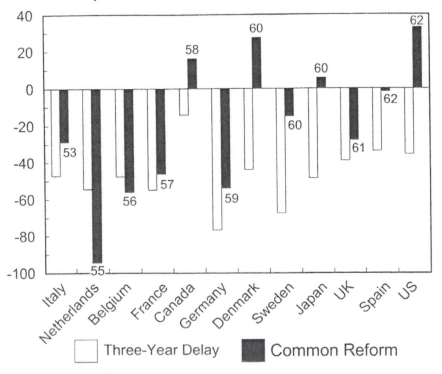

Source: Gruber and Wise (2004, p.34).

Duval (2003) analysed the effects of old-age pension systems and other social transfer programmes on the retirement behaviour of male workers in 22 OECD countries. He confirms the conclusion that implicit taxes on continued work have notable effects on the labour-force exit of older men. Other studies (such as Keenay and Whitehouse, 2003) come to similar conclusions.

The role of economic rewards for retirement is also documented in cross-country studies on the effect of taxes on market work activity. Using a panel of OECD countries, Davis and Henrekson (2004) find large responses to taxes, and estimate that higher tax rates on labour income and consumption expenditures lead to less work in the market sector, an increase in informal household work, a larger underground economy, and lower value added and employment in industries employing low-wage unskilled workers. Similarly, Olovsson (2009) finds a primary role of taxes in explaining differences in labour supply between Sweden and the US. Alesina, Glaeser and Sacerdote (2005), however, conclude that estimated elasticities are not large enough to explain the complete labour-supply differential between Europe and the US.

A final explanation for the employment gap between Americans and Europeans is the larger extent to which home production can be substituted with services available in the labour market (Freeman and Schettkat, 2001). Time-use data show that the total amount of formal plus informal (household) work is the same for Europeans and Americans. The lower tax wedge and wider wage dispersion in the US makes market work relatively more attractive in the US than it is in Europe. Market work is more rewarding for high-wage earners, whose share is larger among women in the US.

On the whole, there is general agreement that financial incentives are crucial for explaining cross-country differences in employment rates and retirement patterns. Studies differ, however, with regard to *which part* of the total observed variation in labour supply they ascribe to purely financial incentives.

3.2 Preferences

Part of the labour-supply literature explains international differences in employment from long-standing cultural norms that form attitudes toward work. Different preferences and social and cultural perceptions imply different utility functions in Europe and the US. Blanchard (2004) argues that with roughly similar productivity levels in the two regions, stronger preferences for more leisure *vis-à-vis* income in Europe might account for the observed reduction in hours worked among European workers over the last 30 years. He finds that most of the reduction in hours worked per capita in Europe in the last decade reflects a decline in hours worked per full-time worker, rather than an increase in unemployment or a fall in labour-force participation. His empirical analysis leads to the conclusion that the decline in hours worked does not represent an effect of taxes, but is mainly due to a preference shift.

3.3   Institutional arrangements

Alesina et al. (2005) explain the decrease in hours worked in Europe and the US-Europe employment gap from labour-market regulations imposed by the power of trade unions in Europe in the 1970s and 1980s. Union density and power is stronger in Europe than it is in the US, which can be explained by political conditions. The importance of unionisation and labour-market regulation increased markedly with the structural shocks of the 1970s and 1980s (Blanchard, 2004). Under the slogan 'work less, work all', unions in declining European industries advocated a policy of work sharing that sought to reduce work hours in order to increase employment (see, for example, Estevão and Sá, 2008, for the French 35-hour workweek reform). Although the influence of powerful unions led to implementation of a number of labour-market regulations, including early retirement and pension laws, these policies did not achieve their goal of lower unemployment (Kapteyn, Kalwij and Zaidi, 2004).

Alesina et al. (2005) find that the effects of taxes on labour supply disappear when they control for unionisation or labour-market regulation. They find a dominant role of labour-market regulation and unionisation in explaining why Europeans today work much less than Americans. Moreover, the policies endorsed by unions in Europe may have had an indirect effect on labour supply through a social multiplier on the marginal utility of leisure (Glaeser, Sacerdote and Scheinkman, 2003). People enjoy leisure more when the leisure of their friends, relatives and social groups increases. This social multiplier may explain why Europeans have a stronger preference for leisure, and prefer to exploit their increased productivity to reduce hours of work.

3.4   Health

Individuals in developed countries are healthier than in the past, as illustrated, for instance, by the secular trends in life expectancy. Simultaneously, there is a tendency to retire at earlier ages, as discussed above. Paradoxically, when asked about reasons for retirement, individuals frequently indicated 'bad health'. Although failing health is a plausible reason to retire, it is hard to believe that work conditions nowadays are more strenuous than they were in the past. Moreover, the improving health in the population would suggest that the role of health in retirement decisions is falling over time.

Several mechanisms can create a link between health and retirement. Failing health may lead one to retire (e.g. Disney, Emmerson and Wakefield,

2006; Kalwij and Vermeulen, 2008). Moreover, retirement may have a (positive or negative) influence on health. It can be beneficial by removing mental stress and physical effort. On the other hand, retirement is a major life event, which in itself creates stress and may reduce health. Existing studies often find a negative effect of retirement on mental health (for example, Bonsang et al., 2007).

Within economics, an obvious framework for studying the relation between health and retirement is the Grossman (1972) model, in which individuals derive utility from consumption and health, but where health also influences earnings. Over the years, several extensions of this model have been proposed. Wolfe (1985) and Galama et al. (2008) extend the basic Grossman model with a retirement decision. Their analysis shows that individuals with lower human capital (and hence lower earnings capacity) have fewer resources to invest in health, so that their health deteriorates faster – implying that health will be positively associated with income and education (Case and Deaton, 2005). A higher earnings capacity also induces people to work longer, as many studies have observed. Yet, secular improvements in health also have an income effect that reduces work effort, explaining trends in early retirement.

4   Research challenges

This section identifies a number of research questions on which progress can, in principle, be made in the near future.

4.1   Pension reforms

Since the early 1990s, many countries have reformed their pension systems, focusing on reducing the generosity of early retirement benefits and changing the incentives to make early retirement less attractive through a system of an actuarially fair compensation for working longer. Reforms have often involved raising the eligibility age (and in many cases equalising eligibility ages for men and women), increasing incentives for continuing work (including bonuses for retirement after the normal pension age), making changes to the benefits calculation (for example, extending the period of earnings measurement or changing qualifying conditions), and replacing defined-benefit (DB) schemes by defined-contribution (DC) or mixed schemes, implementing changes in the indexation of pensions (moving towards partial- or full indexation to prices), and changing pension contribution rates (OECD 2007, 2009).

Country-specific studies have exploited some reforms of state pensions to analyse whether retirement decisions are sensitive to certain financial incentives, such as the level of benefits and the 'accrual' in the benefit level due to additional years of work. Still, substantial progress could be made by rigorously exploiting the variation in a number of countries simultaneously, learning from the variation over time within and between countries. The Gruber and Wise studies referred to above are an interesting attempt in that direction, using similar stylised models to estimate the effects of financial incentives in separate countries.

Lack of accurate data on pension entitlements seems to be the main bottleneck for research on this topic. The availability of new, harmonised data (such as SHARE) creates potential for further progress. While variation across countries could be exploited for studying state pensions, identification improves by exploiting variation over time and across countries with longitudinal data. For occupational pensions, survey questions are not the best way to get accurate measures of workers' future entitlements, and linking to administrative data is a promising way to go (see section 5), particularly when combined with microsimulation models like Euromod (which exploit institutional details to analyse the consequences of country-specific policies; see Sutherland, 2000).

4.2 Gradual retirement

While most employees seem to have a strong aversion to working full-time after the normal retirement age, recent surveys suggest that many might be interested in part-time work. Creating opportunities for gradual retirement may therefore increase labour-force participation of older age groups. Existing research on gradual retirement focuses mainly on the US and is rather descriptive.[3] European policies aimed at stimulating gradual retirement have been described in several papers, but these lack rigorous modelling. The crucial question is whether the positive effect of creating part-time work opportunities on labour-force participation exceeds the negative effect on hours worked of those who would have kept working full-time had they not been offered the possibility of gradual retirement. Particularly since not everyone, yet, has access to gradual retirement (with a combination of part-time earnings and a partial pension), this requires structural modelling, extending the models of Stock and Wise (1990) or Blau (1994), for example.

---

3   See, for example, Rappaport (2008) and Munnell (2008), which reach opposite conclusions in studying the usefulness of gradual retirement to keep people at work and maintain retirement income.

To make this kind of research possible, the main challenge seems to be to disentangle preferences from demand-side factors. If few people take gradual retirement, is this because they do not want it, or because their employer does not offer the opportunity? Answers to this question require additional information, in the form of surveys or survey questions, by creating a database with detailed information on opportunities for gradual retirement offered in each occupational group, or by linking to employer data with information on gradual retirement opportunities.

4.3   Health

The theoretical work of Galama et al. (2008) (extending the Grossman model) to explain the interrelations of health and retirement is promising, offering plausible explanations for various stylised facts. Still, this framework has yet to be developed to its full extent, and needs to be transformed into an empirical model that can be estimated and tested with available data. This is an important challenge that may provide new insights in the often-studied association between health and socioeconomic status (see Smith, 2007; Michaud and van Soest, 2008). To better understand the causal mechanisms at work, it also seems useful to distinguish between mental and physical health, to incorporate job characteristics (such as stress and physical effort), and to make explicit the role of health insurance. Moreover, while the effect of health on labour-market outcomes can be immediate, the reverse pathway (the effect of economic and social events on health) may take much longer to become effective (Van den Berg, Lindeboom and Portrait, 2006). This is an argument for studying health and retirement in a long-term perspective, requiring, for example, data on socioeconomic events and health shocks much earlier in life.

Longitudinal data covering a large part of the life cycle – with rich enough information on socioeconomic status as well as health – are probably the main requirement for further progress in this field. This seems to be a topic where retrospective data could adequately substitute prospective panel data, since major health events and socioeconomic shocks (like unemployment, a change of occupation, or a transition into or out of poverty) can probably be measured reasonably well with retrospective questions. Utilising the life histories that are currently surveyed in SHARE could be an important step towards promising research in this direction. On the other hand, objective measures of physical health and measures of mental health that go beyond diagnosed depression-related diseases cannot be measured retrospectively – so that also for this topic, prospective collection of longitudinal data remains very useful.

4.4   *Economic and non-economic determinants of labour-market position*

Existing studies that have addressed both economic and non-economic determinants typically consider one type without controlling for the other. In principle, this may lead to omitted-variable biases in the results, if the various factors are correlated – and it seems obvious that they often are correlated. Moreover, it seems interesting to analyse the relative contribution of the various factors.

With the availability nowadays of datasets with observed variation in pension entitlements, health, social participation, and family background and circumstances on the same respondents, an analysis that simultaneously incorporates all of the main factors of retirement behaviour becomes feasible. Longitudinal data will help to identify the direction of causality, exploiting the order in which changes take place. Multi-disciplinary data like SHARE seem ideal for these purposes.

4.5   *Demand and supply*

Employer attitudes, and how these are perceived by employees, are important for retirement behaviour, since they affect job characteristics and job satisfaction, which in turn influence the labour-force participation decision. The ideal model should account for these factors by incorporating information on the employer and the firm, by matching firm- and employee data or asking survey questions to employees about their employers.

Progress can be expected, in this respect – although an obvious complication is the fact that the employer is almost always a choice of the employee, and the retirement facilities offered by employers can be a criterion for an employee who chooses an employer. One possibility to address this is to analyse the career choices made during earlier stages of the life cycle. Although retrospective data can help in this respect, it seems preferable here to use prospective longitudinal data that are representative for all people of working ages.

4.6   *Other determinants of retirement and labour supply*

Job characteristics and job satisfaction are important determinants of commitment to work, and affect willingness to keep working (Hayward et al., 1989). The US literature concludes that retirement of couples is based upon joint decision-making. Blau (1997, 1998) and Gustman and Steinmeier (2000, 2004), developed economic models of decision-making of couples, and Henretta et al. (1993) devised a 'Family Organizational Economy' model. More generally, family circumstances and family networks are important, through informal care for elderly parents or grandchildren (Currie and Madrian, 1999; Heitmueller, 2007). The role of non-family social networks as reference groups is discussed in Woittiez and Kapteyn (1998) and Aronsson, Blomquist and Sacklén (1999).

Empirical modelling of these factors (particularly reference-group effects) often requires intricate econometric models and thoughtful consideration of identifying assumptions (Manski, 1999). Similar identification concerns apply to collective models of household decision-making (Vermeulen, 2002). Progress here will doubtless require more specific data than what is typically provided by general socioeconomic surveys – for example, on the basis of field experiments or stated preferences.

4.7   *Alternative exit routes*

There are several ways in which older workers leave the labour market and make the transition from full-time work to complete retirement. Working full-time until some retirement age, and then stopping work to enjoy a retirement pension (also called 'cliff-edge' retirement; see Vickerstaff, Cox and Keen, 2003) is just one of them. Gradual retirement – in the form of either a bridge job at another employer, or in self-employment, or as an employee at the same employer but with reduced hours or an adjusted task requiring less effort – is a second way to make the transition. Labour-market exit may also be non-monotonic, in the sense that retired workers sometimes return to work, or workers in gradual retirement return to their old full-time job ('reverse retirement').

Moreover, workers may spend some time in unemployment or on disability benefits before retiring. Unemployment and disability have been seen as substitutes for early retirement that can be attractive for both employer and employee, although the evidence seems inconclusive. Kerkhofs, Lindeboom and Theeuwes (1999) find some evidence of substitution in the Netherlands, while Riphahn (1999) and Dahl, Nilsen and Vaage (2000) find no substitution in Germany and Norway. Most retirement models, however, ignore alternative exit routes. A promising exception is Heyma (2004), who builds into one structural dynamic model the choice between various exit routes.

With adequate data following the same individuals over a long period of time, and with enough detail on financial incentives, health, and demand-side constraints, the latter strategy seems promising: building models with several exit routes that give more insight in the determinants of labour-force participation and its alternatives. Administrative data, possibly linked to survey data, may be the best way to go here, since asking survey respondents about their perceived opportunities for each exit route seems too much of a burden.

5   Research needs

Research progress in the next five years in the field of ageing, pensions and health will depend importantly on data collection, on the data being made available to the research community, and on knowledge being exchanged between countries and disciplines.

To analyse the effect of economic and social policy measures, it is particularly important to exploit the variation in policy changes over time and across countries, requiring longitudinal data for many countries. Policy diversity captured at a given point in time may be useful, but can only identify the effect of policy measures on the maintained assumption that country-specific effects do not play a role, and that country-specific policies are not affected by the cultural factors or social norms that also drive labour-market behaviour. Under that assumption, country-specific policies are plausibly exogenous, and it is reasonable to interpret the association between labour-market outcomes and national policies in cross-section data as causal effects due to these policies. In many cases, however, this assumption is not plausible, and identification must come from variation over time.

Consider, for example, the association across countries between the minimum age at which men or women are entitled to a state pension and the average retirement age. A positive correlation can be interpreted as a causal effect of the state pension entitlement age on retirement decisions – but this interpretation requires the assumption that the entitlement age is exogenous to retirement behaviour. If differences in preferences or social norms across countries affect not only retirement decisions but also political decisions about entitlement to state benefits, the interpretation may be incorrect.

Longitudinal data offer at least a partial solution to this type of identification problem, if there is enough variation in policies over time. First, looking at the associations between *changes* in labour-market outcomes and *changes* in socioeconomic policies makes it possible to control for country-specific factors that do not vary over time. Thus, in the example above, if changes in the retirement age appear to be correlated with changes in the state-benefit entitlement age, then interpreting this as a causal effect of the entitlement age on the retirement age requires much weaker assumptions than giving the same interpretation to the association in levels – as it is possible to control for time-invariant social norms and cultural factors that can be correlated with both the retirement age and the entitlement age.

Second, in cases where there may plausibly be feedback mechanisms from labour-market outcomes to policy measures, the timing often helps to identify the policy effects and to disentangle them from these reverse effects. For example, it seems likely that if the average retirement age affects the state-benefit entitlement age, this does not happen immediately (since the political process takes time). If we then see an effect of the current entitlement age on the retirement age (controlling for country-fixed effects as well as past entitlement- and retirement ages), we can interpret this as a policy effect. This illustrates that dynamic models with fixed effects can be used to relax the identifying assumptions. Estimating such models, however, requires rich longitudinal data – at least three waves, and with enough variation in policies.

The comparative advantage of the European context is that many policy changes have taken place and will take place in the near future. The systems of state pensions and occupational pensions have been reformed dramatically over time: first by introducing generous early retirement arrangements, and later by removing them and moving towards flexible and actuarially fair systems (Duval, 2003). Many countries have also relaxed the rules for combining pension receipt with earnings, stimulating part-time work at an older age (Belloni et al., 2006). Several countries have also responded to increasing numbers of people drawing disability benefits by limiting access or duration and adjusting benefit levels (OECD, 2003).

5.1 Data collection

Although substantial progress has been made in the past five years, much remains to be done to bring data collection in Europe at the level of that in the US, and to be able to fully exploit the diversity in Europe to help and understand the effects of social policy on labour supply and retirement and all of the related issues discussed above.

SHARE (the Survey of Health Ageing and Retirement In Europe) released in 2004 its first wave of rich individual data on adults aged 50 years and older. The database offers multi-disciplinary data for performing cross-country analyses in Continental Europe, comparable with the US Health and Retirement Study (HRS), a widely used dataset to study issues of older Americans, and the English Longitudinal Study of Ageing (ELSA), a source of data on people aged 50 years and older living in England. The baseline 2004 SHARE study included representative data on 28,517 respondents from 11 countries (Denmark, Sweden, Austria, France, Germany, Switzerland, Belgium, the Netherlands, Spain, Italy and Greece). In 2006-2007, a second wave was fielded in these 11 countries, and SHARE was extended to Poland, the Czech Republic, Ireland and Israel. The longitudinal nature of the data is substantially enriched in the third wave (collected in 2008 and 2009), with retrospective questions on major life events experienced over the life course. SHARE includes objective and subjective measures of physical and mental health, well-being, socioeconomic status (work activity, job characteristics, income, wealth and consumption, housing, education) and social participation (family relations, informal care, volunteer activities).

While SHARE provides a good basis, addressing the research challenges identified in section 4 would benefit from a number of extensions:

- Expanding the longitudinal dimension. This is necessary to identify and understand the causal mechanisms that often take many years to become effective, such as the effect of unemployment or other economic shocks on health and well-being.

  The current project SHARE-Life collects retrospective data on major life events from the SHARE respondents in waves 1 and 2. This is a cost-efficient way to collect data on a long time period. Sociological research has shown that retrospective questions are not error-free, but still provide useful information on the respondents' past, such as their occupational career (see, for example, Ruspini, 2002). While retrospective measures can help in many respects, not everything can easily be measured retrospectively (for instance, job satisfaction, well-being, or cognitive skills) so that genuine panel data with complete information on the same individuals over a long period of time remain important.

- Expanding the international dimension. To learn from international variation in institutions and policies, more variation in institutions and policies would be very useful. For example, the often-used distinction of Esping-Andersen (1990) between liberal, conservative, and social-democratic welfare systems could be used to analyse the consequences of the nature of the welfare system for labour-force participation and well-being – but with only a few countries in each regime, it is hard to disentangle this effect from country-specific factors (see also chapter 5 in this volume).

- Adding more detailed survey information. The multi-disciplinary nature of SHARE has its obvious advantages, but also has the drawback that the room for specific questions on topics of interest is limited. For example, nothing is asked about opportunities or preferences for gradual retirement. Adding modules on specific topics would be useful, perhaps only for specific subsamples, such as employees in a given age group.
- Merging with administrative data. Many variables are inherently hard to measure in surveys. This applies particularly to economic variables such as income from various sources, assets, or pension entitlements, which are often reported with error or not reported at all. In some countries, survey data have been linked to administrative data that originate from, for example, the social security administration or the tax authorities (see, for example, Kapteyn and Ypma, 2007). Merging SHARE records with administrative data can help to substantially improve the quality of the economic variables in the SHARE survey, and thus also improve the accuracy of model estimates and statistical inference using these variables. Another option might involve linking SHARE to other register data – on healthcare utilisation, for example (see Atella et al., 2006).

5.2 Other survey data

SHARE is not the only useful source of data. As discussed in section 4, for many purposes it might be useful to follow people over time during an earlier part of their labour-market career. In that case, a worthwhile option might be to use a panel data study such as the European Community Household Panel (ECHP), giving comparable micro-data in EU countries for all adult age groups.

ECHP (like SHARE) offers the advantage of having a rich questionnaire on socioeconomic status, well-being, family life and social networks. Unfortunately, data collection in ECHP came to an end in 2003. Its successor, SILC, does not have the same multidisciplinary nature. ESS, the European Social Survey, would in principle have had the potential to be a good replacement, but it is quite weak in economic measures. Its core questionnaire has only two income questions – one on the main source of income in the household and a categorical question on household income –, nothing on wealth, portfolio choice, homeownership, individual incomes, wages, and so forth. This is unfortunate, since it renders ESS data inadequate for most economic research. Some individual countries have ongoing panels with rich economic and non-economic information for the complete adult population (GSOEP in Germany, BHPS in the UK, PSID in the US), but this – unfortunately – no longer exists at the European level.

5.3   Laboratory and field experiments

Socioeconomic surveys and administrative data are not the only sources of data for economic research. Laboratory experiments represent an increasingly popular method to generate new insights in labour economics in general (see, for example, Falk and Gächter, 2008). We do not know of any applications to labour supply or retirement behaviour, but experiments could be equally successful there. There are general concerns with the fundamental question whether experimental results in the laboratory say something about the real world, and the answer seems to depend on the nature of the experiment and the adequacy of the experimental design (see, for instance, Levitt and List, 2007); this probably also applies to labour-market behaviour of older workers. Still, more research is needed here.

Field experiments are an alternative to laboratory experiments, with subjects recruited in the field rather than the classroom and exploiting field context rather than having to provide abstract instructions (Harrison and List, 2004). Field experiments can have many formats and may differ from laboratory experiments in many ways. Field experiments involving actual worker decisions have been used successfully in labour economics,[4] but (at least to our knowledge) have not specifically been used to analyse the labour supply of older workers. In principle, this is a promising field of research, which is underdeveloped compared to, for example, field experiments in the analysis of saving for retirement. Combining several types of data also seems a useful trend. All of this requires careful consideration in the context of retirement or labour-supply decisions at an older age.

5.4   Data access

The mere existence of data is not enough – they also need to be used by researchers. In order to achieve this, several conditions need to be satisfied. The monetary cost and the conditions for using the data should not hamper their use, and the data must come in a user-friendly format with adequate documentation. This is particularly important for panel data, which might prove impossible for outsiders to work with, if not organised and documented in an appropriate way. An option might involve workshops in which researchers and programmers familiar with the data teach other researchers how to use the data.

---

4   See, for example, Bellemare and Shearer (2009).

5.5   Exchange of knowledge

Cooperation between researchers in different countries is useful, particularly if we want to learn about the consequences of differences in institutions (such as social security and pension generosity). While it would certainly also be useful to create a database with the main features of national institutional arrangements and their reforms, personal cooperation and the possibility of consultation with national experts will remain necessary.

For many labour-supply issues discussed here, insights from various disciplines can be usefully combined (mainly economics and sociology, and to some extent also psychology and public health). Thus, cooperation between experts from various disciplines certainly has added value. Such cooperation has evolved through the development of SHARE, for example, and it would be worthwhile to create more such networks.

An example of a successful network funded by the EU is the ERA-Age network.[5] This network focuses on ageing, but is very broad – with hardly any focus on economics. It mainly brings together researchers and policymakers in sociology, public health, and biological and medical science.

It seems reasonable to argue that there may also be scope for a research network focusing on ageing from an economic perspective, with input from social sciences and public health. Experts and institutes specialised in economics would then work with non-economists on common topics. It should be clear from the previous sections that many such topics exist. In particular, it would be useful to start a network that combines insights from economics, psychology, sociology – and perhaps even public health – on the role of job characteristics, working conditions, and several features of job satisfaction and their interactions with economic factors in keeping older people at work.

5.6   Priorities in data collection

Survey data collection using face-to-face interviews is not cheap. Still, we have seen that longitudinal data covering a long time period create numerous opportunities for scientific policy-relevant research, and investing in SHARE and other datasets is definitely worthwhile. From a research perspective, adding more countries seems less necessary, although it certainly has its advantages. On the other hand, it may be argued that each additional country itself can benefit from the complete SHARE database.

5   See http://era-age.group.shef.ac.uk/

Making administrative data available for research also requires effort and time – but where the administrative data already exist, it seems relatively cheap compared to new data collection, and it is certainly less burdensome for the respondents. Linking administrative data to survey data has the potential advantages of more accurate data with a lower respondent burden. Possibilities and obstacles need to be explored, and removing legal impediments in some of the European countries should be prioritised. The EU could play an important role in stimulating countries to remove legal barriers and in defining common protocols that harmonise access across member states without jeopardising confidentiality and protection of data at the individual level.

Administrative data can come from various sources, such as tax authorities and social security institutions providing data on income, wealth, and first-pillar state pension entitlements; pension funds with data on second-pillar pension entitlements; banks and insurance companies with data on private savings and life insurance; or healthcare providers and health insurance companies with data on diagnosed health problems and the use of healthcare facilities. A systematic exploration of the diverse opportunities in European countries seems a useful next step.

Administrative data could provide additional information, but would not eliminate the need for survey data. Collecting survey data therefore remains necessary. The rising access to the Internet among all population groups, including the elderly, suggests that data collection may become more efficient in the future, with Internet interviews (for those who can use the Internet) replacing face-to-face interviews. Specific groups (such as the oldest old) could still be interviewed in person. Although more research is needed on differences in answers related to mode of interviewing ("mode effects"), for example, it seems reasonable to expect that Internet interviewing will play a larger role in the future. The HRS already has an Internet equivalent with experimental modules in the off-survey year, and Europe seems to be lagging behind in this respect. There is little doubt that investing in an Internet-interviewing version will be self-financing in the long run.

6   Conclusions

Greater insight into the labour supply of older age groups is scientifically important and policy-relevant. Although academic research has provided many new insights into this topic during the past ten years, much remains to be done. A great many research challenges may fruitfully be addressed in the next five- to ten years, particularly if the infrastructure for socioeconomic research is enriched with innovative and high-quality easily accessible data, and if researchers in several countries and disciplines exchange their knowledge more extensively.

Returning to the policy questions stated in section 1, the following tentative conclusions could be drawn on what is currently known, what might be learned in the near future, and what would be needed in order to move forward.

1.   What is the role of financial incentives in explaining labour-force participation of older age groups and differences that are seen in participation rates across European countries?

A lot is known here, but the models that are currently used could be improved, better data could be used, and non-economic factors could be incorporated. More cooperation would be useful. It could start with producing an overview paper on what the micro-econometric models of Gruber and Wise and their extensions have shown across countries.

2.   Which other factors explain labour-force participation of older age groups? What is the role of health problems that limit ability to work, job characteristics and satisfaction with work, family considerations (such as labour-force participation of the spouse or responsibilities for grandchildren or elderly parents), cultural factors or peer-group effects?

3.   How effective are current pension policies in maximising labour-force participation of older age groups? Can we design additional policies that work through other channels than financial incentives, such as improving work conditions for people with work disabilities?

A lot of progress can be made on these topics, possibly exploiting survey data (SHARE) linked to administrative datasets, and benefiting from the broad expertise in the area. A strong multi-disciplinary research network would help to stimulate this.

4.   How desirable is gradual retirement, from the point of view of workers as well as employers? And how does gradual retirement affect the macro-economic supply of labour?

Current research on this topic focuses on individual countries and specific programmes stimulating part-time work combined with a (partial) pension. Looking at actual behaviour could be helpful, but is hampered by the problem of disentangling demand- and supply effects. Additional information (specific survey modules, for example) seems necessary.

5.   What is the interplay between supply- and demand factors that drive participation, employment, desired hours and actual hours? Are supply policies sufficient, or should we also think of policies focused on employers, on enhancing productivity of older workers, or on creating an appropriate institutional environment?

6.   What are the consequences of increasing the labour supply of older workers for the economy as a whole, for the economic and non-economic well-being of the older population, for poverty among the elderly, and so forth?

Like the previous question, addressing these issues requires additional insights in the demand side. This could come from employer data or, perhaps, from economic theory.

References

Alesina, A, E. Glaeser and B. Sacerdote (2005) 'Work and Leisure in the U.S. and Europe: Why So Different?' NBER working paper 11278.

Aronsson, T., S.N. Blomquist and H. Sacklén (1999) 'Identifying Interdependent Behaviour in an Empirical Model of Labour Supply, *Journal of Applied Econometrics*, 14(6), 607-626.

Atella, V., F. Peracchi, D. Depalo and C. Rossetti (2006) 'Drug Compliance and Health Outcomes: Evidence from a Panel of Italian Patients', *Health Economics*, 15, 875-892.

Bellemare, C. and B. Shearer (2009) 'Gift Giving and Worker Productivity: Evidence from a Firm Level Experiment', *Games and Economic Behavior*, 67, 233-244.

Belloni, M., C. Monticone and S. Trucchi (2006) 'Flexibility in Retirement. A Framework for the Analysis and a Survey of European Countries', Research report commissioned by the European Commission, Turin: CeRP.

Blanchard, O. (2004) 'The Economic Future of Europe', NBER working paper 10310.

Blau, D. (1994) 'Labor Force Dynamics of Older Men', *Econometrica*, 62(1), 117-156.

Blau, D. (1997) 'Social Security and Labour Supply of Older Married Couples', *Labour Economics*, 4, 373-418.

Blau, D. (1998) 'Labor Force Dynamics of Older Married Couples', *Journal of Labor Economics*, 4, 595-629.

Bonsang, E., S. Adam, C. Bay, S. Germain and S. Perelman (2007) 'Retirement and Cognitive Reserve: A Stochastic Frontier approach applied to Survey Data', *CREPP working paper* 2007/04, University of Liege.

Burniaux, J.-M., R. Duval and F. Jaumotte (2004) 'Coping with Ageing: A Dynamic Approach to Quantify the Impact of Alternative Policy Options on Future Labour Supply in OECD Countries', OECD Economics Department working paper 371, Paris: OECD.

Case, A., and A. Deaton (2005) 'Broken Down by Work and Sex: How our Health Declines' in D.A. Wise (ed.), *Analyses in the Economics of Aging* (Chicago: University of Chicago Press), pp. 185-212.

Currie, J. and B. Madrian (1999) 'Health, Health Insurance and the Labor Market' in O. Ashenfelter and D. Card (eds), *Handbook of Labor Economics, Vol. 3A* (Amsterdam: North-Holland), pp. 3309-3416.

Dahl, S.-A, O.A. Nilsen and K. Vaage (2000) 'Work or Retirement? Exit Routes for Norwegian Elderly', *Applied Economics*, 32, 1865-1876.

Davis, S. and M. Henrekson (2004) 'Tax Effects on Work Activity, Industry Mix and Shadow Economy Size: Evidence from Rich-Country Comparisons', NBER working paper 10509.

Disney, R., C. Emmerson and M. Wakefield (2006) 'Ill Health and Retirement in Britain: A Panel data-based analysis', *Journal of Health Economics*, 25, 621-649.

Duval, R. (2003) 'The Retirement Effects of Old-Age Pension and Early Retirement Schemes in OECD Countries', *OECD Economics Department working paper 370*. Paris: OECD.

Esping-Andersen, G. (1990) *The Three Worlds of Welfare State Capitalism* (Cambridge: Cambridge University Press).

Estevão, M. and F. Sá (2008) 'The 35-Hour Workweek in France: Straightjacket or Welfare Improvement?' *Economic Policy*, 23(55), 417-463.

Falk, A. and S. Gächter (2008) 'Experimental Labour Economics' In S.N. Durlauf and L.E. Blume (eds) *The New Palgrave Dictionary of Economics (2nd edition)* (Basingstoke: Palgrave Macmillan).

Freeman, R. and R. Schettkat (2001) 'Marketization of Production and the US-Europe Employment Gap', *Oxford Bulletin of Economics and Statistics*, 63, 647-70.

Galama, T., A. Kapteyn, R. Fonseca and P-C Michaud (2008) 'Health, Saving and Retirement', RAND working paper.

348   *Arthur van Soest*

Glaeser, E., B. Sacerdote and J. Scheinkman (2003) 'The Social Multiplier', *Journal of the European Economic Association*, 1(2-3), 345-353.

Grossman, M. (1972) 'On the Concept of Health Capital and the Demand for Health,' *Journal of Political Economy*, 80(2), 223-255.

Gruber, J. and D. Wise (1999) *Social Security and Retirement around the World* (Chicago: University of Chicago Press).

Gruber, J. and D. Wise (2004) *Social Security Programs and Retirement around the World: Micro-Estimation* (Chicago: University of Chicago Press).

Gruber, J. and D. Wise (2005) 'Social Security Programs and Retirement around the World: Fiscal Implications,' NBER working paper 11290.

Gruber, J. and D. Wise (2007) *Social Security Programs and Retirement around the World: Fiscal Implications of Reform* (Chicago: University of Chicago Press).

Gustman, A. and T. Steinmeier (1986) 'A Structural Retirement Model,' *Econometrica*, 54, 555-584.

Gustman, A. and T. Steinmeier (2000) 'Retirement in Dual-Career Families: A Structural Model,' *Journal of Labor Economics*, 18, 503-545.

Gustman, A. and T. Steinmeier (2004) 'Social Security, Pensions and Retirement Behavior within the Family,' *Journal of Applied Econometrics*, 19, 723.

Harrison, G. and J. List (2004) 'Field Experiments,' *Journal of Economic Literature*, 42, 1009-1055.

Hartlapp, M. and G. Schmid (2008) 'Labour Market Policy for "Active Ageing" in Europe: Expanding the Options for Retirement Transitions,' *Journal of Social Policy*, 37(3), 409-431.

Hayward, M., W. Grady, M. Hardy and D. Sommers (1989) 'Occupational Influences on Retirement, Disability and Death,' *Demography*, 26(3), 393-409.

Heitmueller, A. (2007) 'The Chicken or the Egg? Endogeneity in Labour Market Participation of Informal Carers in England,' *Journal of Health Economics*, 26, 536-559.

Henretta, J., A. O'Rand and C. Chan (1993) 'Joint Role Investments and Synchronization of Retirement: A Sequential Approach to Couples' Retirement,' *Social Forces*, 71(4), 981-1000.

Heyma, A. (2004) 'A Structural Dynamic Analysis of Retirement Behaviour in the Netherlands,' *Journal of Applied Econometrics*, 19(6), 739-759.

Kalwij, A. and F. Vermeulen (2008) 'Health and Labour Force Participation of Older People in Europe: What Do Objective Health Indicators Add to the Analysis?' *Health Economics*, 17, 619-638.

Kantarci, T. and A. van Soest (2008) 'Gradual Retirement: Preferences and Limitations,' *De Economist*, 156(2), 113-144.

Kapteyn, A., A. Kalwij and A. Zaidi (2004) 'The Myth of Worksharing,' *Labour Economics* 11, 293-313.

Kapteyn, A. and J. Ypma (2007) 'Measurement Error and Misclassification: A Comparison of Survey and Administrative Data,' *Journal of Labor Economics*, 25, 513-551.

Kapteyn, A. and T. Andreyeva (2008) 'Retirement Patterns in Europe and the U.S.,' Netspar panel paper 6, Tilburg University.

Keenay, G. and E. Whitehouse. (2003) 'Financial Resources and Retirement in Nine OECD Countries: The Role of the Tax System,' *OECD Social Employment and Migration working paper 8*, Paris: OECD.

Kerkhofs, M., M. Lindeboom and J. Theeuwes (1999) 'Retirement, Financial Incentives and Health,' *Labour Economics*, 6(2), 203-227.

Levitt, S. and J. List (2007) 'Viewpoint: On the Generalizability of Lab Behaviour to the Field,' *Canadian Journal of Economics*, 40(2), 347-370.

Lumsdaine, R. and O. Mitchell (1999) 'New Developments in the Economic Analysis of Retirement, In O. Ashenfelter and D. Card (eds), *Handbook of Labor Economics, Vol. 3C* (Amsterdam: North-Holland), pp. 3261-3307.

Manski, C.F. (1999) *Identification Problems in the Social Sciences* (Cambridge: Harvard University Press).

Michaud, P.C. and A. van Soest (2008) 'Health and Wealth of Elderly Couples: Causality Tests Using Dynamic Panel Data Models,' *Journal of Health Economics*, 27(5), 1312-1325.

Munnell, A. (2008) 'Is Phased Retirement the Path to Retirement Income Security?' paper prepared for conference "The Future of Life Cycle Saving and Investing," Boston University, October 24.

Nickell, S. (2004) 'Employment and Taxes,' CEP discussion paper 634, Center for Economic Performance, London School of Economics and Political Science.

OECD (2003) Transforming Disability into Ability: Policies to Promote Work and Income Security for the Disabled (Paris: OECD).

OECD (2005) Pensions at a Glance: Public Policies Across OECD Countries (Paris: OECD).

OECD (2007) Pensions at a Glance: Public Policies Across OECD Countries (Paris: OECD).

OECD (2009) Pensions at a Glance: Retirement-Income Systems in OECD Countries (Paris: OECD).

Olovsson, C. (2009) 'Why Do Europeans Work So Little?' *International Economic Review*, 50, 39-61.

Rappaport, A. (2008) 'Why Phased Retirement? The Case for Phased Retirement and Some Policy Recommendations,' paper prepared for conference "The Future of Life Cycle Saving and Investing," Boston University, October 24.

Reday-Mulvey, G. (2000) 'Gradual Retirement in Europe,' *Journal of Aging and Social Policy*, 11 (2-3), 49-60.

Riphahn, R. (1999) 'Disability Retirement among German Men in the 1980s,' *Industrial and Labor Relations Review*, 52(4), 628-647.

Ruhm, C. (1990) 'Bridge Jobs and Partial Retirement,' *Journal of Labor Economics*, 8(4), 482-501.

Ruspini, E. (2002) *Introduction to Longitudinal Research* (London: Routledge).

Sigg, R. (2007) 'Extending Working Life: Evidence, Policy Challenges and Successful Responses' in B. Marin and A. Zaidi (eds) *Mainstreaming Ageing: Indicators to Monitor Sustainable Progress and Policies* (Farnham: Ashgate), pp. 447-489.

Smith, J. (2007) 'The Impact of Socioeconomic Status on Health over the Life-Course,' *Journal of Human Resources*, 42(4), 739–764.

Stock, J. and D. Wise (1990) 'Pensions, the Option Value of Work and Retirement', *Econometrica*, 58(5), 1151-1180.

Sutherland, H. (2000) 'Euromod: A Tax-benefit Model for the European Union,' *Transfer: European Review of Labour and Research*, 6, 312-316.

Van den Berg, G., M. Lindeboom and F. Portrait (2006) 'Economic Conditions Early in Life and Individual Mortality,' *American Economic Review*, 96(1), 290-302.

Vermeulen, F.P. (2002) 'Collective Household Models: Principles and Main Results,' *Journal of Economic Surveys*, 16, 533-564.

Vickerstaff, S., J. Cox and L. Keen (2003) 'Employers and the Management of Retirement,' *Social Policy and Administration*, 37(3), 271-287.

Woittiez, I. and A. Kapteyn (1998) 'Social Interactions and Habit Formation in a Model of Female Labour Supply,' *Journal of Public Economics*, 70, 185-205.

Wolfe, J.R. (1985) 'A Model of Declining Health and Retirement,' *Journal of Political Economy*, 93, 1258-1267.

Zaidi, A., M. Markovec and M. Fuchs (2007) 'Transition from Work to Retirement in EU Countries' in B. Marin and A. Zaidi (eds), *Mainstreaming Ageing: Indicators to Monitor Sustainable Progress and Policies* (Farnham: Ashgate), pp. 395-419.

Labour Supply and Employment
of Older Workers

Comments by Maarten Lindeboom

Introduction

This chapter sketches research challenges and research needs in the area of labour supply and employment of older workers. The topic is one of the key policy questions in the economics of ageing, and it is difficult not to agree with most of the points made by the author. The remainder of this discussion centres on section 4, which presents the research challenges. My comments will also include suggestions for some other potentially important avenues for future research.

Research challenges

The challenges identified in section 4 are all important – and indeed both the academic literature and policymakers can and will benefit from more research in these areas.

Interactions between subprojects

While some of the challenges are seen as separate issues, quite a lot can, presumably, be learned from projects that deal with the interaction of the issues. For instance, section 4 mentions pensions as a separate subject. It is argued that cross-national variation as well as time variation in pension systems could be exploited to better measure the incentive effects of pension wealth on retirement behaviour. While I do not disagree with this, I think that changes in the pension system could also be relevant for other issues mentioned in section 4.

For instance, there is a long-standing debate about the relative importance of health- and financial incentives in explaining retirement behaviour. Independent (from health) variation in pension wealth induced by the reforms could be used to learn more about the relative importance of both factors.

Furthermore, the retirement decision can be seen as an individual optimisation problem, where the agent considers the optimal timing of retirement, given the available retirement options. These alternative retirement options concern, for instance, applying to a Disability Insurance (DI) scheme. Changes in the reward of one of the options (pensions) will affect the relative rewards of other options (DI and work). This will affect behavioural choices. Indeed, as mentioned in the chapter and noted in previous papers, alternative exit routes are important – but I think that institutional changes (such as changes in the pension system) should be exploited to learn more about this.

Health

Health is an important issue, and a lot of what the author describes in subsection 4 on the role of health makes a great deal of sense. Particularly promising is the life-cycle perspective in explaining later retirement and health outcomes. In this connection, it is good to mention that health problems do not arrive randomly at later ages; already at relatively young ages a substantial fraction of workers report having health problems – and this is strongly associated with socioeconomic problems (Smith, 2004). Also note that health and employment status are strongly correlated – and that much of the association between health and income disappears as soon as one conditions on labour-market status (Case and Deaton, 2003). The picture below, for the Netherlands, describes employment patterns for workers of different cohorts by different health status. The thin lines refer to those who report being in poor health, the thick lines to those who report being in good health (defined as not poor). The figure clearly shows that employment patterns of those who report being in good health are close to unity up to the ages at which early retirement becomes relevant. Prime-aged workers (35-50) who are in poor health have employment levels that are 30 to 40 percentage points lower. A group of workers of this size cannot be ignored. About 10% of the 30 year-olds report being in poor health – and this fraction grows to almost 30% at the age of 50.

As a consequence of the importance of health in determining/influencing employment among the older population, policies aimed at improving participation rates of older workers should also focus on the causes and consequences of poor health prior to retirement age. Research and policy should also focus on ways to keep these workers in the labour force. Or, phrased as a question: how can we avoid premature exit from the labour force, and how many years of labour might be gained by intervention?

Figure 1:   Employment patterns for workers of different cohorts by different health status

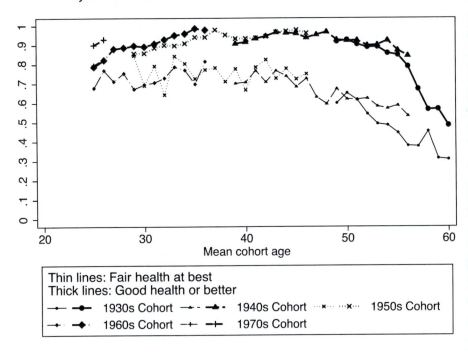

Thin lines: Fair health at best
Thick lines: Good health or better

1930s Cohort   1940s Cohort   1950s Cohort
1960s Cohort   1970s Cohort

References

Case, A. and A. Deaton (2005) 'Broken Down by Sex and Work, How our Health Declines', chapter 6 in D.A. Wise (ed.), *Analyses in the Economics of Aging* (Chicago: Chicago University Press).

Labour Supply and Employment
of Older Workers

Comments by Ruud de Mooij

In this chapter, Arthur van Soest offers an excellent review of the current state of knowledge regarding labour supply of older workers. Since the Gruber and Wise project a decade ago (and the various follow-up papers), we know quite a bit about the role of financial incentives for elderly labour supply. In particular, actuarially unfair early retirement schemes have in many countries imposed high implicit tax rates on participation at older ages. This has tremendously depressed the labour supply of older persons – especially in continental European countries, where participation rates dropped to very low levels. Indeed, early retirement schemes have induced a culture in which it is normal to stop working long before the official retirement age. But Van Soest argues that financial incentives are not the only reason for this shift. Non-financial aspects (such as health, preferences, demand restrictions, restrictions to work part-time, job characteristics) and certain institutions (such as trade unions) have also played a role. To properly understand the culture of early retirement, we need to know more about the interplay between financial and non-financial aspects. Van Soest proposes a multidisciplinary approach to attain this knowledge, with an emphasis on empirical research using micro data. His chapter offers helpful suggestions for improving data and an ambitious, sensible research agenda for the future.

I would add to this the imperative for better theory in order to understand the interplay between financial and non-financial factors. This is vital for identifying the key policy challenges that governments are facing today. For instance, a normative question that is relevant for policymakers is how institutions should be designed optimally, from a social welfare perspective.

Thereby, governments face trade-offs. For instance, many institutions, such as social insurance, employment protection legislation, labour contracts and health policy, offer insurance against various kinds of risk during the life cycle of individuals. This is valuable for people and, *ceteris paribus*, increases welfare. Insurance, however, is inevitably ac-companied by moral hazard in the form of reduced incentives to work, learn and search for a new job. The policy challenge, therefore, is to find institutions that achieve an optimal trade-off between insurance and incentives. This is a daunting task – especially in a world with heterogeneous risks and preferences, and limited information for governments. Yet, theory, by sharply identifying trade-offs, could offer empirical researchers proper guidance in looking for the right relationships, specifications and parameters that subsequently could help policymakers to make sound decisions.

An example may illustrate the importance of a proper conceptual analysis for designing good policy. Economic theory emphasises that maximising participation of older workers is not the true policy goal. Indeed, policies regarding elderly participation should follow from a broader goal of social welfare maximisation. This does not necessarily coincide with maximisation of elderly participation. Yet, Dutch policy during the past ten years has aimed at just that. The result is that policy might currently be overshooting. Let me explain this in some more detail.

During the 1980s and early 1990s, labour-market participation of older persons was very low in the Netherlands. This was partly due to poor incentives imposed by early retirement schemes. Indeed, implicit tax rates on postponed retirement ranged from 80 to 100%. These policies were deemed unsustainable in the long run, especially in light of the ageing of the population. Since the 1990s, the government has implemented a series of reforms (early retirement schemes were made actuarially fair, for example) to encourage participation of older workers. As a result, participation rates have been rising steadily since the mid-1990s.

Despite this positive trend, elderly participation remained high on the Dutch policy agenda. Indeed, the implicit tax on working at an older age has been transformed into a substantial implicit subsidy. This holds for three reasons. The first is due to the Dutch pension system. Both the contribution rate of occupational pensions and the accrual of pension benefits are uniform and independent of age. Yet, the pension savings of the young have a much longer maturation period. Accordingly, young workers pay an implicit tax of approximately 9% because their pension contributions exceed their pension accrual.

For older workers, the uniform pension premium and accrual imply an implicit subsidy on delayed retirement of approximately 6% (Bonenkamp, 2009). Second, wage profiles in the Netherlands are steep, and keep increasing even beyond the age of 50. This most likely reflects a deferred-payment scheme, whereby young workers receive a wage below their productivity and older workers a wage above their productivity (Van Vuuren and De Hek, 2009). Again, this deferred payment can be interpreted as an implicit tax on the young and an implicit subsidy for the old. Third, the government has introduced various kinds of age-specific tax relief for people working beyond the age of 62. In particular, workers beyond the age of 62 are granted an earned-income tax credit that is 770 euro higher than that of younger workers. Moreover, workers receive a special bonus that increases from almost 2300 euro per year for a 62 year-old to almost 4600 euro for a 64 year-old. Together with the implicit subsidies, these allowances render the average tax burden negative for most workers beyond age 62.

Does this make sense from a welfare perspective? Is stimulating labour-market participation of older people (still) attractive if it requires society to pay a net subsidy? Economic theory offers counterarguments. Cremer et al. (2008) argue that a small implicit tax (as opposed to an implicit subsidy) on elderly workers is most likely to be optimal. Moreover, subsidising older workers may oppose equity objectives. Indeed, participation at old age is positively correlated with good health, high ability and high lifetime income. Subsidies for this group thus have perverse distributional effects (involving redistribution from the lifetime poor to the lifetime rich).

The emphasis on the supply of older workers might also be the wrong focus of policy in the Netherlands. Now that elderly labour participation rates are rapidly increasing, it is becoming increasingly clear how poorly the Dutch labour market for elderly workers actually functions (Euwals et al., 2009). Firms are reluctant to hire older workers, wages are rapidly rising with tenure, and strict employment protection is forging 'golden chains' for elderly workers. Job-to-job mobility is thus low, tenures are long, unemployment duration is exceptionally high and unemployed older individuals face a very low probability of finding a job. Consequently, one can hardly speak of a 'labour market', since almost no transactions take place at the wages that firms are willing to offer. Without future reform, the Netherlands runs the risk of becoming a country of people with extended working lives, but with older workers that are increasingly working in the wrong jobs and with too low productivity.

The approach suggested by Van Soest recognises that it is not just financial incentives for participation that matter for public policy. Indeed, his chapter emphasises the importance of demand factors and demand restrictions for labour participation of older persons. This is to be applauded from a policy perspective. His approach is just what we need to make research not only innovative from a scientific perspective, but also relevant for public policy around the world.

References

Bonenkamp, J. (2009) 'Measuring Lifetime Redistribution in Dutch Occupational Pensions', *De Economist*, 157(1), 49-77.

Cremer, H., J.M. Lozachmeur and P. Pestieau (2008) 'Social Security and Retirement Decision: A Positive and Normative Approach', *Journal of Economic Surveys*, 22(2), 213-233.

Euwals, R., R. de Mooij and D. van Vuuren (eds.) (2009) *Rethinking Retirement*, CPB special publication 80.

Vuuren, D. van and P. de Hek (2009) 'Firms, Workers, and Life-cycle Wage Profiles' in R. Euwals, R. de Mooij and D. van Vuuren (eds), *Rethinking Retirement*, CPB special publication 80.

12

Policy and Research Challenges for Ageing, Health and Pensions in Europe

Arthur van Soest / Lans Bovenberg / Asghar Zaidi

1 Introduction

Background and motivation

The combination of rising life expectancy and a declining fertility rate has led to a rising share of older persons in the populations of European and many other advanced economies. This demographic transition, referred to as population ageing, is undoubtedly one of the key economic and social developments shaping the 21st century in European countries. In meeting the challenges arising from population ageing, many European countries have already been making important institutional and policy changes, particularly through reforms of pension systems. However, further and more widespread reforms may still be required in many of these countries, especially in the labour market and health and social care policies, so as to avoid an adverse impact of this demographic transition on potential economic growth and on public social spending.

Understanding the impact of population ageing on our societies and designing suitable policy responses require a good deal of research and empirical evidence as well as forward planning. Within the context of European societies, insights can be drawn from a cross-national comparative point of view, since comparing different policy approaches and exploring the consequences of various policy changes in different institutional and economic cycle environments will help greatly in learning about the type of policy reforms required.

This concluding chapter draws insights from the work presented in ten chapters of this volume, Chapters 2-11. In so doing, the chapter lays out a research agenda that exploits the diversity of European pension- and health systems to study causal links between institutional arrangements, individual decision-making, labour force transitions, financial security and well-being of older age groups, including health outcomes. One of the key aims of this volume has been to integrate economic, psychological, sociological and epidemiological approaches to gain insight about individual decision-making related to health and pensions. This concluding chapter discusses the implications of this multidisciplinary information for financial and labour markets, financial institutions and public policy. The chapter thereby stimulates future scientific research on ageing-related issues.

Setting the context

The evidence on the extent of population ageing and its associated budgetary challenges in European countries will be used to set the context. Results drawn from the 2009 Ageing Report of the European Commission, presented in Figure 12.1, show that the old-age dependency ratio (people aged 65 or above relative to the working-age population aged 15-64) in EU Member States is expected to rise from 25.4% to 53.5% over the period 2008-2060. This implies that the EU, on average, would move from having four working-age people for every person aged over 65 years to a ratio of two to one. The EU of today already had the highest old-age dependency ratio in the world in 1950, and during the period 2000 to 2050 the largest increases are projected to take place in Japan (by close to 50 %-points), China and the EU27 (by almost 30 %-points).[1]

Ageing (and its impact on social systems) is listed as one of the five major trends in society in the METRIS report (European Commission, 2009b). It will generate pressure on the labour markets (to absorb a greater fraction of labour force in the wake of a shrinking working-age population) as well as on the well-being of European citizens (partly through displacement of other social benefits in favour of ageing-related benefits, and partly because the resources available for ageing-related benefits will be spread over a higher number of people in older age cohorts). Results on the budgetary consequences of population ageing in EU27 countries are available in the 2009 Ageing Report, and summary results appear in Figure 12.2.

---

1   These conclusions are drawn from the estimates based on the UN World Population Prospects: The 2008 Revision, as referred to in European Commission (2009a: 48).

Figure 12.1: Population ageing in EU countries, 2008-2060, as given by the old-age dependency ratio (number of 65+/number of 15-64)

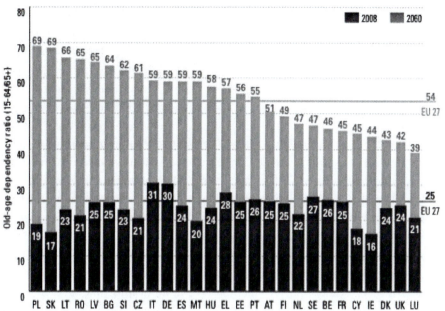

Source: European Commission (2009a), data originally from Eurostat, EUROPOP2008.

Figure 12.2: Age-related government expenditures, 2007-2060, % of GDP

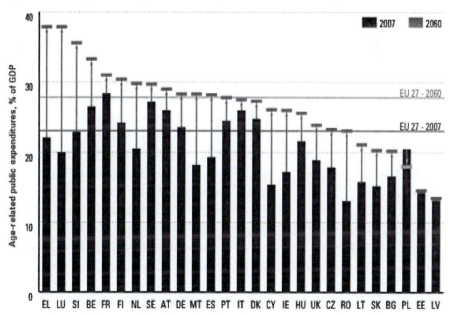

Source: European Commission (2009a).

It shows that the fiscal impact of ageing is projected to be substantial in almost all member states: on the basis of current policies, age-related public expenditure will increase on average by about 4.7 %-points of GDP by 2060 in the EU. Most of the projected increases in public spending over the period 2007-2060 will be on pensions (+2.4 %-points of GDP), followed by healthcare (+1.5 %-points of GDP) and long-term care (+1.1 %-points of GDP).[2]

Other considerations

The Lisbon agenda defined an ambitious programme to improve Europe's lagging growth performance – despite the fact that empirical evidence suggests that population ageing could weaken Europe's economic growth prospects and welfare. Clearly, Europe will have to devise innovative policies and institutional solutions that will allow its societies to benefit longer from the talents of older people. For example, the labour market should be able to benefit from the fact that lower birth rates and improved education enable women to have a higher and fuller participation in the formal labour market. However, innovative initiatives may be necessary to facilitate formal and informal childcare. The longer working lives of women can be exploited to allow a more successful combination of career and motherhood. The life-cycle perspective needs to be adopted to allow human resources to be maintained better during the entire life course of both men and women – which benefits both income security and well-being in old age.

Many studies have analysed the impact of ageing on adequate provision of health- and long-term care and pensions. These studies vary in their focus: some analyse individual behaviour and individual well-being, while others explore the changing roles of institutions (banks, pension funds and insurance companies), the consequences for public finances and the macroeconomic consequences of ageing in an international context (including spill-over effects from other countries with different institutions and demographic compositions). Moreover, while economic decisions are central to our approach, it is by now well-established that economic decisions cannot be isolated from decision- and environmental factors at other levels and in other domains, implying a large added value for an approach that uses insights from other social science disciplines. A society that achieves successful ageing is one in which the well-being of older people is optimised and where older people make the maximum contribution to society as a whole. Additionally, the notion of social sustainability in societies experiencing population ageing requires enabling

2   For details, see European Commission (2009a: 24).

practical, mutually beneficial and satisfying relationships between older and younger generations.[3]

This book combines insights from many disciplines in economics and the social sciences. It includes microeconomics of individual behaviour, sociological and psychological aspects related to the economics of ageing, the economics of health and healthcare, financial markets and institutions in an ageing society and the public policy implications of population ageing.

Three themes on ageing-related policy and research issues

This chapter is organised along the lines of three broad policy and research themes followed throughout the book: income security of the elderly, (wider) well-being of the elderly and the labour market and older workers.

1.   The first theme, income security of the elderly, is quite wide-ranging, including everything related to the design of pension systems, individual decisions on pensions and other retirement savings, consumption patterns before and after retirement, decisions to annuitize or not, financial knowledge, psychological factors that may lead to decisions that are not in the households' own long-run interest, and so forth. It addresses important questions such as how should governments, individuals and financial institutions share the responsibilities for financial security during old age – in particular, for those who belong to typically vulnerable groups (immigrants, women, or low-skilled and low-educated workers). The disciplines providing insight into these questions are mainly microeconomics, finance, macroeconomics – but also public policy plays a role, through the implications of pension systems and pension reforms for income inequality and poverty. Health comes into consideration through the financial risks of health shocks and health (and work disability) insurance.

2.   The second theme, well-being of the elderly, goes beyond economic status and includes factors such as psychological well-being, family contacts and other social networks and inter- and intra-generational transfers, time use and satisfaction with daily activities (including paid work and volunteer work), social exclusion, physical and mental health (and health behaviour and prevention), availability of formal and informal long-term care (LTC) and other aspects of the healthcare system. It explores what might be an appropriate European model for the organisation and financing of

---

3   For further discussion on the issue of what policy challenges with respect to intergenerational solidarity are identified in the international policy agenda, and how European societies are responding to them, see Zaidi, Gasior and Sidorenko (2010).

LTC, and how social productivity (volunteering, participating in social organisations, informal care tasks) of older people might be augmented. Insights are combined from health, sociology, psychology, microeconomics and public policy.

3.   The third theme involves a key policy concern of European policymakers: how to raise labour-force participation of older workers? Retirement and labour supply decisions of older workers are core issues of microeconomic research, but also important are factors such as quality of work, social networks and peer-group behaviour. Health status and its dynamics play a major role (through work disability, morbidity or expected longevity). Public policy determines the economic environment under which decisions are made by affecting the incentives to work and maintaining human capital. The demand side of the labour market is equally important, particularly when labour supply is being increased. How can the productivity of older workers be maintained and how can wage rises be kept in line with productivity in older age? How can existing stereotypes and prejudice towards older workers be eliminated? How effective are laws against age discrimination or public campaigns to promote the image of older workers? Which demand-side adjustments will be necessary to accommodate an increased supply of older workers (with regard to hiring policies, reducing wage costs, training older workers, using alternative exit routes, accommodating workers with health problems and facilitating gradual retirement)? These adjustments may require changes involving substantial macroeconomic implications in labour-market institutions.

Structure of the chapter

For each of the three themes, this chapter discusses the most important policy questions and also the progress in our knowledge until now in addressing these policy issues. It then identifies the gaps in current scientific knowledge and the challenges for research in the next five to ten years. All of these are discussed in respective sections for each theme, in sections 2-4. Section 5 sketches a road map of European research that can fill many of the gaps identified in the previous sections, emphasising the feasibility of addressing the identified research challenges, the data requirements and the need for research cooperation at the European level – and the steps needed to make progress.

2 Theme I: Income security of the elderly

Financial security in old age can be provided in many different ways. Many European countries seek to achieve this objective through a government system of old-age social security provision (first-pillar pensions), occupational pensions through private pension funds (the second pillar) and voluntary household savings for retirement (for example, in the form of life insurance; the third pillar). General household savings that can be used for retirement but also for other purposes are sometimes seen as a fourth pillar. A specific example is investing in owner-occupied housing, where a reverse mortgage or the proceeds of selling the house can be used to finance (part of) the cost of living in old age. Financial support by family members often also plays an important role. The four pillars are often used in combination.

2.1 Policy questions

The important policy questions raised on this theme can be summarised as follows.

- *Should Pay-As-You-Go systems be supplemented by funded pensions?* How has the recent financial and economic crisis of 2008-2009, during which the performance of returns from funded systems has taken a beating, raised doubts whether the move towards funded pension schemes is the right strategy?

- *What are the advantages and disadvantages of collective pensions against individual ones?* What are the relative advantages and drawbacks of collective models?

- *Which regulations must be imposed on pension funds and other financial institutions?* The relevance of this policy question became clear during the recent financial and economic crisis, by revealing serious agency problems: interests of financial managers do not always coincide with those of the participants and the firms for whom they invest.

- *Can we give guidelines for portfolio allocation of a funded pension system in a given institutional context?* How does a pension fund's optimal portfolio of risky and safe assets depend on, for example, flexibility of the retirement age, contribution rates, standard of living and the housing market?

- *How can people be helped to make life-cycle saving- and investment decisions that guarantee financial security in old age and maximise lifetime welfare?*

- *Should portfolio risk be reduced at older age?* How does answering this question involve taking account of the relationships between labour income and asset returns and risks in the long- and short run? How should these considerations change when retirement becomes more flexible (through gradual retirement or the opportunity to delay or postpone retirement)?
- *Defined-benefit or defined-contribution pensions?* How can the two options of defined-benefit- and defined-contribution plans be combined in new pension products?
- *Annuities or liquidation of household wealth?* Should annuitization be made mandatory to ensure income security in old age (as is done in many European countries)? Should reverse mortgages be stimulated by public policy and by financial institutions?
- *Which policies can improve financial planning and retirement savings of groups that do not adequately prepare for retirement?* What specific help may be required by certain groups (the low-skilled, self-employed and immigrants) to avoid sub-optimal retirement savings decisions due to lack of information or status quo bias?

2.2 Major progress in understanding

Adequacy of savings from the perspective of the pension system

Differences in European retirement provisions are reflected in differences in the age-saving profiles (Börsch-Supan and Lusardi, 2003). While other driving forces (such as the stringency of borrowing constraints) may explain part of these differences, some of the variation is due to different pension set-ups. The more generous social security in Italy and Germany, for example, reduces the need to save for retirement during working age. For the adequacy of pensions, as important as benefit levels *per se* is the government's role in providing a good *ex ante* allocation and diversification of risks.

Adequacy from the individual perspective

The empirical evidence, implicitly or explicitly based on the life-cycle model is mixed (and largely concentrated on US data). Most of the more recent studies using more sophisticated models (such as Scholz, Seshadri and Khitatrakun, 2006) find that saving is adequate for most of the population, and that *under-saving* is concentrated among households with low socioeconomic status.

Wealth liquidity is another important element of saving adequacy, particularly the illiquidity of housing wealth – a major component of wealth in old age. Financial markets have developed instruments (reverse mortgages) to extract equity from a home and transform it into more liquid forms. These (still under-utilised) instruments offer greater opportunity for households to finance a preferred consumption path (Muellbauer, 2007). These new instruments, however, are still rarely used in practice.

The availability of *annuities* is another essential aspect of retirement saving adequacy. Simulation exercises show that annuities are quite valuable (Geanakoplos, Mitchell and Zeldes, 2000), even when the optimal consumption trajectories differ substantially from the time paths of annuity payouts. In practice, annuity markets are thin, as many problems limit individuals' propensity to annuitize. Models of fully rational economic agents fail to explain the lack of demand for annuities. Psychological factors (for example, preference for lump sums *as such)* and other forms of 'irrational' or bounded rational behaviour (such as hyperbolic discounting) are likely to be at work (Brown, 2009).

Retirement planning, information about pensions and financial literacy

A growing literature has documented significant departures from the model of a fully rational economic agent and pointed to behavioural and psychological factors that limit individual planning ability. Lusardi (1999) found that one-third of the age group 50-60 in the US had not thought about retirement at all. Lack of financial planning is concentrated among specific groups of the population (those with low educational attainment, African-Americans, Hispanics and women). Several US studies show that workers were also poorly informed about their pensions and the characteristics of their pension plans. One reason individuals do not engage in planning or are not knowledgeable about pensions is that they lack financial literacy. Lusardi and Mitchell (2007) show that financial literacy and pension knowledge are positively associated with financial planning and retirement savings.

Risk and portfolio choices in individual and collective pension plans

In a static framework, the portfolio-choice problem is a compromise between risk and performance, formalised by maximising expected utility with a concave utility function on final consumption. The optimum depends on the degree of concavity of the utility, which determines the degree of absolute risk aversion and the distribution of returns to financial assets.

Key progress has been made since the late 1960s by introducing time into portfolio-choice strategies, exploring how the length of the time horizon affects the optimal structure of the portfolio. In the benchmark model, the only decision to be made at each point in time regards asset allocation. Real-world households also control their labour supply or the intensity of their saving effort. Gollier (2001) shows that this greatly affects the relationship between the investor's time horizon and the share of wealth invested in risky assets.

When the model includes human capital wealth, total wealth (consisting of financial wealth and human capital wealth) can be invested in either stocks or bonds. Human capital is not tradable. Young agents should thus invest more of their *financial* wealth in risky assets than old agents do, since the young are less dependent on financial wealth for their consumption – as they have labour income as an alternative risk-less income source.

A rapidly developing literature analyses to what extent the core conclusions of this benchmark model are robust to alternative assumptions with respect to human capital, the properties of financial markets and the specification of individual preferences. Cocco, Gomes and Maenhout (2005) find that young investors choose portfolios that are less tilted towards equity than middle-aged investors do. The desire to hold a safe, liquid stock of precautionary savings may explain both the 'equity premium puzzle' (the large rewards for risk-taking that are hard to explain on the basis of the benchmark model) and why young households do not participate in the stock market (Constantinides and Duffie, 1996).

Conclusions

In conclusion, theoretical models of financial behaviour, financial markets and preparing for retirement seem to be relatively well developed. Still, these models miss several real-life features, the most important of which may be the assumption that individuals behave like the traditional *homo economicus* with rational expectations and optimal decisions. Recent empirical work emphasises lack of financial literacy and pension knowledge and suboptimal decision-making based upon rules of thumb, default choices and the influences of peer-group decisions. Pure economics is not enough – abundant evidence suggests that psychological and sociological factors play a vital role. The way forward should involve incorporating these empirical 'anomalies' into rigorous theoretical and empirical models of retirement saving and life-cycle labour supply and consumption.

3   Theme II: Well-being of the elderly

Well-being of older Europeans has many dimensions, among them the economic aspects covered in the previous section. The second theme considers other dimensions and how they relate to the economic dimension. For analytical purposes, a distinction should be made between various age groups. The younger age group among the elderly (which may extend until around the age of 70) will constitute a larger fraction of the population in the next decades; they are also expected to be relatively healthier, more affluent, increasingly well educated and active. They will play a major role in formal work and in other socially productive activities. Good working conditions and opportunities to remain socially productive after (early) retirement can contribute to higher well-being of this subgroup. Both intrinsic and extrinsic rewards for these activities are crucial.

For older groups (say, those above 70 years of age), physical and mental health problems often become a growing concern, leading to increasing reliance on formal and informal health and social care. Mobility tends to diminish, increasing the importance of suitable housing and living arrangements. Social networks play an important role at all ages – but their nature changes over time, as they become smaller and more family-centred at older ages. The role of family networks also changes with age, from giving support to parents and children and grandchildren to receiving support from younger generations. The role of networks also depends on the national context, such as the intergenerational balance of income and wealth, and the nature of the housing market.

3.1   Policy questions

The theme on the wider measures of well-being of older people covers quite diverse areas. The policy issues are also quite varied, as outlined below.

- *How can satisfaction from formal work and well-being in early old age be promoted?* What can be done to organise work within the firm and in society to keep people motivated – and thus keep as many older people as possible in employment? How can the well-being, particularly health, of middle-aged to early old-aged workers be better protected and improved?
- *How can retirees be equipped and motivated to continue or initiate socially productive activities, such as volunteering, coaching, or caring for a sick or disabled person?* How can we augment the proportion of socially productive older people? What are the costs and benefits of extending opportunities and incentives?

- *Should the government or the family cover life-course risks?* What is the optimal role of the government, financial transfers and social support by the family?

- *How can public policy encourage people to have children while being employed?* How to reconcile parenthood and employment? What are the consequences of combining different ways of organising and funding formal childcare with informal childcare?

- *How to provide care to the dependent elderly?* Can voluntary intergenerational transfers in a wider community take over part of the role of the family in providing social care for the elderly?

- *How can access to the Internet and e-networking improve family and social contacts in old age?* Should government interventions stimulate Internet access of the older age groups (subsidies, training)?

- *How can public policies promote subjective well-being (SWB) of older people?* Which life circumstances affect SWB of older adults? How can policies at the local, regional, national or supranational levels help to promote SWB of older people? To what extent does high versus low SWB of older adults produce positive versus negative individual and societal outcomes?

- *How to project and deal with the costs of healthcare in ageing societies?* How should the healthcare system be organised? Should personal care budgets be used to give more choice and responsibility to the consumers of healthcare?

- *What is an appropriate European model for the long-term care (LTC) system?* What is a financially feasible system for LTC finance and organisation in Europe? Should policy measures stimulate investments in labour supply of LTC staff? Can and should public policy stimulate informal care as an alternative to formal care? Can housing adjustments and housing market flexibility reduce the need for care?

3.2 Major progress in understanding

A rich literature focuses on factors that determine well-being of older age groups, building on theoretical arguments from sociology, psychology, economics and epidemiology. The empirical evidence is mainly descriptive, confirming the associations predicted by the various theories. Often, however, the same associations can be explained by different causal mechanisms or by confounding factors. Substantial progress could be made in this respect, particularly when longitudinal data become available.

Links between socially productive activities and well-being

Socially productive activities range from continuation of gainful work to voluntary work. The long-term neglect of socially productive activities of older people has been redressed in the literature (Künemund, 2001). Social participation within and beyond the family has impressive benefits, including increased emotional well-being, better health and even increased longevity.

Role of formal and informal work for well-being in early old age

The association of employment duration and well-being is bi-directional: poor health and disability are powerful predictors of early exit from the labour market, but leaving the job prematurely also reduces well-being. Poor quality of work is not only more prevalent among workers with low socioeconomic status, but also exerts direct effects on health and well-being. An adverse psychosocial work environment negatively affects health among older workers by increasing their probability of long sickness absence, reduced performance and productivity, and forced early retirement due to illness or disability.

Social networks

Social networks are important throughout the life cycle, but their nature changes with age (particularly from early old age to later old age). The fact that network size decreases with age is partly explained by the increasing probability of the death of parents, partners, siblings and friends. The theory of socioemotional selectivity (Carstensen, 1991) explains these changes from changes in individual time perspective: people select with whom to spend their remaining time, and they drop less-important relations.

The literature has established that intergenerational family bonds usually remain strong throughout adulthood and old age, and may even become more important for well-being over the life course than nuclear family ties (Kohli, 1999). Financial transfers and social support are frequent and substantial, occur mostly in the generational lineage, and their net flow is mostly from parents to children; financial transfers *inter vivos* are complemented by bequests.

The tension between the time spent to help parents and formal labour supply for both women and men is especially acute for those in the position of the 'sandwich generation' (Künemund, 2006), that has a double obligation of care for dependent parents and children. With the rising labour force participation of women and the extension of working life through a later retirement age, the potential time crunch is likely to become harsher.

A crucial policy question is whether the available family potential can be activated in times of need. Studies disagree on the importance of the burden due to competing demands from parents, children and the labour market and the consequences for well-being (see Künemund, 2006). Evidence suggests that welfare-state provisions, instead of crowding out family support, enable the family to provide new intergenerational support and transfers (Kohli, 1999).

Subjective well-being of older age groups

Research on SWB among older adults is conducted in several disciplines: social-scientific survey research, personality psychology, social psychology, life-span developmental psychology and gerontology. Measurement in many existing studies focuses on one component: life satisfaction based on the survey question "How do you feel about your life as a whole?" Kahneman et al. (2004) propose alternative approaches such as the Day Reconstruction Method (DRM): respondents keep a diary recording events of the previous day and assess how they felt during each event on selected dimensions of well-being.

Pinquart and Sörensen (2001) analysed 286 studies on the association of socioeconomic status, social network and competence with SWB in older people. Their results stress the importance of life circumstances and social indicators for SWB in older adults: low income, low educational status and functional deficits accompany reduced life satisfaction, low happiness and reduced self-esteem in old age.

A few *prospective longitudinal studies* show that, depending on the type of event, the adaptation of SWB (that is, return to pre-event level) may be slow and incomplete. For instance, adaptation to widowhood took about seven years, on average. SWB decline in response to severe disability was even more pronounced and adaptation took either longer or never happened at all.

Health and socioeconomic status

Trying to understand the relationship between health and socioeconomic status (SES) is one of the most challenging research issues in the economics of ageing – particularly the direction of causality. Research shows that the strong relationship holds for a variety of health variables (most illnesses, mortality, self-rated health status, psychological well-being, biomarkers) and alternative measures of SES (wealth, education, occupation, income, level of social integration). In the case of wealth, this association is known as the *health-wealth gradient*. One direction of causality is from wealth to health because individuals with more wealth can afford better medical care, live in healthier environments, and so forth. Another is from health to wealth. Healthier individuals may be able to work more than those who are ill, enabling them to accumulate more wealth. Finally, wealth and health status may be simultaneously determined by possibly unobserved common factors. For policy to improve welfare, health and well-being, these explanations of the health-wealth gradient must be distinguished in the context of an ageing population. Several publications deal with this issue using longitudinal data in the US. Adams et al. (2003) find evidence of causal effects of health on wealth – and much less in the other direction. Whitehouse and Zaidi (2008) review more than 50 past studies to show that there are large socioeconomic differences in mortality and life expectancy (especially for men), which appear to have become larger over time. Still, the causality debate remains an open issue, and additional research is required, particularly in European countries.

Ageing and healthcare expenditure

The data show that for a given year, health expenditure increases with age – except for the very old. The claim is often made that population ageing will result in an acceleration of health expenditure. But this argument seems to confuse the notion of correlation with that of causality. Its validity deserves to be examined more closely by studying the role of other possible factors of health expenditure growth.

Various econometric studies show that advances in medical technology – and its diffusion – are leading predictors of the increase in healthcare expenditure. NBER studies analysing the growth of in-patient Medicare costs showed that the price that Medicare pays for admissions has declined over time, but that the technological intensity of the treatment has increased (Cutler and McClellan, 1998). An explanatory factor of the evolution of healthcare expenditure is obviously not death as such – but rather medical care aimed at treating various pathologies, some of which lead to the death of patients. This approach strongly suggests a change of paradigm, in which population ageing and the increase in longevity have little impact on the growth of healthcare expenditure, whereas the larger part of this increase is due to the evolution of medical technology.

The future cost and organisation of healthcare in ageing societies of Europe is a core topic in health economics. Researchers in this area may want to build on the methods and type of data used in Goldman et al. (2004), which projects healthcare costs of older age groups in the US using a micro-simulation demographic and economic model, the Future Elderly Model (FEM). The model uses Medicare claims records linked to actual medical care use and costs over time. A similar exercise would be very useful to carry out for European countries, combining information provided by health insurers (claims data) and medical data in a single database.

Ageing and long-term care (LTC)

LTC expenditure growth is expected to accelerate in the future, mainly as a result of larger numbers of older persons and a steep increase in the numbers of the oldest old. LTC issues are thus becoming increasingly important on the health and social policy agendas of developed countries. Norton (2000) (and recently Huber et al., 2009) provides a detailed survey of the issues. Informal care provided at home by family members, friends or voluntary organisations is the most important source of LTC in all OECD countries. Presently, women aged over 45 provide the bulk of informal care. Men are more likely to assume the role of caregiver for their spouses than for other family members.

Most demographically derived predictions of future health and LTC spending focus on number of people, assuming their average health status and healthcare needs remain constant. Demographers are divided in their opinions on the extent to which life expectancy will be further prolonged in the future, and the factors driving the decline of mortality are poorly understood. Clearly, future LTC demand will be driven not just by the rising number of very old people, but also by their health and healthcare needs. Although some studies agree that favourable disability trends in the future could have a substantial mitigating effect on future demand for LTC, the fast-growing number of oldest old is nonetheless expected to increase care needs (and related spending) substantially in the future.

4   Theme III: The labour market position of older workers

Keeping older workers employed is at the heart of the public policy debate on the economic consequences of ageing and the sustainability of pension systems.

Many existing studies focus on the consequences of financial incentives and other factors for labour supply of older workers, so this seems relatively well understood. There appears to be common agreement that financial incentives strongly influence whether older people are willing to work, provided that other factors (like health and job characteristics) do not keep an individual from working. Much less empirical research is performed on the demand side of the labour market, which becomes particularly relevant if policy measures stimulating supply become effective.

4.1 Policy questions

The main policy questions concerning labour supply, labour demand, the productivity and skills of older workers are the following:

- *Which economic and non-economic factors drive labour supply of older workers?* How important are factors such as health, job characteristics and job satisfaction, family considerations and cultural or peer-group effects? The quantitative importance of these other factors is still unclear and policies focusing on them are not as yet forthcoming.
- *What are the advantages and disadvantages of gradual or phased retirement, for workers, employers and the macro-economy?*
- *What are the consequences of working longer for well-being at an older age?* What about the consequences for health and cognitive skills?
- *How to eliminate negative attitudes towards older workers among employers?* How effective are legislation against age discrimination and public campaigns promoting a more positive view of older workers?
- *What determines labour demand for older workers, and how effective are specific policies to increase this?*
- *How can life-long learning be promoted in an effective way?*
- *How is labour market flexibility related to the position of older workers?* Would removing institutional restrictions and increasing flexibility improve the situation?
- *Does the economic crisis affect the relevance of existing policy recommendations?* Is increasing youth unemployment reducing the political pressure on increasing participation of older workers?
- *Is education the key to the European employment problem in the long run?* The lower education levels of older cohorts in Europe compared to younger cohorts and to the US could explain the employment gap of older workers between Europe and the US.

4.2 Major progress in understanding

Labour supply of older workers

At least three explanations for the large variation in retirement patterns across countries have been given (see Kapteyn and Andreyeva, 2008). The first is financial incentives, emphasised by a group of researchers led by Gruber and Wise, who performed country-specific studies using a harmonised methodology (see Gruber and Wise, 2004), and many other studies that evaluate the link between provisions of social security programmes and national retirement patterns.

Another explanation for the employment gap between Americans and Europeans is the greater extent to which home production can be substituted with services available in the labour market (Freeman and Schettkat, 2001). Alternative explanations point to differences in preferences or culture (Blanchard, 2004) and the role of institutions, particularly the power of unions (Alesina, Glaeser and Sacerdote, 2005).

A discussion of retirement patterns must also consider the role of health. Failing health may lead to retirement (Kalwij and Vermeulen, 2008). If this were the only mechanism at play, then one would expect the increase in population health to be accompanied by an upward trend in retirement ages. Retirement status may also have a (positive or negative) influence on health: it can remove mental stress and physical work effort, but is also a major life event, which in itself creates stress and may reduce health. Existing studies often find a negative effect of retirement on, in particular, mental health (see Bonsang et al., 2007).

The demand for older workers

The demand for older workers can be analysed in terms of the factors that influence job creation and job destruction (the 'equilibrium models'), or in terms of the factors that drive the hiring decisions of individual firms. Hetze and Ochsen (2005) build on traditional equilibrium models and incorporate the impact of age-based heterogeneity in the labour force on the formation of search equilibrium. They show how ageing affects job creation and job destruction: if older workers are less productive than their younger counterparts, ageing will reduce the number of vacancies (which would reduce the company's revenues).

Of the models that predict the demand for older workers from the firm's perspective, the best-known is Lazear's 'delayed compensation contracts' model (Lazear, 1979). Lazear argues that a work contract where the worker is paid less than the value of his/her marginal product at younger ages, and more at later ages, has advantages for both workers and employers. For workers, this type of contract increases their lifetime wealth. Employers, while forced to bear the higher fixed costs associated with delayed compensation, gain from improvements in performance and stronger employee commitment that are induced by the workers' fear of losing delayed compensation. Mandatory retirement is required as a way of terminating the contractual work, as the worker would not voluntarily retire due to the high wage. In this model, firms avoid hiring older workers, as this reduces the possible benefits of delayed compensation.

Productivity and wages

Skirbekk (2003) summarises the literature relating age to productivity, presenting evidence suggesting a decline in several aspects of physical and mental functioning from around age 50. Other studies, however, argue that older workers often rely on their professional experience to adapt and compensate for the decline in physical and mental ability, and find no significant difference between the job performance of older and younger workers. Recent studies using matched employer-employee data (see Crépon and Aubert, 2003) suggest that individual productivity declines in some dimensions with age, but experience, personal aids and suitable workplace adjustments can partly compensate for this.

Employer attitudes and age discrimination in recruitment

Some insight into attitudes and beliefs among the general public is provided by a special Eurobarometer survey (European Commission, 2007). It shows that almost half of the surveyed population feels that a candidate's age is, together with the onset of a disability, one of the most important criteria that might put the candidate at a disadvantage when competing for a job against someone with the same qualifications. In a survey of 500 large employers in the UK, Taylor and Walker (1994) show that a sizable group of respondents had negative stereotypes of older workers, especially with regard to their openness to training and their ability to adapt to new technologies. Laboratory and field experiments seem to confirm that employers discriminate against older applicants (see Riach and Rich, 2002). More recent studies show mixed results. For example, McNair, Flynn and Dutton (2007) found that age stereotypes and attitudes tend to favour older workers in general (emphasising more skills, life experience, reliability, loyalty and ability to cope with pressure or deal with others, especially customers) although negative views exist as well (such as lack of willingness to adopt new work methods).

Labour market regulation, recruitment and retention

Cheron, Hairault and Langot (2008), extending the standard model of equilibrium unemployment with an explicit role of age, argue that a standard retirement age explains the lower employment rate amongst the older population: reducing the willingness of firms to recruit older workers, as the expected returns will be lower. The shorter expected job duration, moreover, reduces the tendency of older workers to invest in job search activities.

Several empirical studies explore the impact of employment protection legislation (EPL). Bassanini and Duval (2006), using cross-country / time-series data from 21 OECD countries over the period 1982-2003, found a positive relation between the level of EPL and the employment rate of workers aged 55 to 64. OECD (2004), in contrast, finds a negative relation between employment protection and the hiring rate of men aged 50 and over (but this is compensated by a decrease in firings).

Qualitative research using cross-country comparisons suggests that a global age-management approach with a 'coordinated and comprehensive package of age friendly employment measures and policies' leads to the best results (Sigg, 2007), but much more work is necessary to determine what characterises optimal age management, considering economic as well as non-economic factors.

On-the-job training (OJT)

With regard to OJT, the literature offers several policy implications: promoting life-long learning or later retirement will not be effective if strong disincentives caused by labour market institutions, early retirement schemes and incentives for pension savings remain in place. Moreover, promoting private savings for old age may inadvertently create implicit taxes on skill formation and indirectly stimulate early retirement, thereby worsening the ageing problems. In a nutshell, any policy reform should take into account the dynamic interactions of OJT investment, retirement and pension saving.

Empirical work analysing OJT of older workers is scarce. A major empirical problem in the training literature is that OJT investment is difficult to measure. Heckman (2000) finds that most training is informal rather than formal, limiting the applicability of commonly employed training measures, which are often based on subjective data (from firms or employees), on formal OJT investment. Moreover, firms and employees seem to have different views on the participation intensity of training. See also Leuven (2005) for an elaborate review. Not only the costs (of OJT investment, for example), but also the returns (future wages) are difficult to measure empirically: earnings are not equal to labour productivity (even if labour markets are perfectly competitive), since OJT time investment drives a wedge between gross labour productivity and gross labour earnings. This is often overlooked. Clearly, time costs are the most important ingredient of investment in human capital, and worker productivity cannot be inferred from labour earnings.

Heckman, Lochner and Taber (1998) do obtain estimates, however, by identifying skill prices per unit of human capital from the earnings of the older workers who are in the final years of their careers. Indeed, human capital investments would approximately be zero for these workers, so that labour earnings indeed reflect productivity.

5 A roadmap for research on ageing, pensions and health in Europe

5.1 *European research networks on ageing, pensions and health*

For many of the priorities identified above, cooperation between experts from various disciplines will certainly add value. Such cooperation has started up in several networks and infrastructural projects, many of which are EU-funded. Below, we refer to some important networks, although the list is not exhaustive.

- An example of a successful EU-funded network is the ERA-Age network coordinated by Alan Walker at the University of Sheffield.[4] ERA-Age aims to promote the development of a European strategy for research on ageing, thereby enabling Europe to gain maximum added value from investment in this field. The network focuses on ageing but is very broad, with scant focus on economics. It brings together researchers and policymakers in sociology, public health, and biological and medical science.
- A second example is the family of EU-funded research networks and infrastructural projects focusing on collecting and using the SHARE data, mainly coordinated by the Mannheim Economics of Ageing Institute

---

4 See http://era-age.group.shef.ac.uk/. ERA-AGE has 12 partner countries; Romania is the only one in Central/Eastern Europe.

(MEA). Here, economists play a leading role. The main drawback is that not all EU countries are represented – and particularly most countries in Central and Eastern Europe are lacking.[5]

- A third relevant network is the ESA Research Network on Ageing in Europe.[6] This network (part of the European Sociological Association) focuses on ageing from a sociological point of view, but the topics of interest overlap substantially with those of economists (costs of pensions and healthcare, changing patterns of consumption and lifestyle, intergenerational transfers and so forth).

- An example of a more specific network is the AAMEE project,[7] which focuses on the promotion of active ageing and social, cultural and economic integration of older migrants and old-age groups in ethnic minorities, emphasising volunteer activities and the emergence of new culturally sensitive products and services in the fields of, for instance, housing, care, education, leisure, culture and marketing.

- The European Network of Economic Policy Research Institutes (ENEPRI)[8] brings together 24 leading national economic policy research institutes from most of the EU-27 countries, to foster the international diffusion of existing research, coordinate research plans, conduct joint research and increase public awareness of the European dimension of national economic policy issues. Several projects related to the socioeconomics of ageing conducted by ENEPRI members receive EC funding, such as AHEAD (Ageing, Health Status and Determinants of Health Expenditure) and AIM (Adequacy and Sustainability of Old-age Income Maintenance).

- A network specifically focusing on ageing in Central and Eastern Europe is the EAST (Eastern European Ageing Societies in Transition) network of the Oxford Institute of Ageing. While the primary focus is gerontological and geriatric research, EAST also targets research on social policy, poverty, unemployment, social security, and so forth.[9]

---

5   Current projects are SHARE LIFE, SHARE LEAP and SHARE-PREP. The website lists 16 participating countries, with the Czech Republic and Poland representing Central and Eastern Europe. See http://www.share-project.org/.

6   See http://www.ageing-in-europe.org/.

7   Active Ageing of Migrant Elders across Europe. See http://www.aamee.eu/.

8   See http://www.enepri.org/.

9   See http://www.ageing.ox.ac.uk/. The network currently involves academics from Albania, Belarus, Bulgaria, Croatia, Czech Republic, Estonia, (East) Germany, Hungary, Latvia, Lithuania, Poland, Romania, Russia, Serbia, Slovakia, Slovenia and Ukraine.

- An example of a network that does not focus particularly on ageing but in which ageing is one of the main themes is the Generations and Gender Programme (see Vikat et al., 2007)[10], which organises and maintains a system of national Generations and Gender Surveys (GGSs) and contextual databases concerning European and some non-European countries. Its main goal is to improve understanding of demographic and social developments and of the factors that influence these developments, with particular attention on relationships between children and parents (generations) and between partners (gender), through the use of GGS data.
- Another network is formed by the project 'Mainstreaming Ageing: Indicators to Monitor Implementation (MA:IMI)',[11] sponsored by the UNECE and carried out by the European Centre Vienna. The network started in 2003 and is now in its second stage. The project pursues the goal of exchanging experiences in the field of ageing-related policies, carries out and fosters data collection, research and analysis on ageing-related issues and maintains a network of organisations, national authorities, and other concerned bodies and individuals active in the field of ageing.

It can be argued that there is ample scope for a network that focuses on research on ageing from an economic perspective with input from social sciences and public health to the extent that they relate and can contribute to the economics of ageing.[12] The core would be experts and institutes specialised in economics, but also non-economists working on adjacent topics would be included. Institutes from Central and Eastern Europe that could be involved are the Warsaw School of Economics, the School of Political and Administrative Studies in Bucharest, the Center for Economic Research and Graduate Education in Prague, the Central European University in Budapest and the European Centre Vienna. Such a network will stimulate policy-relevant studies at the research frontier by allocating funds for research grants, organising meetings, facilitating international data collection and data access, and so forth.

---

10   See http://www.unece.org/pau/_docs/ggp/GGP_2006_SumDescr.pdf. Eastern and Central Europe are reasonably well represented; Hungary is in the core group of the programme and data are collected in nine Eastern and Central European countries.

11   For more details, see http://www.euro.centre.org/detail.php?xml_id=81

12   In the US, such networks do exist, funding research projects and organising meetings, e.g., the retirement research centres (at NBER, Boston College and Michigan University) funded by the Social Security Administration, and the NBER programme on Ageing. Broad European networks similar to NBER like CEPR and CESIfo do not have a structured programme on ageing.

5.2 Objectives: Three research priorities

Based upon the analysis in this chapter (and in other chapters in this volume), we can identify three important and promising areas for future European research on the economics and social science of ageing. The three priorities consist of many research questions; some of them could be fruitfully addressed in the next five years; others require new data, making a five-year horizon unrealistic.

I.   The first is *employability of individuals at early old age*, with emphasis on employer attitudes, productivity of older workers and institutional factors, which are studied less often (and understood less well) than are workers' retirement decisions. While pension reforms have stimulated older workers to remain in the labour force longer, keeping them there requires that certain demand conditions be satisfied: jobs and working conditions must be attractive to workers, who in turn must remain attractive to firms. This involves not only economic factors (skills, productivity and labour costs) but also perceptions and attitudes of employers and co-workers towards older workers, which vary across countries and can be influenced by age-management policies. The effects of employment protection and other features of labour-market flexibility and opportunities for job mobility and self-employment after a career job need to be better understood. The variation across Europe in institutions and institutional reforms needs to be better exploited, to improve our understanding of the mechanisms at work and the relationships between these economic and non-economic factors.

II. The second research priority concerns *collective versus individual responsibility for retirement* saving, to guarantee financial security in the years after retirement and its maintenance until late old age. How can we reconcile the tension between paternalism and collective risk sharing and freedom of choice and heterogeneity of preferences? The consequences of financial illiteracy and sub-optimal decisions need to be better understood in rigorous modelling frameworks supported by empirical research. The institutional diversity in Europe, with its many pension reforms involving various combinations of (minimum) state pensions, occupational pensions, and voluntary contributions to retirement savings plans can be exploited to a greater extent.

III. The third priority concerns the long-run- and short-term determinants of *health, healthcare use, and well-being* in later old age. What mechanisms link conditions earlier in life (including socioeconomic conditions) to health and well-being in later life? The variety in European healthcare systems and reforms of these systems can be exploited to better understand how various systems benefit different socioeconomic groups, within reasonable bounds of cost-effectiveness. Particular attention should be given to analysing how diverse socioeconomic groups, with varying levels of morbidity and mortality outcomes, draw differential benefits from welfare regimes of distinct types. The relationships between economic welfare, housing conditions and living arrangements and environments, social networks, cognitive skills, healthcare demand, and mental and physical health need to be studied in different cultures, involving distinct family networks and cultural and historical norms – in addition to different healthcare arrangements and other institutions.

To make European research on these priorities feasible and successful, the most important needs identified in this book involve various sorts of data: administrative and survey data at the individual- and the firm level, as well as collection and dissemination of knowledge on the institutions in European countries. Moreover, we need better access to and documentation of these data (and to already-existing data), and the forging of greater coordination in the European research agenda on these priorities.

European-level coordination would be useful to apprise researchers across Europe of the data available (particularly in countries other than their own), improve documentation for international users and simplify access procedures. Such a European network could be an important step forward to achieve this goal. This might be done in cooperation with national data archiving organisations and with CESSDA (the Council of European Social Science Data Archive). Workshops in which researchers and programmers familiar with specific data sources teach researchers how to use the data are also useful. Such workshops are already organised, for example, for the German Socioeconomic Panel (GSOEP).

SHARE (the Survey of Health, Ageing and Retirement in Europe) is a general survey covering adults aged 50 and older, offering multidisciplinary micro data in a number of European countries. It is selected as one of the large-scale European infrastructural projects that deserve further development in the ESFRI Roadmap of the EC. The measures in the SHARE dataset are also comparable with those from the US Health and Retirement Study (HRS), and the English Longitudinal Study of Ageing (ELSA). SHARE includes many objective and subjective measures of physical and mental health, psychological conditions, socioeconomic status, and social participation. To address the research challenges identified above, useful extensions of SHARE would be required (see summary in Box 12.1).

Box 12.1:   Further extensions required in SHARE, the Survey of Health, Ageing and Retirement in Europe

Expanding the longitudinal dimension
This expansion is needed to identify and understand the causal mechanisms that often take years to become effective, such as the effect of unemployment or other economic shocks on health and well-being in old age. While retrospective measures can help in many respects, genuine panel data with complete information on the same individuals over a long period of time remain important.
Expanding the European coverage
To learn as much as possible from international variation in reforms of labour market institutions, pension systems and healthcare policies, it would be useful to include more countries. Central and Eastern European countries are underrepresented. Also the use of anchoring vignettes should be improved to enhance the comparability of subjective measures of well-being, which until now have been administered only experimentally to small sub-samples in selected countries.
Including additional detailed information
Detailed survey information should be added on, job characteristics, OJT, employer attitudes, gradual retirement opportunities and preferences and job satisfaction, for example. This could be done through interviews of SHARE respondents on specific topics in years when there is no core survey, a strategy also pursued in the US for the HRS. Internet interviews, along the lines of the HRS in the US, would be cost-effective. Specific groups such as the oldest old can still be interviewed face to face.
Merging with administrative sources of data
The collected data could be merged with administrative data (on income, wealth and, in particular, pension entitlements). These variables are inherently hard to measure in surveys, and merging with administrative data can improve the quality of the economic content of SHARE. Another avenue might be to link SHARE to register data (available in several countries) on healthcare utilisation. To study labour demand, it would be useful to merge the SHARE data to an employer survey that interviews the employers of the SHARE respondents. This would provide a more direct way to analyse employer attitudes and firm- and workplace characteristics than asking the employees. Matched employer-employee data are often used in labour economics, but not in the specific context of older workers.

5.3   Actions: What steps are needed, and in which order?

A first priority is to set up a European centre of expertise with institutional knowledge on labour market institutions, financial, pension and social security institutions and healthcare systems. This centre should provide information on which institutional arrangement applies to which individual at a given point in time. In addition, a broad European network should be established aimed at international exchange of country-specific survey- and administrative data in the field of ageing, health and pensions. One major set of tasks of such a network could be related to SHARE (see Box 12.1 for our recommendations on further extensions required in SHARE). Some special topics (relating to the three research priorities identified in this chapter) that could be covered in SHARE:

- Among respondents in early old age: detailed interview on job charac-teristics, job satisfaction, employer accommodation, gradual retirement opportunities, interest in self-employment and other bridge jobs, and employer accommodation of and attitudes towards older workers.
- Among respondents of all ages: detailed interview on time preference, future expectations, risk attitudes and, for the younger groups, retirement saving behaviour. Experiments with hypothetical or real payments could be part of this.
- Among the older age groups (possibly using proxy interviews with a close relative): detailed interview on expectations concerning LTC and other healthcare use, preferences for healthcare use, social networks, housing and living arrangements and well-being (using the existing extensive modules on health and healthcare use).

Concrete actions must be taken to build on existing informal networks and the contacts established through the Forward Look project, which produced the material included in this book. Actions are required to implement and coordinate the European research agenda on the economics and social sciences of ageing by establishing a European research network structure. Existing national and international centres of expertise like Netspar, CeRP and the European Centre Vienna can be asked to take the lead. This will facilitate a better integration of researchers from Central and Eastern Europe into European networks. It will provide facilities to organise regular meetings and make research grants available to activate and expand the network, focusing on the three research priorities identified above.

Steps must be taken to create a European forum for exchange of knowl-edge between academics, business and public policy. Academic research can contribute to the most urgent policy questions and obtain access to relevant expertise and data only if researchers interact with the private pension industry (insurance companies, pension funds, banks) and government policymakers. Hence, a European network for research on ageing, health and retirement pro-vision that is set up and run jointly by knowledge institutions, national and European government institutions and the pension industry will be a mutually advantageous endeavour for all stakeholders.

References

Adams, P., M.D. Hurd, D. McFadden, A. Merrill, and T. Ribiero (2003) 'Healthy, Wealthy and Wise? Tests for Direct Causal Paths between Health and Socioeconomic Status.' *Journal of Econometrics*, 112, 3-56.

Alesina, A., E. Glaeser and B. Sacerdote (2005) 'Work and Leisure in the U.S. and Europe: Why so Different?' NBER working paper 11278.

Bassanini, A. and R. Duval (2006) 'The Determinants of Unemployment across OECD Countries: Reassessing the Role of Policies and Institutions', *OECD Economic Studies*, 42 (Paris: OECD).

Blanchard, O. (2004) 'The Economic Future of Europe', NBER working paper 10310.

Bonsang, E., S. Adam, C. Bay, S. Germain and S. Perelman (2007) 'Retirement and Cognitive Reserve: A Stochastic Frontier Approach Applied to Survey Data', *CREPP working paper* 2007/04.

Börsch-Supan, A. and A. Lusardi (2003) 'Saving: Cross-national Perspective' in A. Börsch-Supan (ed.) *Life-Cycle Savings and Public Policy: A Cross-National Study in Six Countries* (Amsterdam: Elsevier).

Brown, J. (2009) 'Understanding the Role of Annuities in Retirement Planning' in A. Lusardi (ed.) *Overcoming the Saving Slump: How to Increase the Effectiveness of Financial Education and Saving Programs* (Chicago: University of Chicago Press), forthcoming.

Carstensen, L.L. (1991) Socioemotional Selectivity Theory: Social Activity in Life-span Context', *Annual Review of Gerontology and Geriatrics*, 11, 195-217.

Cheron, A., J.-O. Hairault and F. Langot (2008) 'Age-dependent Employment Protection', IZA discussion paper 3851.

Cocco, J.F., F.J. Gomes and P.J. Maenhout (2005) 'Consumption and Portfolio Choice over the Life Cycle', *The Review of Financial Studies*, 18, 492-533.

Constantinides, G.M. and D. Duffie (1996) 'Asset Pricing with Heterogeneous Consumers', *Journal of Political Economy*, 105, 219-240.

Crépon, B. and P. Aubert (2003) 'Productivité et Salaire des Travailleurs âgés', *Economie et Statistique*, 368, 157-185.

Cutler, D.M. and M.B. McClellan (1998) 'What is Technological Change?' in D.A. Wise (ed.) *Inquiries in the Economics of Aging* (Chicago: University of Chicago Press), pp. 51-81.

European Commission (2007) *Discrimination in the European Union. Special Eurobarometer 263* (Brussels: European Commission).

European Commission (2009a) *2009 Ageing Report: Economic and Budgetary Projections for the EU-27 Member States (2008-2060)* (Brussels: European Commission).

European Commission (2009b) *The METRIS Report: Emerging Trends in Socio-economic Sciences and Humanities in Europe* (Brussels: European Commission).

Freeman, R. and R. Schettkat (2001) 'Marketization of Production and the US-Europe Employment Gap', *Oxford Bulletin of Economics and Statistics*, 63, 647-70.

Geanakoplos, J., O. Mitchell and S. Zeldes (2000) 'Social Security Money's Worth', NBER working paper 6722.

Goldman, D.P., B. Shang, J. Bhattacharya, A.M. Garber, M. Hurd, G.F. Joyce, D.N. Lakdawalla, C. Panis and P.G. Shekelle (2004) *Health Status and Medical Treatment of the Future Elderly: Final Report* (Santa Monica: RAND Corporation).

Gollier, C. (2001) *The Economics of Risk and Time* (Cambridge: MIT Press).

Gruber, J. and D. Wise (2004) *Social Security Programs and Retirement around the World: Micro-Estimation* (Chicago: University of Chicago Press).

Heckman, J. (2000) 'Policies to Foster Human Capital', *Research in Economics*, 54, 3-56.

Heckman, J., L. Lochner and C. Taber (1998) 'Explaining Rising Wage Inequality: Explorations with a Dynamic General Equilibrium Model of Labor Earnings with Heterogeneous Agents', *Review of Economic Dynamics*, 1, 1-58.

Hetze, P. and C. Ochsen (2005) 'How Aging of the Labor Force Affects Equilibrium Unemployment', Thünen Series of Applied Economic Theory working paper 57.

Huber, M., R. Rodrigues, F. Hoffmann, K. Gasior and B. Marin (2009) 'Facts and Figures on Long-term Care. Europe and North America', European Centre for Social Welfare Policy and Research.

Kahneman, D., A.B. Krueger, D.A. Schkade, N. Schwarz and A.A. Stone (2004) 'A Survey Method for Characterizing Daily Life Experience: The Day Reconstruction Method', *Science*, 306, 1776-1780.

Kalwij, A. and F. Vermeulen (2008) 'Health and Labour Force Participation of Older People in Europe: What Do Objective Health Indicators Add to the Analysis?' *Health Economics*, 17, 619-638.

Kapteyn, A. and T. Andreyeva (2008) 'Retirement Patterns in Europe and the U.S.', Netspar panel paper 6.

Kohli, M. (1999) 'Private and Public Transfers between Generations: Linking the Family and the State', *European Societies*, 1, 81-104.

Künemund, H. (2001) *Gesellschaftliche Partizipation und Engagement in der zweiten Lebenshälfte. Empirische Befunde zu Tätigkeitsformen im Alter und Prognosen ihrer zukünftigen Entwicklung* (Berlin: Weißensee Verlag).

Künemund, H. (2006) 'Changing Welfare States and the 'Sandwich Generation'– Increasing Burden for the Next Generation?' *International Journal of Ageing and Later Life*, 1, 11-30.

Lazear, E. (1979) 'Why is There Mandatory Retirement?' *Journal of Political Economy*, 87, 1261-1284.

Leuven, E. (2005) 'The Economics of Private-sector Training: A Review of the Literature', *Journal of Economic Surveys*, 19, 91-111.

Lusardi, A. (1999) 'Information, Expectations, and Savings for Retirement' in H. Aaron (ed.), *Behavioral Dimensions of Retirement Economics* (Washington, DC: Brookings Institution and Russell Sage Foundation), pp. 81-115.

Lusardi, A. and O.S. Mitchell (2007) 'Baby Boomer Retirement Security: The Role of Planning, Financial Literacy and Housing Wealth', *Journal of Monetary Economics*, 54, 205-224.

McNair, S., M. Flynn and N. Dutton (2007) 'Employer Responses to an Ageing Workforce: A Qualitative Study', UK Department for Work and Pensions research report 455.

Muellbauer, J. (2007) 'Housing, Credit and Consumer Expenditure', Kansas Federal Reserve Symposium paper.

Norton, E.C. (2000) 'Long-term Care' in A.J. Culyer and J.P. Newhouse (eds), *Handbook of Health Economics, vol. 1* (Amsterdam: Elsevier).

OECD (2004) *OECD Employment Outlook* (Paris: OECD).

Pinquart, M. and S. Sörensen (2001) 'How Effective are Psychotherapeutic and Other Psychosocial Interventions with Older Adults? A Meta-Analysis', *Journal of Mental Health and Aging*, 7, 207-243.

Riach, P. and J. Rich (2002) 'Field Experiments of Discrimination in the Market Place', *Economic Journal*, 112, 480-518.

Scholz, J.K., A. Seshadri and S. Khitatrakun, (2006) 'Are Americans Saving 'Optimally' for Retirement?' *Journal of Political Economy*, 14, 607-643.

Sigg, R. (2007) Extending Working Life: Evidence, Policy Challenges and Successful Responses' in B. Marin and A. Zaidi (eds) *Mainstreaming Ageing: Indicators to Monitor Sustainable Progress and Policies* (Farnham: Ashgate), pp. 447-489.

Skirbekk, V. (2003) 'Age and Individual Productivity: A Literature Survey', Max Planck Institute for Demographic Research working paper 2003-028.

Taylor, P. and A. Walker (1994) 'The Ageing Workforce: Employers' Attitudes Towards Older People', *Work, Employment and Society*, 8, 569-591.

Vikat, A., Z. Spéder, G. Beets, F.C. Billari, C. Bühler, A. Désesquelles, T. Fokkema, J.M. Hoem, A. MacDonald, G. Neyer, A. Pailhé, A. Pinnelli and A. Solaz (2007) 'Generations and Gender Survey (GGS): Towards a Better Understanding of Relationships and Processes in the Life Course', *Demographic Research*, 17, 389-440.

Whitehouse, E.R. and A. Zaidi (2008) 'Socio-Economic Differences in Mortality: Implications for Pensions Policy', OECD Social, Employment and Migration working paper 71.

Zaidi, A., K. Gasior and A. Sidorenko (2010) 'Intergenerational Solidarity: Policy Challenges and Societal Responses', European Centre for Social Welfare Policy and Research policy brief, July.

List of Contributors

THOMAS BOLL is a Researcher and Lecturer at the Research Unit INSIDE, University of Luxembourg, Walferdange, Luxembourg

LANS BOVENBERG is Professor of Economics at Tilburg University, The Netherlands

FRANK DE JONG is Professor of Financial Markets and Risk Management at Tilburg University, The Netherlands

RUUD DE MOOIJ is Head of Sector 'Labour Market and Welfare State' at CPB (Netherlands Bureau for Economic Policy Analysis), and Professor in Public Economics, Erasmus University Rotterdam, The Netherlands

MICHAEL DEWEY is Professor of Statistical Epidemiology at the Institute of Psychiatry, Department of Health Services and Population Research, King's College London, The United Kingdom

JOHN DOLING is Professor of Housing Policy at the Institute of Applied Social Studies, University of Birmingham, The United Kingdom

ROB EUWALS is Programme Leader Labour Market at CPB (Netherlands Bureau for Economic Policy Analysis), The Hague, The Netherlands

DIETER FERRING is Professor of Psychology and Director of the Research Unit INSIDE at the University of Luxembourg, Walferdange, Luxembourg

ANTONIO M. FÓNSECA is Assistant Professor at the Catholic University, Faculty of Education and Psychology, Porto, Portugal

ELSA FORNERO is Professor of Economics at University of Turin, and Scientific Coordinator of CeRP (Centre for Research on Pensions and Welfare Policies), Turin, Italy

ERIC FRENCH is a Senior Researcher in economic research at the Federal Reserve Bank of Chicago, USA

CHRISTIAN GOLLIER is Director of Toulouse School of Economics (TSE) and Research Director at the Institut d'Economie Industrielle (IDEI), Toulouse, France

DANIEL HALLBERG is a Researcher at the Institute for Future Studies, Stockholm, and at the Department of Economics, Uppsala University, Sweden

ALBERTO HOLLY is Professor Emeritus at Department of Economics, University of Lausanne, Switzerland

BAS JACOBS is Professor of Economics and Public Finance, Department of Economics, Erasmus University Rotterdam, The Netherlands

ARIE KAPTEYN is Director Labor and Population, RAND, Santa Monica, USA

MARTIN KOHLI is Professor of Sociology at the Department of Social and Political Sciences, European University Institute, San Domenico di Fiesole, Italy

HARALD KÜNEMUND is University Professor, Ageing and Empirical Research Methods, University of Vechta, Germany

MAARTEN LINDEBOOM is Professor of Economics at the Free University of Amsterdam, The Netherlands

ANAMARIA LUSARDI is the Joel Z. and Susan Hyatt Professor of Economics at Dartmouth College, Dartmouth, a Research Associate at the NBER and also Director of the Financial Literacy Center, a joint consortium of Dartmouth, RAND, and the Wharton School, USA

CHIARA MONTICONE is a Researcher at CeRP (Centre for Research on Pensions and Welfare Policies), Turin, Italy

AMILCAR MOREIRA is a Post-Doctoral Research Fellow at Oslo University College, Norway

JØRGEN MORTENSEN is Associate Senior Research Fellow at the Centre for European Policy Studies, Brussels, Belgium

THEO NIJMAN is the Van Lanschot Professor in Investment Theory at Tilburg University, The Netherlands

JIM OGG is Associate Research Fellow, Directorate of Research on Ageing, at CNAV (Caisse Nationale d'Assurance Vieillesse), Paris, France

CONSTANÇA PAÚL is Professor at University of Porto, Porto, Portugal

ANTOON PELSSER is Professor of Finance and Actuarial Science, at the Department of Finance, Maastricht University, The Netherlands

JANNEKE PLANTENGA is Director Graduate School of Economics, Utrecht University, The Netherlands

HENRIËTTE PRAST is Professor Personal Financial Planning at Tilburg University and a Board Member of the Scientific Council for Government Policy (WRR), The Netherlands

JOHANNES SIEGRIST is Professor at the Institute of Medical Sociology, Heinrich Heine University, Düsseldorf, Germany

LOU SPOOR is Senior Manager Strategy, Eureko Group, The Netherlands

EDDY VAN DOORSLAER is Professor of Health Economics at Erasmus School of Economics, Rotterdam, The Netherlands

MAARTEN VAN ROOIJ is Senior Economist at the Research Department of De Nederlandsche Bank, The Netherlands

ARTHUR VAN SOEST is Professor of Econometrics at Tilburg University, The Netherlands

MORTEN WAHRENDORF is an Associate at the Institute of Medical Sociology, Heinrich Heine University, Düsseldorf (Germany), and Researcher at the ESRC International Centre for Life Course Studies in Society and Health, London, The United Kingdom

BRENDAN WHELAN is Research Director of the Irish Longitudinal Study of Ageing at Trinity College Dublin, Ireland

ASGHAR ZAIDI is Director Research and also one of the heads of the project 'Mainstreaming Ageing: Indicators to Monitor Implementation (MA:IMI)', at the European Centre for Social Welfare Policy and Research, Vienna, Austria.

Index

DATE DUE